Micro- and Nanocomposites

This book aims to explore the field of tribology in micro- and nano-reinforced composites to enhance material performance across a range of applications. Researchers and engineers are actively researching different reinforcement configurations and material combinations to develop composites with superior tribological properties. This research will lead to more efficient and reliable engineering solutions.

Micro- and Nanocomposites: A Tribological Viewpoint will include a brief examination of previous reinforcement techniques and their impact on composite materials. It will also analyze how changes in tribological parameters affect material surface behavior. Understanding the role of tribology is crucial in assessing the performance of reinforced composite materials, especially in sustainable applications. A key focus of the book will be the significance of reinforcement selection from a tribological perspective in studying the environmental impact of modern industrial practices. Case studies are presented at the end of the book to illustrate the key principles discussed throughout.

This book is suitable for researchers, professionals, corporations, and individuals in higher education courses. It can also serve as a valuable reference for those studying advanced composite materials.

Multi-Scale and Multi-Functional Materials: Design, Development, and Applications

Series Editors: Chander Prakash and Alokesh Pramanik

The demand for Multi-Scale and Multi-Functional polymers, metal alloys, and composites for various engineering applications such as biomedical, aerospace, electronics, tissue engineering, healthcare, and energy storage applications has been increasing day by day. These materials are a unique combination of compositional and microstructural to satisfy specific sets of performance requirements of thermal, structural, mechanical, and biological. This book series covers the concurrent design of materials via multi-scale modeling and a multi-level robust design approach and includes the most recent advances in artificial intelligence-based techniques for (a) accelerating material discovery, (b) improving/accelerating synthesis protocols, (c) enhancing existing characterization methods, and (d) developing robust, efficient, and accurate materials models, as well as simulation methods at multiple length scales; advances in machine learning/artificial intelligence algorithms targeted towards materials, chemistry, and applications will also be discussed. Books will cover processes such as precision machining, processing, coating, deposition, cladding, surface finishing, and emerging advanced manufacturing technologies to enhance the performance of materials to reduce friction coefficients, lower wear rates, increase corrosion resistance, improve fatigue performance, and a range of multi-functional features. Therefore, this book series is aimed to provide a globalized platform to share innovative manufacturing methodologies for the synthesis and development of Multi-Scale and Multi-Functional materials.

Artificial Intelligence and Machine Learning in the Thermal Spray Industry: Practices, Implementation, and Challenges
Lalit Thakur, Hitesh Vasudev, and Gaurav Prashar

Micro- and Nanocomposites: A Tribological Viewpoint
Edited by Abhijit Bhowmik, Jitendra Kumar Katiyar, Chander Prakash, Alokesh Pramanik, and Animesh Basak

Micro- and Nanocomposites
A Tribological Viewpoint

Edited by
Abhijit Bhowmik, Jitendra Kumar Katiyar,
Chander Prakash, Alokesh Pramanik,
and Animesh Basak

CRC Press
Taylor & Francis Group
Boca Raton London New York

CRC Press is an imprint of the
Taylor & Francis Group, an **informa** business

Designed cover image: Shutterstock

First edition published 2026
by CRC Press
2385 NW Executive Center Drive, Suite 320, Boca Raton FL 33431

and by CRC Press
4 Park Square, Milton Park, Abingdon, Oxon, OX14 4RN

CRC Press is an imprint of Taylor & Francis Group, LLC

ISBN: 978-1-032-87016-8 (hbk)
ISBN: 978-1-032-87096-0 (pbk)
ISBN: 978-1-003-53090-9 (ebk)

DOI: 10.1201/9781003530909

Typeset in Times LT Std
by KnowledgeWorks Global Ltd.

Contents

Preface

Tribology is an important factor that helps to determine the mechanical performance, durability, and overall efficiency of reinforced composites. Tribology is also important in developing micro- and nano-component-reinforced composites. Reinforced composites are materials made by mixing a matrix (typically a polymer, metal, or ceramic) with reinforcing fibers or particles (such as carbon fibers, glass fibers, micro- or nanoparticles). The resistance of the material to wear and the frictional behavior during sliding contact with other surfaces may be impacted by the kind of reinforcements and the orientation and concentration of those reinforcements. For instance, composites with high-strength fibers oriented in the direction of sliding may demonstrate reduced wear rates due to their capacity to distribute the load properly. Variations in tribological qualities are possible when using various kinds of reinforcements.

Nowadays, research on reinforced composites often focuses on investigating the lubrication's impact on lowering friction and wear levels. The wear and friction properties of the composite may be affected by the hardness, roughness, and chemical composition of the surface it is rubbing against. Temperature, as well as the circumstances of the surrounding environment, may have a significant impact on the tribological performance of reinforced composites. The wear processes may be influenced by high temperatures, high humidity, or harsh chemicals, all of which can impair the qualities of the composite. The tribological attributes of a material are often modified to suit a particular purpose better. For instance, reinforced composites may be used in the aircraft or automobile sectors to minimize friction and wear in crucial components. This would result in higher fuel economy and increased durability.

This book aims to study the tribology of micro- and nano-reinforced composites to maximize the performance of materials in various applications. Researchers and engineers continue investigating reinforcing configurations and material combinations to produce composites with improved tribological qualities. This will ultimately result in engineering solutions that are more effective and dependable. The proposed book will incorporate a brief study of reinforcement used earlier and its various effects on composite materials. The material surface's behavior under the tribological parameter change will also be evaluated. The impact of tribology plays a vital role in understanding the performance of reinforced composite material in sustainable applications. One of the main topics of this book will continue to be the critical role played by reinforcement selection with a tribological perspective in studying the environmental effects of modern industrial activities. Case studies will be provided at the end of the book to clarify the principles addressed.

About the Editors

Abhijit Bhowmik is currently the head of the department and assistant professor under the Mechanical Engineering Department at Dream Institute of Technology, Kolkata, India. Dr. Bhowmik obtained a B.E. in mechanical engineering from the Tripura Institute of Technology, India, in 2012 and an M. Tech degree in mechanical engineering, specializing in thermal science and engineering from the National Institute of Technology Agartala, India, in 2014. He has served as an assistant professor in the Mechanical Engineering Department at Dream Institute of Technology, Kolkata, since 2015. In 2023, he completed his Ph.D. from the National Institute of Technology Agartala, India. His research work has been published in 7 patents, 50+ SCI-indexed journals, and 110+ Scopus-indexed journal papers. His research interests span composite materials, tribology, mechanical properties, fracture analysis, welding, microstructural evaluation, and optimization techniques. Dr. Bhowmik is also a lead guest editor of the renowned publishers Sage, De Gruyter Brill, and Bentham Science Publishers. He is an associate member of the Institution of Engineers, India.

Jitendra Kumar Katiyar is an associate professor at the Department of Mechanical and Industrial Engineering, Manipal Institute of Technology, Manipal Academy of Higher Education, Manipal, India. He has 10 years of teaching and research experience. Dr. Katiyar's research interests include modern manufacturing techniques, tribology of carbon materials, polymer composites, self-lubricating polymers, lubrication tribology, and coatings for advanced technologies. He obtained his bachelor's degree from UPTU Lucknow, India, with honors in 2007, his master's from the Indian Institute of Technology Kanpur in 2010, and his Ph.D. from the same institution in 2017. He is a life member of the Tribology Society of India; the Malaysian Society of Tribology; the Institute of Engineers, India; the Indian Society for Technical Education (ISTE); the American Society of Mechanical Engineers; and was elected executive committee member of the Tribology Society of India (2024–2026). Dr. Katiyar is also an editor and guest editor of multiple esteemed journals and publishers.

Chander Prakash is Vice Chancellor (Officiating) of Sanskaram University, Jhajjar, Haryana, India. He is an accomplished academic leader with extensive experience in academic administration, research, and global university rankings. A visionary professional, he has driven transformative initiatives in higher education, focusing on elevating institutional reputation and cultivating a thriving research ecosystem. He is a passionate researcher with diversified research interests, such as developing materials for biomedical and healthcare applications and additive manufacturing and developing and exploring new cost-effective manufacturing technologies for biomedical industries. He has published over 550 (450 SCI/Scopus) scientific articles in peer-reviewed, reputable, top-notch journals, conferences, and books. As a prolific author, Prof. Prakash has edited or authored 35 books, served as a series editor for 3 book series, and contributed as a guest editor for leading journals. He is also an editorial board member of highly reputed journals. Beyond publications, Prof. Prakash has significantly contributed to intellectual property rights (IPRs), with 18 patents filed; 6 granted; and has successfully raised $1.2 million in research grants from bodies like DSTSERB, UKIERI-DST, and the Ministry of Science and Technology, Government of India. His outstanding contributions have earned him the Research Excellence Award (2019, 2020, 2021). He has actively contributed to institutional accreditation as a NAAC assessor and chairperson, ensuring academic quality and excellence.

Alokesh Pramanik is a distinguished researcher and academic in the field of mechanical and aerospace engineering, currently serving as a senior lecturer at Curtin University, Australia. With over 20 years of research experience, his expertise spans manufacturing processes, composite materials, and advanced alloys. Dr. Pramanik has authored more than 200 research publications, including books, book chapters, and peer-reviewed journal articles. Dr. Pramanik is actively involved in the academic community as an associate editor and editorial board member of several prestigious international journals. He is also a sought-after keynote speaker at global conferences, contributing to the advancement of knowledge in his field. Dr. Pramanik's academic contributions continue to influence both theoretical and applied aspects of mechanical engineering, making him a respected figure in the global research community.

Animesh K. Basak is a well-established academic and researcher in the broad field of materials science and engineering. Upon the completion of his master's and Ph.D. in materials science and engineering from Katholieke Universiteit Leuven (KULeuven), Belgium, Dr. Basak worked as an academic at various Australian universities and is currently working at Adelaide Microscopy, Adelaide University, Adelaide, South Australia, Australia. Dr. Basak has authored more than 300 research publications, including books, book chapters, and peer-reviewed journal articles in his areas of research. This includes nanostructured materials, high-resolution electron microscopy and probe analysis, nanoscale fabrication of materials, deformation mechanisms of novel materials, nano-mechanics, tribology and tribo-corrosion, and electrochemical measurement techniques. Dr. Basak is actively serving the academic community as journal editor, editorial board member, and reviewer of journals and funding agencies. Dr. Basak also disseminates his research work at global conferences as an invited keynote speaker.

Contributors

Abayomi Adewale Akinwande
Department of Metallurgical and Materials Engineering
Federal University of Technology
Akure, Ondo State, Nigeria

Abdulaziz AlHazaa
Department of Physics and Astronomy
College of Science
King Saud University
Riyadh, Saudi Arabia

E. R. Babu
Department of Mechanical Engineering
Bangalore Institute of Technology
Bengaluru, Karnataka, India

Devakant Baviskar
Department of Mechanical Engineering
Saraswati College of Engineering
Navi Mumbai, Maharashtra, India

K. Benarji
Department of Mechanical Engineering
Shri Vishnu Engineering College for Women
Bhimavaram, India

Mohd Nadeem Bhat
National Institute of Technology Srinagar
Hazratbal, Srinagar, India

Abhijit Bhowmik
Department of Mechanical Engineering
Dream Institute of Technology
Kolkata, India
Division of Research and Development
Lovely Professional University
Phagwara, Punjab, India

Bharathkumar B. R.
Department of Mechanical Engineering
Jain College of Engineering and Technology
Hubballi, India

Vishwanath B. R.
Department of Mechanical Engineering, BNMIT,
Bengaluru, India

Jayesh Subhash Chordiya
Department of Mechanical Engineering
Pimpri Chinchwad College of Engineering
Pune, India

Payel Deb
Department of Mechanical Engineering
PIET, Parul University
Vadodara, Gujarat, India

Nellaiah Hariharan
Shelcel Biozeen Technologies Pvt. Limited
Bangalore, India

Katika Harikrishna
Department of Mechanical Engineering
Shri Vishnu Engineering College for Women
Bhimavaram, India

Anand K. Hosamani
Department of Mechanical Engineering
Jain College of Engineering
Belagavi, India

Md. Farid Hossain
Department of Mechanical Engineering
Dream Institute of Technology
Kolkata, India

Arun Prakash J.
Department of Aeronautical Engineering
Nehru Institute of Engineering and Technology
Coimbatore, Tamilnadu, India

Hemalata Jena
School of Mechanical Engineering
KIIT (deemed to be University)
Bhubaneswar, Odisha, India

Hemanth K.
School of Mechanical Engineering
REVA University
Bengaluru, Karnataka, India

Prashant D. Kamble
Department of Mechanical Engineering
Yeshwantrao Chavan College of Engineering
Wanadongri, Maharashtra, India

Nazrul Islam Khan
Department of Mechanical Engineering
Netaji Subhas University of Technology
New Delhi, India

Yugesh A. Kharche
Department of Mechanical Engineering
Padmashri Dr. V. B. Kolte College of Engineering
Malkapur, Maharashtra, India

Manoj Kumar
Department of Mechanical Engineering
ABES Engineering College
Ghaziabad, India

Naveen Kumar
Department of Mechanical Engineering
United College of Engineering & Research
Naini, Prayagraj, Uttar Pradesh, India

P. Satish Kumar
Department of Mechanical Engineering
Rathinam Technical Campus
Coimbatore, Tamilnadu, India

Raman Kumar
Department of Mechanical and Production Engineering
Guru Nanak Dev Engineering College
Ludhiana, Punjab, India
Jadara Research Center
Jadara University
Irbid, Jordan

Sujeet Kumar
Department of Mechanical Engineering
Bonam Venkata Chalamayya Engineering College
Odalarevu, Andhra Pradesh, India

Sumit Das Lala
Department of Mechanical Engineering
PIET, Parul University
Vadodara, Gujarat, India

Nagoor Meeran M.
Department of Chemistry
Vel Tech Rangarajan Dr. Sagunthala R&D Institute of Science and Technology
Chennai, Tamilnadu, India

Harshit Majumder
Department of Mechanical Engineering
PIET, Parul University
Vadodara, Gujarat, India

Guttikonda Manohar
Department of Mechanical
CVR College of Engineering
Hyderabad, India

Rajesh Kamalkishor Mantri
Department of Mechanical Engineering
Government Polytechnic
Maharashtra, India

Upendra Maurya
Department of Mechanical Engineering
Pimpri Chinchwad College of Engineering
Pune, India

Shalini Mohan
Biomaterials and Product Development Laboratory
Department of Biotechnology
Kalasalingam Academy of Research and Education
Krishnankoil, India

Ashokkumar Mohankumar
Centre for Sustainable Materials and Surface Metamorphosis
Chennai Institute of Technology
Chennai, India

Shuhaib Mushtaq
Department of Mechanical Engineering
Islamic University of Science and Technology
Awantipora, Kashmir, India

Lakshmanan Muthulakshmi
Biomaterials and Product Development Laboratory
Department of Biotechnology
Kalasalingam Academy of Research and Education
Krishnankoil, India

G. J. Naveen
Department of Mechanical Engineering
Visvesvaraya Technological University
Belagavi, Karnataka, India

Abeyram Nithin
Department of Mechanical Engineering
Sreenidhi Institute of Science and Technology
Hyderabad, India

Vignesh Packkirisamy
Centre for Sustainable Materials and Surface Metamorphosis
Chennai Institute of Technology
Chennai, India

Mohammad Arif Parray
Department of Mechanical Engineering
Islamic University of Science and Technology
Awantipora, Kashmir, India

Rajdeep Paul
Department of Mechanical Engineering
Techno College of Engineering
Agartala, Tripura, India

Suyash Pawar
Department of Mechanical Engineering
MVPS's KBT College of Engineering
Nashik, Maharashtra, India

Sharath P. C.
Faculty of Engineering and Technology
Jain (deemed to be University)
Bengaluru, Karnataka, India

Priyabrata Pradhan
Department of Mechanical Engineering
Synergy Institute of Engineering and Technology
Odisha, India

Abhilash Purohit
Department of Mechanical Engineering
Galgotias University
Greater Noida, Uttar Pradesh, India

Nagaraja C. Reddy
Department of Mechanical Engineering
Bangalore Institute of Technology
Bengaluru, Karnataka, India

Valentin Romanovski
Department of Materials Science and Engineering
University of Virginia
Charlottesville, VA, USA

Shilpa B. Sahare
Department of Mechanical Engineering
Yeshwantrao Chavan College of Engineering
Wanadongri, Maharashtra, India

Vinay Saini
Department of Chemical and Petroleum Engineering
University of Calgary
Calgary, Canada

Rajasekaran Saminathan
Department of Mechanical Engineering
College of Engineering and Computer Science
Jazan University
Saudi Arabia

P. Sampathkumaran
Department of Mechanical Engineering
Sambhram Institute of Technology
Bengaluru, Karnataka, India

Amith S. C.
Centre for Sustainable Materials and Surface Metamorphosis
Chennai Institute of Technology
Chennai, India

S. Seetharamu
National Design and Research Forum
Bengaluru, Karnataka, India

Santosh R. Shekokar
Department of Mechanical Engineering
Padmashri Dr. V. B. Kolte College of Engineering
Malkapur, Maharashtra, India

Sehijpal Singh
Department of Mechanical and Production Engineering
Guru Nanak Dev Engineering College
Ludhiana, Punjab, India
Department of Mechanical Engineering
Graphic Era (deemed to be University)
Dehradun, India

Shrishail B. Sollapur
Department of IIAEM
Faculty of Engineering and Technology
JAIN (deemed to be University)
Bengaluru, Karnataka, India

Arunkumar Thirugnanasambandam
Centre for Sustainable Materials and Surface Metamorphosis
Chennai Institute of Technology
Chennai, India

N. R. Thyagaraj
Department of Mechanical Engineering
SJC Institute of Technology
Chickballapur, Karnataka, India

Vaishali
Department of Mechanical Engineering
Government Engineering College
Samastipur, Bihar, India

V. S. S. Venkatesh
School of Mechanical Sciences
Indian Institute of Technology
Bhubaneswar, Odisha, India

Section I

Fundamentals of Composites

1 Fundamentals and History of Composites for Engineering Applications

Lakshmanan Muthulakshmi, Shalini Mohan, and Nellaiah Hariharan

1.1 INTRODUCTION

Composites are defined as materials that are composed of constituents with diverse properties, combined together as layers or phases dispersed within a matrix (Donald, 2010). Composite materials have been used in almost all industries because of their special properties arising from being constituted of several materials that make them versatile (Knight et al., 2003). In the 1200s AD, the Mongol army remained powerful for a long time because of the utilisation of composite technology in the production of the first composite bows using a combination of bone, animal glue, and woody materials, along with birch bark. The accuracy and military superiority of these bows helped Genghis Khan have hegemony over the world till the advent of gunpowder (Randall, 2016). Compression moulding techniques like sheet moulding and bulk moulding facilitated rapid advancements in the automobile age through the manufacture of multiple models of cars and other vehicles (Mathijsen, 2017). The compound annual growth rate (CAGR) of composite materials between 2016 and 2021 was pegged at 7.04%, exemplifying their demand in the global marketplace (Godara et al., 2021). Composites are materials of two or more components with different characteristics that give rise to boundaries with interwoven properties, leading to the creation of newer materials not attainable by conventional means. Natural materials that have evolved over a long period of time may be treated as composite materials (Vasiliev & Morozov, 2013). Their versatile properties have widened the scope of composite materials for use across a number of applications for the development of novel solutions to improve the quality and attractiveness of newer markets (with no exclusive privileges) for defence, aerospace, or production of materials of high value (Todor et al., 2018). The polymer matrices used in composites are generally made up of thermosetting or thermoplastic resins, usually reinforced with fibres, for improving the mechanical properties of the polymer composite as a whole (Park & Seo, 2011). Thermoplastics could be used both in the crystalline and amorphous states, making them applicable in many prominent matrices. Appreciable fibre–matrix bonding could be obtained by the use of thermoset polymers (Schulte

DOI: 10.1201/9781003530909-2

& Von Lacroix, 2000). Polymer matrix composites have improved toughness and strength, wear resistance, stability at elevated temperatures, and strength-to-weight ratio (Singh et al., 2021). Ceramic matrix composites have a matrix made up of ceramic with fibres embedded in them (Marimuthu et al., 2023). Metal matrix composites consist of light metals like titanium, aluminium, and magnesium reinforced with fibres (Kandola & Kandare, 2008). Reinforced polymers are one of the most widely used polymer composites in recent times. Reinforcements tend to improve the mechanical properties of the polymer matrix. Common ways of reinforcing polymer matrices are incorporation of grainy material, fibres, and sheets/layers (Miskolczi, 2013). Composite materials are used in various activities beneficial to humans, such as construction of bridges and other structures; design of automobile bodies; development of materials of high strength and low density for the aeronautics industry; manufacturing of industrial parts and items of housing facilities like storage tanks, bathtubs, shower stalls, and washing sinks; and developing biomaterials (Campbell, 2004; Egbo, 2021; Sharma et al., 2018).

This chapter provides a brief account of the historical evolution of composite materials and their use by the human race, defines composites and polymers, outlines classification of composite materials, and mentions the types of reinforcements used in composites.

1.2 HISTORY OF COMPOSITE MATERIALS

Composites are defined as materials that are made up of constituents with different properties, combined together either as layers or phases that are dispersed within a matrix (Donald, 2010). These materials have been used in almost all industry fields because of their special properties arising from being constituted of several materials that make them versatile (Knight et al., 2003). The earliest use of composite materials dates back to 1500 BC when the Egyptians and Mesopotamians used a mixture of straw and mud as a composite material for building strong and durable structures (Nagavally, 2017). Similarly, weapons like the Japanese-made samurai swords and Damacus gun barrels were made using innovative composite technology since ancient times (Sivasubramanian et al., 2021). In the 1200s AD, the Mongol army remained powerful for a long time because of the utilisation of composite technology in the production of the first composite bows using a combination of bone, animal glue, and woody materials, along with birch bark. The accuracy and military superiority of these bows helped Genghis Khan have hegemony over the world till the advent of gunpowder (Randall, 2016). The lightness and other advantageous properties of the composite material Bakelite, which was produced in the early 1900s, made it a crucial material for military purposes during World War II (Geyer et al., 2017). Along with this, fibre-reinforced polymers (FRP) were developed, with more than 7 million pounds of glass fibre produced in 1945 for several military applications. Around 1947, a complete composite-frame vehicle was developed using FRP (Godara et al., 2021). The chronological usage of composite materials by mankind is presented in Figure 1.1.

Compression moulding techniques like sheet moulding and bulk moulding facilitated rapid advancements in the automobile age through the manufacture of multiple

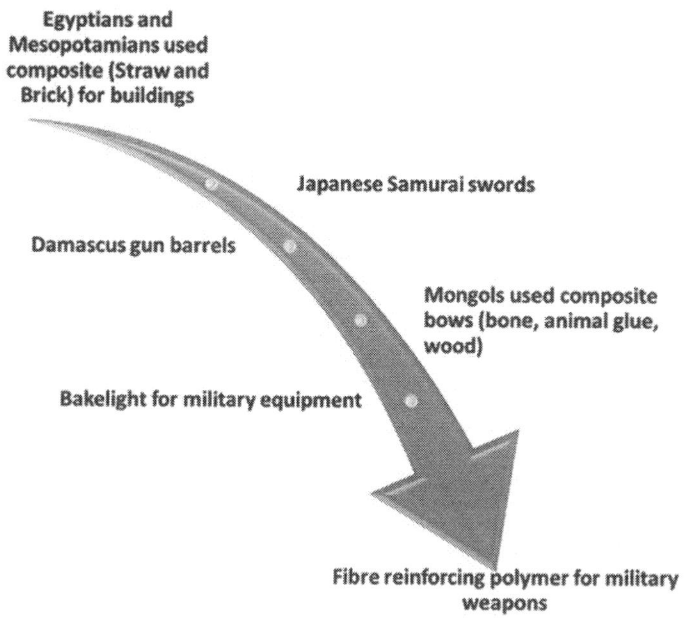

FIGURE 1.1 Historical events in the development of composite materials.

models of cars and other vehicles (Mathijsen, 2017). The marine market utilised humongous amounts of composite materials in the 1960s, which was only surpassed by the automobile industry in the 1970s. The first invented carbon fibre was patented around 1961 and was commercially available for a longer period till then. Composites were used for infrastructure applications as in FRP-reinforced concrete bridges in Europe, America, and Asia in the 1980s (Ngo, 2020; Šerifi et al., 2018). During 1990–2000, the use of composite materials became widespread for everyday items of use like handles and doorways. In the mid-2000s, the Dreamliner Boeing 787 for inflexible and high-energy applications was developed (Giurgiutiu, 2022). The CAGR of composite materials between 2016 and 2021 was predicted to be 7.04%, exemplifying their worldwide demand on the global marketplace (Godara et al., 2021).

1.3 INTRODUCTION TO COMPOSITE MATERIALS

Composites are assemblies of two or more materials on a macroscopic scale to form a novel hybrid material with enhanced properties compared to its separate constituents. The fibre and resin possess advantages like improved strength, resistance to corrosion, wear, fatigue and damage, and thermal resistance as a whole which the individual constituents do not possess (Carey, 2017). The constituents of a composite are shown in Figure 1.2.

In other words, composites are materials that are made up of two or more components with different properties and interwoven boundaries between them for the creation of newer materials that are not attainable by conventional means. Natural

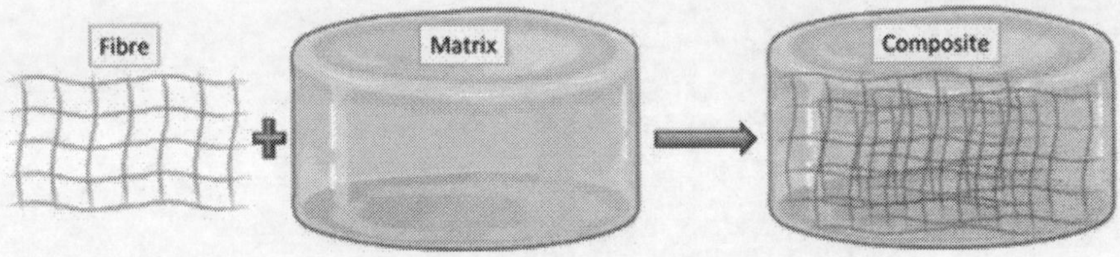

FIGURE 1.2 Constituents of a composite material.

materials that have evolved over a long period of time may be treated as composite materials (Vasiliev & Morozov, 2013). Materials such as steel do not have the properties of composite materials. Some of the most common properties of the composites are a reduced weight-to-strength ratio, lower electrical and thermal conductivity, increased impact strength, and a higher corrosion, fatigue, and creep resistance (Egbo, 2021). Their versatile properties have widened the scope of composite materials for use across a number of applications for the development of novel solutions to improve the quality and attractiveness of newer markets (with no exclusive privileges) for defence, aerospace, or production of materials of high value (Todor et al., 2018). Composites are classified based on the type of reinforcement used for fabricating them, such as short-fibre reinforcement, long-fibre reinforcement, particulate reinforcement, filler reinforcement, and flake reinforcement. Composites are seen to consist of distinct phases, separated by an interface, upon microscopic examination (Park & Seo, 2011). Fibre-reinforced composites are materials composed of polymeric or ceramic matrices in which fibres of high strength and modulus are embedded or bonded to (Derradji et al., 2019). Particulate composites are made up of particles suspended in a matrix by uniform scattering to improve the isotropic properties like toughness (Satankar et al., 2023).

1.4 INTRODUCTION TO POLYMERS

A polymer is a large macromolecule which is made up of smaller units joined together by bonded interactions. Polymers may possess a wide range of desirable properties, making them ubiquitous in almost every commodity used in our daily lives. The word "polymer" is derived from two Greek words "polos" and "meros" meaning "many" and "parts", respectively (Patil et al., 2018). Polymeric materials are made up of either similar or different units joined together. Polymers could also be defined as the following:

- Component made up of structural units joined together by polymerisation reaction.
- Higher molecular weight compounds obtained by the combination of lower molecular weight units.
- Combinations of two or more polymeric components.
- Products of polymerisation reactions.
- Macromolecular compounds (Bakri et al., 2021).

Polymers can be categorised as synthetic and natural based on their origin. Both of these types are employed in a wide range of industries due to their unique properties. They are used either alone or in combination with other additives for improving their properties (Tajeddin & Arabkhedri, 2020). The starting materials for synthesis of polymer are monomeric units which are joined together by polymerisation reactions, leading to combinations of molecules with or without the elimination of smaller groups like amine or water (Chundawat et al., 2019). Organic and inorganic polymers have found applications in the development of elastomers, coating materials, adhesives, blends, fibres, ceramics, plastics, composites, and caulks. Polymeric materials have accelerated the progress of material science through promoting interdisciplinary research for widening their applications (Namazi, 2017). Many materials of biological origin like DNA, tissue, skin, hair, and many more components are made up of biological polymers, also known as biomolecules. These biological polymers play a vital role in the functioning of biochemical pathways, promoting their functioning in a sequestered manner (Bergstrom, 2015). Another classification of polymers includes thermosets, thermoplastics, and elastomers based on their inherent properties. These materials are of lower density, implying their ease of processibility, and resistant to corrosion (Loos, 2014).

1.5 TYPES OF POLYMER MATRIX

The polymer matrices used in composites are generally made up of thermosetting or thermoplastic resins, usually reinforced with fibres for improved mechanical properties of the polymer composite as a whole (Park & Seo, 2011). The polymer matrix is generally composed of polymerising organic compounds that are capable of bonding together for increasing the stiffness, strength, resistance to corrosion, lightness, and feasibility. The four major categories of polymer matrix include semi-crystalline thermoplastics and thermosets and amorphous thermoplastics and thermosets (Parvej et al., 2022). The polymer matrix acts as a continuous phase that is capable of fixing reinforcing agents at uniform positions, resulting in composites with the appropriate physical and chemical properties. These polymer matrices are made up of non-conducting resins and embedded with metallic or conducting fillers (Paul et al., 2022). Thermoplastics are capable of reversible chemical reactions upon changes in temperature whereas thermosets are found to be difficult to reprocess. The recyclability of thermoplastics is attributed to weaker van der Waals forces that aid in their softening when subjected to a temperature higher than their glass-transition temperature (Adeniran et al., 2022). Thermoplastics may be used both in the crystalline and amorphous states, making them suitable for many prominent matrices. Appreciable fibre–matrix bonding could be obtained by the use of thermoset polymers (Schulte & Von Lacroix, 2000). The differences between thermoplastics and thermosets are listed in Table 1.1.

Some of the advantages of thermoplastics over thermosets include the non-requirement of curing reactions, thereby accelerating their fabrication process, clean production, infinite shelf life of the precursor materials used, a wide selection of tougher matrix materials, and their recyclability (Gibson, 2000).

TABLE 1.1
Differences between Thermoplastic and Thermoset Polymers

Sl. No.	Thermoplastic	Thermoset
1.	Weak interactions between polymer chains.	Strong covalent bonds between polymer chains.
2.	Softens on heating.	Remains hard on heating.
3.	Can be reshaped.	Cannot be reshaped.
4.	Low tensile strength.	Higher tensile strength.

Thermoplastic matrices may be reprocessed, reshaped, and require shorter manufacturing cycles; yet, they are not suitable for the synthesis of structures by liquid moulding techniques for the production of complex geometries, which is generally done under extreme operating conditions. Low-viscosity thermoplastic could be used as a replacement for thermosets that have improved mechanical properties (Zhang et al., 2020). Monomers in thermoset polymer matrices have more than two reactive ends that are capable of crosslinking in three dimensions. These monomers tend to form stronger physical bonds that are not reversible upon heat treatment or shaping. Thermosets that are moulded possess higher resistance to elevated working temperatures, chemical attack, and corrosion (Tawiah et al., 2023). Thermoset polymers are found to show a reduction in their elastic modulus and composite strength upon continuous exposure to water absorption (Bakri et al., 2022). Examples of thermoplastic-based composites include polypropylene, polyethylene, polystyrene, polycarbonate, and polyvinyl chloride; thermoset-based composites include polyesters and epoxy- and polyurethane-based composites (Yashas Gowda et al., 2018).

1.6 POLYMER-BASED COMPOSITE MATERIALS

Some composite materials classified on the basis of their polymer matrix (Rajak et al., 2019) are described here, and their classification is shown in Figure 1.3.

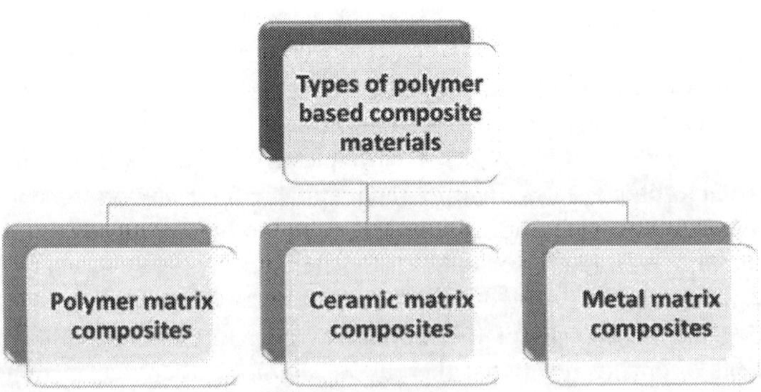

FIGURE 1.3 Types of polymer-based composite materials.

1.6.1 Polymer Matrix Composites

Polymer matrix composites have attracted the attention of the scientific community by virtue of their wide applications in engineering and technology, attributed to their superior properties like higher specific strength and stiffness. These materials are therefore regarded as advanced composites in a variety of industries (Kangishwar et al., 2023). Polymer matrices reinforced with natural fibres have higher interfacial bonds and resistance that help in maintaining their mechanical and chemical identities (Ashori, 2008; Cao & Wu, 2008). Polymer matrix composites have improved toughness and strength, increased wear resistance, enhanced stability at elevated temperatures, and higher strength-to-weight ratios (Singh et al., 2021). A variety of polymeric materials are used for reinforcement in the fabrication of polymer matrix composites for achieving desired properties (Ravichandran et al., 2020).

1.6.2 Ceramic Matrix Composites (CMCs)

Ceramic matrix composites (CMCs) have a matrix of ceramic with fibres embedded in the matrix. Ceramic materials like Al_2O_3, SiC, AlN, and Si_3N_4 are used in the preparation of these composites, which have a higher crack resistance due to the integration of fibres into the matrix (Marimuthu et al., 2023). CMCs are referred to as inverse composites because of their lower failure strain of the matrix than the fibres. The higher toughness of CMCs could be achieved by the use of both oxide and non-oxide fibre coatings in the interface of the composites (Paladugu et al., 2022). Examples of oxide CMCs include alumina and aluminosilicate, while non-oxide CMCs are represented by carbon, silicon-carbide, titanium carbide, and tungsten carbide. The steps involved in the production of CMCs are selection of the source material, treatment of fibres, fabrication, deposition of interphase, impregnation of matrix, drying, thermal processing, shaping and machining, coating, quality control, and final finishing (Shrivastava et al., 2024). Some notable advantages of CMCs are higher fracture toughness, improved resistance to catastrophic failure, lightness, high strength, lower thermal expansion, higher resistance to oxidation, and ability to withstand higher temperatures (Rana & Fangueiro, 2016).

1.6.3 Metal Matrix Composites (MMCs)

Metal matrix composites (MMCs) are materials of light metals like titanium, aluminium, and magnesium reinforced with fibres (Kandola & Kandare, 2008). The market value of MMCs has exploded due to their large-scale applications in transportation hardware, thermal management, industry, aerospace, recreational, and other infrastructural systems (Miracle, 2005). MMCs are lightweight materials commonly used in a number of aerospace entities like spacecraft, aircraft, and helicopters. Superalloys from nickel could be used as matrix material for certain high-temperature applications (Mouritz, 2012). Some important properties of MMCs include lower densities, higher strength-to-weight ratios, improved strength of retention at higher temperature, good creep, and wear and fatigue resistance.

These are potential replacements for materials like cast iron that are used in engines, brakes, connecting rods, pistons, pins, and drive shafts (Kumar et al., 2020; Mallick, 2012).

1.7 TYPES OF POLYMER REINFORCEMENTS

Reinforced polymers are one of the most widely used polymer composites in recent times. Reinforcements tend to improve the mechanical properties of the polymer matrix. Common ways of reinforcing a polymer matrix are incorporation of the reinforcing material as grains, fibres and sheets/layers (Miskolczi, 2013). The types of composite reinforcements are shown in Figure 1.4.

1.7.1 Fibre-Reinforced Composites (FRCs)

Fibre-reinforced composites (FRCs) are made up of polymeric, metallic or ceramic matrices which have fibrous materials of higher modulus and tensile strength embedded in them (Derradji et al., 2018). FRCs are used for many industrial applications due to their synthesis techniques that involve a wide range of characteristics of the fibre and matrix materials. The selection of these starter materials depends on their applications and desired properties (Khandelwal & Rhee, 2020). Common applications of FRCs are in the production of sporting goods, cars, and the wings and fuselages of aeroplanes, which are made of carbon fibre–reinforced composites (Valorosi et al., 2020).

1.7.2 Particle-Reinforced Composites (PRCs)

Particle-reinforced composites (PRCs) are materials whose matrices are reinforced with either spherical or irregularly shaped particles (Mouritz, 2012). PRCs find applications in many wear-resistance applications like road surfacing, where the hardness of the cement is increased by the addition of gravel as a reinforcing filler material

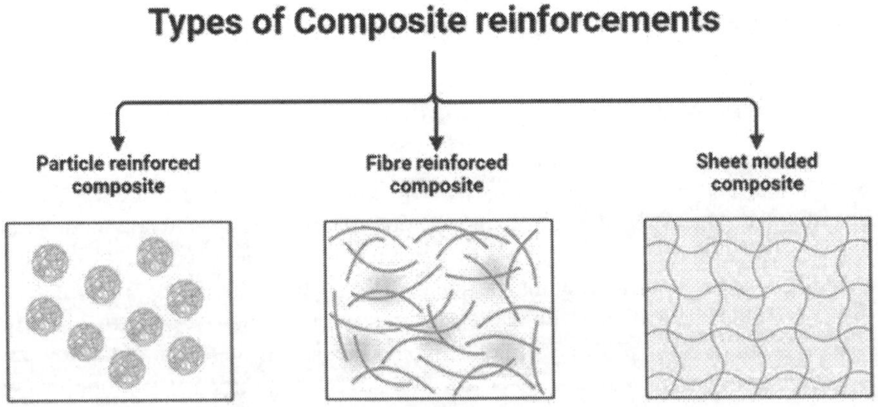

FIGURE 1.4 Types of composite reinforcements.

(Rajak et al., 2019). The major advantage of PRCs over FRCs is their reduced cost, ease in production, and their matrix-forming capacity (Tanzi et al., 2019). Typical reinforcements used in PRCs are nitrides, carbides, and borides, based on their applications and chemical reactivity between the matrix and the particles to be embedded in it (Dieringa et al., 2013).

1.7.3 Sheet Moulding–Reinforced Composites

Sheet moulding composites are materials made up of polymer resins, inert fillers, fibre reinforcements, stabilisers, thickeners, release agents, and catalysts (Ilyas et al., 2021). Sheet moulding is a fibre-reinforced polymer material that is produced by compression moulding. These compounds are produced as flat sheets with thermosets reinforced with chopped glass or hybrid fibres (Gogoi et al., 2023). A sheet moulding composite is made up of a thermoset resin that is reinforced with fibres manufactured in two phases, namely, compounding and maturation (Park & Lee, 2012).

1.8 APPLICATIONS OF COMPOSITES

Composite materials are used in various activities beneficial to humans such as construction of bridges and other structures; design of automobile bodies; development of materials of high strength and low density for the aeronautics industry; manufacturing of industrial parts and items of housing facilities like storage tanks, bathtubs, shower stalls, and washing sinks; and developing biomaterials (Campbell, 2004; Egbo, 2021; Sharma et al., 2018). Some examples of composite materials reported earlier are listed in Table 1.2.

TABLE 1.2
Examples of Composites Reported Previously

Sl. No.	Composite	Type	Properties	Reference
1.	Silicon carbide fibre	Ceramic matrix composite	• Good mechanical properties. • Better fracture toughness and high temperature creep rate.	Brennan and Prewo (1982)
2.	Cobalt boride composite	Metal matrix composite	• Better elastic and hardness properties. • Good modulus at higher temperature.	Rinaldi et al. (2010)
3.	High-density polythene composite	Polymer matrix composite	• Agro-residues used as reinforcements. • Less flexural strength.	Panthapulakkal and Sain (2007)
4.	Boron nitride-reinforced polythene composite	Particle-reinforced composite	• Improved thermal conductivity. • Variation in conductivity based on particle size.	Zhou et al. (2007)

(Continued)

TABLE 1.2 *(Continued)*
Examples of Composites Reported Previously

Sl. No.	Composite	Type	Properties	Reference
5.	Sugarcane fibre/polyester composite	Fibre-reinforced composite	• Improved resistance to friction and wear. • Fibre bonding with polymer good.	El-Tayeb (2008)
6.	Carbon nanotube/ epoxy composites	Polymer-reinforced composite	• Improved mechanical properties. • Surface functionalisation of improved reinforcement of nanotubes.	Sun et al. (2008)
7.	Graphite oxide/ polyurethane composite	Polymer-reinforced composite	• Higher resistance to scratching. • Improved Young's modulus and hardness.	Cai et al. (2009)
8.	Pineapple leaf fibre/ polycarbonate composite	Fibre-reinforced composite	• Higher tensile strength and impact strength. • Improved Young's modulus.	Threepopnatkul et al. (2009)
9.	Ceramic/ carbon nanotube composite	Ceramic matrix composite	• Better electrical and thermal properties. • Less brittle than plain ceramic material.	Cho et al. (2009)
10.	Aluminium metal/graphite composite	Metal matrix composite	• Lower wear and friction coefficient.	Sharma et al. (2018)

ACKNOWLEDGEMENT

Ms. Shalini Mohan, Junior Research Fellow acknowledges the Department of Biotechnology (DBT), Government of India for the financial grant offered with the fellowship ID DBT/2022-23/KARE/2059. Dr. Lakshmanan Muthulakshmi acknowledges the Department of Scientific and Industrial Research for the financial grant (DSIR/TDUPW-1101/3/2021-IRD (SC)/-DSIR and Kalasalingam Academy of Research and Education for seed money grant in the year 2021–2022.

REFERENCES

Adeniran, O., Cong, W., & Aremu, A. (2022). Material design factors in the additive manufacturing of carbon fiber reinforced plastic composites: A state-of-the-art review. Advances in Industrial and Manufacturing Engineering, 5, 100100.

Ashori, A. (2008). Wood–plastic composites as promising green-composites for automotive industries!. Bioresource Technology, 99(11), 4661–4667.

Bakri, M. K. B., Rahman, M. R., Khui, P. L. N., Jayamani, E., & Khan, A. (2021). Use of sustainable polymers to make green composites. In Advances in Sustainable Polymer Composites (pp. 109–129). Woodhead Publishing.

Bakri, M. K. B., Rahman, M. R., & Matin, M. M. (2022). Cellulose reinforcement in thermoset composites. In M. R. Rahman (Ed.), Fundamentals and Recent Advances in Nanocomposites Based on Polymers and Nanocellulose (pp. 127–142). Elsevier.

Bergstrom, J. S. (2015). Mechanics of Solid Polymers: Theory and Computational Modeling. William Andrew.

Brennan, J. J. & Prewo, K. M. (1982). Silicon carbide fibre reinforced glass-ceramic matrix composites exhibiting high strength and toughness. Journal of Materials Science, 17, 2371–2383.

Cai, D., Yusoh, K., & Song, M. (2009). The mechanical properties and morphology of a graphite oxide nanoplatelet/polyurethane composite. Nanotechnology, 20(8), 085712.

Campbell, F. C. (2004). Introduction to composite materials and processes: Unique materials that require unique processes. In Manufacturing Processes for Advanced Composites (pp. 1–37). Elsevier.

Cao, Y. & Wu, Y. Q. (2008). Evaluation of statistical strength of bamboo fiber and mechanical properties of fiber reinforced green composites. Journal of Central South University of Technology, 15(Suppl 1), 564–567.

Carey, J. P. (2017). Introduction to braided composites. In J. P. Carey (Ed.), Handbook of Advances in Braided Composite Materials (pp. 1–21). Woodhead Publishing.

Cho, J., Boccaccini, A. R., & Shaffer, M. S. (2009). Ceramic matrix composites containing carbon nanotubes. Journal of Materials Science, 44, 1934–1951.

Chundawat, N. S., Rathore, B. S., & Meghwal, K. (2019). Phenolic and epoxy-based copolymers and terpolymers. In M. Mozafari & N. P. S. Chauhan (Eds.), Advanced Functional Polymers for Biomedical Applications (pp. 143–165). Elsevier.

Derradji, M., Jun, W., & Wenbin, L. (2018). Fiber-reinforced phthalonitrile composites. In M. Derradji, W. Jun, & L. Wenbin (Eds.), Phthalonitrile Resins and Composites: Properties and Applications (pp. 241–294). William Andrew.

Derradji, M., Zegaoui, A., Medjahed, A., Dayo, A. Q., Wang, J., Arse, Y. B., … & Liu, Y. G. (2019). Cost effective surface-modified basalt fibers-reinforced phthalonitrile composites with improved mechanical properties and advanced nuclear shielding efficiency. Polymer Composites, 40(S1), E912–E919.

Dieringa, H., Kainer, K., & Kaya, A. A. (2013). Applications: Magnesium-based metal matrix composites (MMCs). Fundamentals of Magnesium Alloy Metallurgy, 317, 341.

Donald, M. M. (2010). Atomistic Film Growth and Some Growth-Related Film Properties. 333–398.

Egbo, M. K. (2021). A fundamental review on composite materials and some of their applications in biomedical engineering. Journal of King Saud University-Engineering Sciences, 33(8), 557–568.

El-Tayeb, N. S. M. (2008). A study on the potential of sugarcane fibers/polyester composite for tribological applications. Wear, 265(1–2), 223–235.

Geyer, R., Jambeck, J. R., & Law, K. L. (2017). Production, use, and fate of all plastics ever made. Science Advances, 3(7), e1700782.

Gibson, A. G. (2000). Continuous molding of thermoplastic composites. In Comprehensive Composite Materials (pp. 979–998).Elsevier.

Giurgiutiu, V. (2022). Stress, Vibration, and Wave Analysis in Aerospace Composites: SHM and NDE Applications. Academic Press.

Godara, S. S., Yadav, A., Goswami, B., & Rana, R. S. (2021). Review on history and characterization of polymer composite materials. Materials Today: Proceedings, 44, 2674–2677.

Gogoi, G., Mandal, M., & Maji, T. K. (2023). Animal fiber–based green composites. In T. Altalhi, & Inamuddin (Eds.), Green Sustainable Process for Chemical and Environmental Engineering and Science (pp. 305–346). Elsevier.

Ilyas, R. A., Sapuan, S. M., Nurazzi, N. M., Norrahim, M. N. F., Ibrahim, R., Atikah, M. S. N., Huzaifah, M. R. M., Radzi, A. M., Izwan, S., Azammi, A. N., & Hassan, C. S. (2021). Macro to nanoscale natural fiber composites for automotive components: Research, development, and application. In S.M. Sapuan, & R.A. Ilyas (Eds.), Biocomposite and Synthetic Composites for Automotive Applications (pp. 51–105). Elsevier.

Kandola, B. K. & Kandare, E. (2008). Composites having improved fire resistance. In A. R. Horrocks, & D. Price (Eds.), Advances in Fire Retardant Materials (pp. 398–442). CRC Press.

Kangishwar, S., Radhika, N., Sheik, A. A., Chavali, A., & Hariharan, S. (2023). A comprehensive review on polymer matrix composites: Material selection, fabrication, and application. Polymer Bulletin, 80(1), 47–87.

Khandelwal, S. & Rhee, K. Y. (2020). Recent advances in basalt-fiber-reinforced composites: Tailoring the fiber-matrix interface. Composites Part B: Engineering, 192, 108011.

Knight, M. G., Wrobel, L. C., & Henshall, J. L. (2003). Micromechanical response of fibre-reinforced materials using the boundary element technique. Composite structures, 62(3–4), 341–352.

Kumar, S., Singh, R., & Hashmi, M. S. J. (2020). Metal matrix composite: A methodological review. Advances in Materials and Processing Technologies, 6(1), 13–24.

Loos, M. (2014). Carbon Nanotube Reinforced Composites: CNT Polymer Science and Technology. Elsevier.

Mallick, P. K. (2012). Advanced materials for automotive applications: An overview. In J. Rowe (Ed.), Advanced Materials in Automotive Engineering (pp. 5–27). Woodhead publishing.

Marimuthu, S., Malathi, A. C. J., Raghavan, V., & Grace, A. N. (2023). Processing of ceramics. In P. Cao, Z. Chen, & Z. Xia (Eds.), Advanced Ceramics for Energy Storage, Thermoelectrics and Photonics (pp. 19–39). Elsevier.

Mathijsen, D. (2017). What does the future hold for composites in transportation markets?. Reinforced Plastics, 61(1), 41–46.

Miracle, D. B. (2005). Metal matrix composites–from science to technological significance. Composites Science and Technology, 65(15–16), 2526–2540.

Miskolczi, N. (2013). Polyester resins as a matrix material in advanced fibre-reinforced polymer (FRP) composites. In J. Bai (Ed.), Advanced Fibre-Reinforced Polymer (FRP) Composites for Structural Applications (pp. 44–68). Woodhead Publishing.

Mouritz, A. P. (2012). Fibre–polymer composites for aerospace structures and engines. In A. P. Mouritz (Ed.), Introduction to Aerospace Materials Woodhead Publishing (pp. 338–393).

Mouritz, A. P. (2012). Metal matrix, fibre–metal and ceramic matrix composites for aerospace applications. Introduction to Aerospace Materials, 16, 394–410.

Nagavally, R. R. (2017). Composite materials-history, types, fabrication techniques, advantages, and applications. International Journal of Current Research, 5(9), 82–87.

Namazi, H. (2017). Polymers in our daily life. BioImpacts: BI, 7(2), 73.

Ngo, T. D. (Ed.). (2020). Composite and Nanocomposite Materials: From Knowledge to Industrial Applications. IntechOpen.

Paladugu, S. R. M., Sreekanth, P. R., Sahu, S. K., Naresh, K., Karthick, S. A., Venkateshwaran, N., Ramoni, M., Mensah, R. A., Das, O., & Shanmugam, R. (2022). A comprehensive review of self-healing polymer, metal, and ceramic matrix composites and their modeling aspects for aerospace applications. Materials, 15(23), 8521.

Panthapulakkal, S. & Sain, M. (2007). Agro-residue reinforced high-density polyethylene composites: Fiber characterization and analysis of composite properties. Composites Part A: Applied Science and Manufacturing, 38(6), 1445–1454.

Park, C. H. & Lee, W. I. (2012). Compression molding in polymer matrix composites. In K.-T. Hsiao, & S. G Advani (Eds.), Manufacturing Techniques for Polymer Matrix Composites (PMCs) (pp. 47–94). Woodhead Publishing.

Park, S. J. & Seo, M. K. (2011). Types of composites. In Interface Science and Technology (Vol. 18, pp. 501–629). Elsevier.

Parvej, M. S., Khan, M. I., & Hossain, M. K. (2022). Preparation of nanoparticle-based polymer composites. In Nanoparticle-Based Polymer Composites (pp. 55–94). Woodhead Publishing.

Patil, S. V., Shelake, S. S., & Patil, S. S. (2018). Polymeric materials for targeted delivery of bioactive agents and drugs. In P. Balakrishnan, M.S. Sreekala, & S Thomas (Eds.), Fundamental Biomaterials: Polymers (pp. 249–266). Woodhead Publishing.

Paul, R., Parvathy, P. A., Vijayan, S. P., John, B., & Sahoo, S. K. (2022). Biobased conducting polymer composites for electromagnetic interference shielding and electronics applications. In V. Kumar, K. Sharma, R. Sehgal, & S. Kalia (Eds.), Conjugated Polymers for Next-Generation Applications (pp. 215–237). Elsevier.

Rajak, D. K., Pagar, D. D., Kumar, R., & Pruncu, C. I. (2019). Recent progress of reinforcement materials: A comprehensive overview of composite materials. Journal of Materials Research and Technology, 8(6), 6354–6374.

Rana, S. & Fangueiro, R. (2016). Advanced composites in aerospace engineering. In S. Rana, & R. Fangueiro (Eds.), Advanced Composite Materials for Aerospace Engineering (pp. 1–15). Woodhead Publishing.

Randall, K. C. (2016). Origins and comparative performance of the composite bow (Doctoral dissertation, University of South Africa).

Ravichandran, M., Balasubramanian, M., Chairman, C. A., Pritima, D., Dhinakaran, V., & Stalin, B. (2020, December). Recent developments in polymer matrix composites–a review. In IOP Conference Series: Materials Science and Engineering (Vol. 988, No. 1, p. 012096). IOP Publishing.

Rinaldi, A., Correa-Duarte, M. A., Salgueirino-Maceira, V., Licoccia, S., Traversa, E., Dávila-Ibáñez, A. B., Peralta, P., & Sieradzki, K. (2010). Elastic properties of hard cobalt boride composite nanoparticles. Acta Materialia, 58(19), 6474–6486.

Satankar, R. K., Sharma, N., Katariya, P. V., Kumar, V., Dewangan, H. C., Pal, A., & Panda, S. K. (2023). Computational modelling and analysis of thermoacoustic behaviour of carbon nanotube-reinforced plant fibre epoxy composite – An extensive review. Materials Today Communications, 36, 106717.

Schulte, K. & Von Lacroix, F. R. A. N. K. (2000). High-density polyethylene fiber/polyethylene matrix composites. In Comprehensive Composite Materials (pp. 231–248).

Šerifi, V., Tarić, M., Jevtić, D., Ristovski, A., & Isović, M. Š. (2018). Historical development of composite materials. The Annals of the University of Oradea. Economic Sciences, XXVII (XVII), 1–8.

Sharma, S., Sudhakara, P., Nijjar, S., Saini, S., & Singh, G. (2018). Recent progress of composite materials in various novel engineering applications. Materials Today: Proceedings, 5(14), 28195–28202.

Sharma, V. K., Singh, R. C., & Chaudhary, R. (2018). Wear and friction behaviour of aluminium metal composite reinforced with graphite particles. International Journal of Surface Science and Engineering, 12(5–6), 419–432.

Shrivastava, S., Rajak, D. K., Joshi, T., Singh, D. K., & Mondal, D. P. (2024). Ceramic Matrix Composites: Classifications, Manufacturing, Properties, and Applications. Ceramics, 7(2), 652–679.

Singh, S., Ghorai, M. K., & Kar, K. K. (2021). Fly ash-reinforced epoxy composites. In K. K. Kar (Ed.), Handbook of Fly Ash (pp. 335–356). Elsevier.

Sivasubramanian, P., Mayandi, K., Arumugaprabu, V., Rajini, N., & Rajesh, S. (2021). History of composites and polymers. In Polymer-Based Composites (pp. 1–21). CRC Press.

Sun, L., Warren, G. L., O'reilly, J. Y., Everett, W. N., Lee, S. M., Davis, D., Lagoudas, D., & Sue, H. J. (2008). Mechanical properties of surface-functionalized SWCNT/epoxy composites. Carbon, 46(2), 320–328.

Tajeddin, B. & Arabkhedri, M. (2020). Polymers and food packaging. In Polymer Science and Innovative Applications (pp. 525–543). Elsevier.

Tanzi, M. C., Farè, S., & Candiani, G. (2019). Organization, structure, and properties of materials. Foundations of Biomaterials Engineering, 3, 3–103.

Tawiah, B., Frimpong, C., & Seidu, R. K. (2023). Tribology of hybrid nanofiller/polymer nanocomposites. In S. C. George, J. T. Haponiuk, S. Thomas, R. Reghunath, & S. P. S. (Eds.), Tribology of Polymers, Polymer Composites, and Polymer Nanocomposites (pp. 265–296). Elsevier.

Threepopnatkul, P., Kaerkitcha, N., & Athipongarporn, N. (2009). Effect of surface treatment on performance of pineapple leaf fiber–polycarbonate composites. Composites Part B: Engineering, 40(7), 628–632.

Todor, M. P., Bulei, C., Heput, T., & Kiss, I. (2018, January). Researches on the development of new composite materials complete/partially biodegradable using natural textile fibers of new vegetable origin and those recovered from textile waste. In IOP Conference Series: Materials Science and Engineering (Vol. 294, p. 012021). IOP Publishing.

Valorosi, F., De Meo, E., Blanco-Varela, T., Martorana, B., Veca, A., Pugno, N., Kinloch, I. A., Anagnostopoulos, G., Galiotis, C., Bertocchi, F., & Palermo, V. (2020). Graphene and related materials in hierarchical fiber composites: Production techniques and key industrial benefits. Composites Science and Technology, 185, 107848.

Vasiliev, V. V. & Morozov, E. V. (2013). Advanced Mechanics of Composite Materials and Structural Elements. Newnes.

Yashas Gowda, T. G., Sanjay, M. R., Subrahmanya Bhat, K., Madhu, P., Senthamaraikannan, P., & Yogesha, B. (2018). Polymer matrix-natural fiber composites: An overview. Cogent Engineering, 5(1), 1446667.

Zhang, J., Chevali, V. S., Wang, H., & Wang, C. H. (2020). Current status of carbon fibre and carbon fibre composites recycling. Composites Part B: Engineering, 193, 108053.

Zhou, W., Qi, S., An, Q., Zhao, H., & Liu, N. (2007). Thermal conductivity of boron nitride reinforced polyethylene composites. Materials Research Bulletin, 42(10), 1863–1873.

2 Atomic-Scale Interface of Matrix and Reinforcement for Tribological Applications

Upendra Maurya, Jayesh Subhash Chordiya, and Vinay Saini

2.1 INTRODUCTION

The field of space tribology has demonstrated that solid lubricating materials are the optimal solution for addressing lubrication challenges in extreme working conditions. This preference is due to their unique properties, such as a low evaporation rate, a broad operational temperature range, and resistance to atomic radiation and corrosion. Beyond aerospace applications, these materials find use in various industries, including metal forming, automotive, and microelectronics. As a result, a fundamental understanding of friction and wear at the atomic scale has become increasingly important. Solid lubricants encompass a range of materials, including carbon-based compounds, soft metals, polymers, ceramics, and layered structures like hexagonal boron nitride (*h*-BN), boric acid (H_3BO_3), and transition metal dichalcogenides (TMDs). However, most atomic-scale tribological research has focused on carbon-based materials and TMDs.

Among carbon-based materials, the unique valence orbitals of carbon atoms allow for diverse crystalline and disordered structures. Examples include the sp^3-hybridized three-dimensional diamond, the sp^2-hybridized planar graphite, and amorphous carbon (*a*-C), which combines sp^3 and sp^2 bonding. Diamond's exceptional hardness, wear resistance, and corrosion resistance have made it a focal point of scientific and technological interest, particularly for tribological applications. Diamond-like carbon (DLC), an amorphous carbon form, stands out as a highly promising solid lubricant. Its ability to reduce friction and prevent adhesive transfer of plastics to tools makes it invaluable in the molding and automotive industries. DLC exists in four forms: non-hydrogenated DLC (*a*-C), tetrahedral amorphous carbon (*ta*-C), hydrogenated DLC (*a*-C:H), and tetrahedral amorphous carbon with hydrogen (*ta*-C:H). Graphite, with its lamellar structure, has long been recognized as an effective solid lubricant. Its weak interlayer interactions enable exfoliation into single layers, leading to the discovery of graphene. This monolayer of carbon atoms, arranged in a two-dimensional honeycomb lattice, has garnered significant attention in tribology due to its extraordinary physical properties.

DOI: 10.1201/9781003530909-3

In addition to carbon-based materials, molybdenum disulfide (MoS_2) from the TMD family has proven to be one of the most successful solid lubricants in tribological applications. Its versatility and effectiveness further solidify the importance of exploring atomic-scale interfaces for advanced tribological solutions. Among the transition metal dichalcogenides (TMD) family, apart from the carbon-based materials previously discussed, molybdenum disulfide (MoS_2) stands out for its extensively validated applications in tribological domains ???. Renowned for its efficiency, MoS_2 is predominantly utilized as a micron-scale thin film in solid lubrication systems. However, despite its widespread application, the fundamental mechanisms that dictate friction at the atomic scale remain a subject of intense investigation and debate.

Understanding friction at the atomic level is crucial for elucidating the fundamental principles underpinning macroscopic friction. Such insights are invaluable for advancing fields requiring high-precision machinery, including nanotechnology and microfabrication. Furthermore, the exploration of atomic-scale friction mechanisms is expected to facilitate the discovery of novel solid lubricants with superior performance characteristics. The advent of advanced methodologies such as molecular dynamics simulations, first-principle calculations, and the emergence of atomic force microscopy (AFM) has significantly enhanced scientific understanding of atomic-scale tribological phenomena. These tools have enabled substantial progress in deciphering the intricate interactions between solid lubricating materials and metallic interfaces.

This chapter endeavors to review recent theoretical advancements in atomic-scale tribology, focusing on the frictional properties of solid sheets. The review is structured around key approaches, including first-principle calculations, molecular dynamics simulations, and experimental findings derived from AFM studies, thereby providing a comprehensive understanding of the field's current landscape.

2.2 ATOMIC-SCALE TRIBOLOGICAL STUDIES

The field of atomic-scale tribology has witnessed remarkable advancements over the years, with researchers delving into the fundamental mechanisms governing friction, wear, and lubrication at nanoscale dimensions. The reviewed papers collectively provide a nuanced understanding of the interplay between materials, computational methods, and experimental techniques, illuminating the path toward enhanced tribological performance across diverse applications. This chapter synthesizes key contributions, methodologies, and findings from the selected research papers, highlighting their implications and future directions.

Han et al. (2019) explored the friction and wear behaviors of Ti_2AlN / TiAl composites through molecular dynamics (MD) simulations. Their findings underscore the role of atomic-scale interactions in influencing macroscopic tribological properties. The study revealed that the intricate bonding and structural stability of these composites contribute to their superior wear resistance. The computational approach employed provided a granular understanding of deformation mechanisms, bridging the gap between theoretical predictions and practical applications in aerospace and automotive sectors.

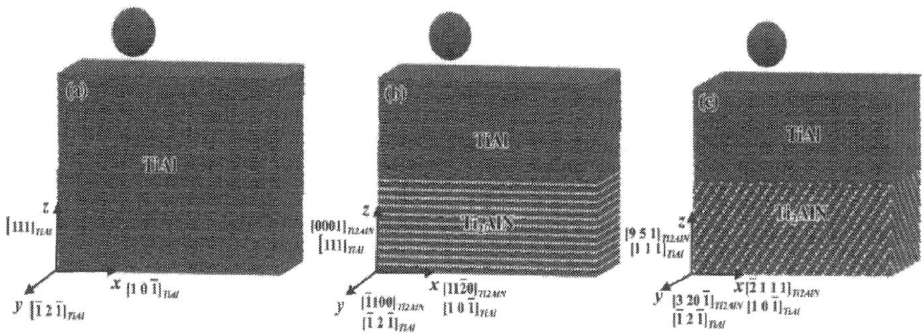

FIGURE 2.1 MD simulation models for friction and wear: (a) Pure TiAl; (b) TiAlN/TiAl coherent interface; (c) TiAlN/TiAl incoherent interface. (Reproduced with permission [Han et al., 2019].)

Figure 2.1 illustrates three MD simulation models designed for the analysis of friction and wear behaviors. These include:

1. A pure TiAl model, where Ti and Al atoms form an ordered face-centered tetragonal structure.
2. A coherent Ti_2AlN/TiAl composite model, featuring a Ti_2AlN (0001)/TiAl (111) interface with excellent atomic alignment, promoting dislocation blocking.
3. An incoherent Ti_2AlN/TiAl composite model with a Ti_2AlN $(10\bar{1}3)$/TiAl (111) interface, characterized by misalignment at the atomic level.

Figure 2.2 showcases the surfaces of scratched TiAl and Ti_2AlN/TiAl composites post-wear analysis, highlighting pile-up formations. It compares:

1. The pure TiAl model, demonstrating the most extensive pile-up due to unimpeded dislocation movement.
2. The coherent Ti_2AlN/TiAl composite, showing moderate pile-up resulting from dislocation blocking at the interface.
3. The incoherent Ti_2AlN/TiAl composite, displaying the least pile-up as the interface facilitates partial dislocation annihilation and transmission.

The review by Qiu et al. (2024) elucidated the structural modeling and mechanical behavior of graphene and carbon nanotube-reinforced metal matrix composites. Employing atomic-scale simulations, the authors demonstrated the synergistic effects of graphene's remarkable strength and carbon nanotubes' flexibility. The paper emphasized the transformative potential of these composites in enhancing mechanical properties such as tensile strength, thermal stability, and tribological performance, positioning them as promising candidates for high-performance engineering applications. Pan et al. (2022) presented a comprehensive analysis of metal

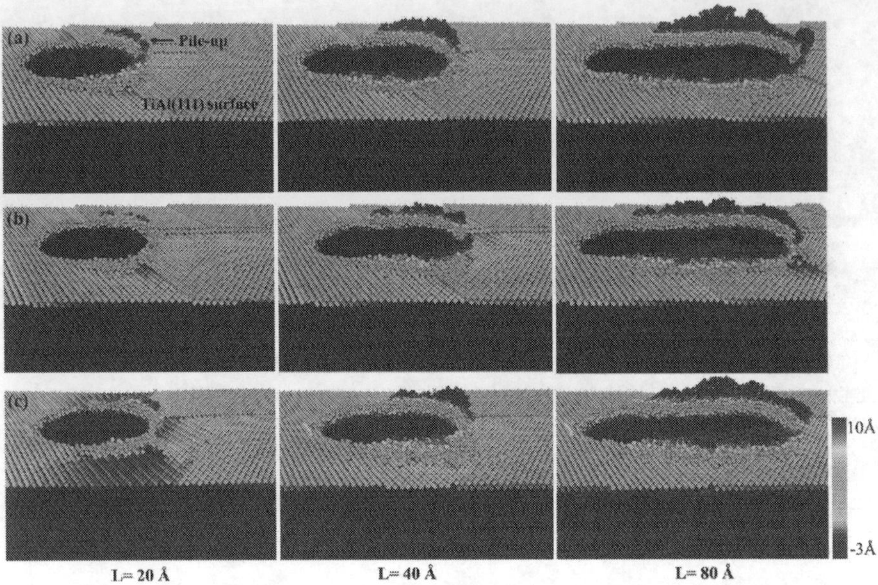

FIGURE 2.2 Surface scratches and atom pile-up: (a) TiAl (b) TiAlN/TiAl with a coherent interface, and (c) TiAlN/TiAl with an incoherent interface. Atom colors represent their height. (Reproduced with permission [Han et al., 2019].)

matrix nanocomposites (MMNCs) in tribology, focusing on manufacturing processes, performance metrics, and underlying mechanisms. Their review synthesized experimental and computational studies, highlighting the pivotal role of nanoparticle reinforcements in reducing wear rates and friction coefficients. The paper also addressed challenges in large-scale fabrication, underscoring the necessity for further exploration of cost-effective synthesis techniques.

The research by Qiu et al. (2021) offered atomic-scale insights into the tribological mechanisms of graphene and fullerene, complemented by machine learning predictions of material properties. The study bridged theoretical modeling and data-driven approaches, unveiling the frictional behavior of these materials under various loading conditions. This innovative integration of computational and machine learning methods demonstrated the potential for predictive modeling in accelerating the discovery of advanced lubricants. Manu et al. (2021) reviewed the tribological properties of two-dimensional (2D) materials and their composites, providing an extensive overview of recent advancements. The authors highlighted the versatility of 2D materials such as graphene, hexagonal boron nitride, and transition metal dichalcogenides in achieving low friction and wear. The paper also identified critical challenges, including scalability and environmental stability, which remain barriers to widespread industrial adoption.

Li et al. (2019) delved into the enhancement of mechanical and tribological properties of polymer composites reinforced with carbon nanotubes and graphene sheets. Through molecular dynamics simulations, the study elucidated the mechanisms by which these reinforcements enhance wear resistance and thermal conductivity. The findings provided valuable insights into the design of lightweight, high-performance polymers for aerospace and automotive industries. Hou et al. (2022) examined intermolecular and surface forces in atomic-scale manufacturing, shedding light on the role of van der Waals forces, hydrogen bonding, and electronic interactions in shaping tribological behavior. The paper demonstrated how these fundamental forces influence adhesion, friction, and wear at the nanoscale, offering a theoretical foundation for the development of precision manufacturing techniques.

Wang et al. (2024) emphasized the multifaceted contributions of carbon materials in tribology, ranging from lubrication to wear mechanisms. The study provided a holistic view of how materials such as graphene, graphite, and carbon nanotubes interact with different substrates to achieve superior tribological performance. By integrating experimental data and theoretical insights, the paper proposed strategies for optimizing carbon-based lubricants. Moghadam et al. (2015) reviewed the mechanical and tribological properties of self-lubricating MMNCs reinforced with carbon nanotubes and graphene. The paper highlighted the synergistic effects of these reinforcements in enhancing load-bearing capacity and reducing friction. The authors underscored the potential of MMNCs in applications requiring high durability and minimal maintenance. Sose et al. (2023) discussed the integration of machine learning in tribology, exploring its applications in material discovery, property prediction, and process optimization. The paper showcased how data-driven approaches could complement traditional methodologies, offering predictive insights into tribological behavior under varying conditions. The review called for further advancements in machine learning algorithms to tackle complex, non-linear phenomena in tribology. Zheng et al. (2024) investigated the atomic-scale wear behavior of two-phase TiAl alloys using vibrational horizontal friction techniques. Their findings revealed the interplay between alloy composition and wear resistance, providing a molecular-level understanding of deformation and adhesion processes. The study offered practical implications for the design of high-performance alloys in extreme environments.

Zhang et al. (2024) explored the robustness of carbon nanotubes in tribology, transitioning from nanoscale to macroscale applications. The study underscored the challenges in scaling up the properties observed at the nanoscale while maintaining their exceptional mechanical and tribological characteristics. The authors proposed innovative strategies for bridging this gap, paving the way for broader adoption of carbon nanotube-based technologies. Doan et al. (2024) provided atomic-scale insights into the mechanical enhancement and frictional properties of amorphous/graphene multilayers. The study utilized MD simulations to explore interfacial dynamics, revealing the potential of these multilayers in reducing wear and enhancing load transfer. The findings hold promise for the development of advanced coatings for high-stress environments. Guo et al. (2024) investigated the multi-scale structure and reinforcement mechanisms of multi-element composite ceramic coatings. Their research demonstrated how the hierarchical organization of these coatings contributes to superior wear resistance and thermal stability. The paper highlighted the potential of ceramic coatings in extending the service life of components in high-temperature applications.

Zhang et al. (2020) examined vulcanization modeling and mechanisms for improving the tribological performance of styrene-butadiene rubber at the atomic scale. The study provided a molecular-level understanding of the cross-linking processes that enhance wear resistance and elasticity. The findings have direct implications for the design of high-performance elastomers in dynamic loading conditions. Teng et al. (2021) conducted an in-depth MD simulation to investigate the enhanced tribological properties of vulcanized natural rubber composites through the incorporation of carbon nanotubes (CNTs). Their study illustrates how CNTs significantly improve wear resistance and reduce the coefficient of friction (COF). By analyzing atomic-scale interactions, the research provides a nuanced understanding of the reinforcement mechanism, laying a foundation for designing high-performance rubber composites for industrial applications.

Wang et al. (2021) utilized MD simulations to explore the thermomechanical and tribological behavior of graphene-reinforced natural rubber nanocomposites. Their findings underscore graphene's effectiveness in enhancing mechanical stability, thermal conductivity, and tribological performance. The study meticulously examines the interfacial dynamics, heat dissipation mechanisms, and frictional behavior, offering valuable insights into the development of advanced rubber-based materials for demanding applications. Liu et al. (2021) present a comprehensive analysis of nitrile rubber composites reinforced with nano-SiO_2, utilizing MD simulations to elucidate the mechanical and tribological enhancements. Their work demonstrates how the dispersion of nano-SiO_2 particles improves wear resistance and modulates interfacial adhesion, leading to a reduction in COF. This study offers profound insights into the structural reinforcement of elastomeric materials. Jiang et al. (2024) introduce a novel approach combining reinforcement learning with graph convolutional neural networks to design TiAl/TiAlN atomic-scale interfaces. This computational framework advances the predictive capability for optimizing mechanical and tribological properties, demonstrating its potential for the efficient design of complex material systems. Xu et al. (2014) investigated the tribological behavior of TiAl matrix composites reinforced with multilayer graphene. Their findings reveal that graphene significantly enhances wear resistance and reduces friction, attributing these improvements to graphene's unique lubrication properties and high structural stability under stress.

Ain et al. (2021) provide a broad overview of polymer nanocomposites, focusing on their mechanical and tribological characteristics. The study emphasizes the role of nanofillers such as carbon nanotubes and graphene in modulating frictional and wear behavior. The conference paper synthesizes recent advancements, bridging the gap between material design and practical applications. Li et al. (2019) delve into the wear and COF variations of vulcanized rubber under dynamic loads and sliding velocities using atomic-scale simulations. Their findings elucidate the molecular mechanisms driving wear resistance and friction modulation, contributing to a deeper understanding of elastomer behavior in variable operational environments. Barry (2009) offers a molecular dynamics perspective on the atomic-scale mechanisms underlying polytetrafluoroethylene (PTFE) tribology. The study identifies key interfacial phenomena, such as adhesion and shear, which govern the material's low-friction properties, making it a staple in tribological applications. Mohammadi et al.

(2020) explore the role of nanofiller geometry in determining the tribological performance of Al-based nanocomposites. Their findings demonstrate how filler shape and orientation influence interfacial strength and wear resistance, offering guidelines for optimizing composite materials.

Hossain et al. (2023) review the advancements in chemically modified CNTs for tribological applications. The study highlights how functionalization improves CNT dispersion and interfacial adhesion, resulting in enhanced frictional and wear properties across diverse material systems. Li et al. (2022) conducted MD simulations to study the reinforcement of polymer composites with graphene. The results show significant improvements in mechanical strength and tribological behavior, attributing these enhancements to the strong interfacial interactions between graphene and the polymer matrix. Xie et al. (2022) investigate the atomic-scale origins of anti-friction and wear-resisting properties in graphene-coated high-entropy alloys. Their findings highlight graphene's ability to reduce adhesive wear and improve load-bearing capacity, demonstrating its efficacy in enhancing alloy performance.

Scharf et al. (2008) examine the atomic-scale structure and composition of tungsten carbide reinforced DLC films. The study reveals how the reinforcement enhances wear resistance and reduces COF, providing insights into designing advanced coating materials for tribological applications. Chhattal et al. (2023) review the tribological potential of MXenes, emphasizing their unique structural and chemical properties. The study outlines current advancements and identifies future research directions, highlighting MXenes' versatility in tribological applications. Wang et al. (2024) provide an extensive review of interface mechanics in carbon nanomaterials-based nanocomposites. Their work focuses on the interfacial bonding mechanisms and their impact on mechanical and tribological performance, offering a roadmap for designing next-generation composites.

The reviewed papers collectively underscore the transformative potential of atomic-scale tribological research in revolutionizing material design and application. While significant progress has been made, challenges such as scalability, environmental stability, and cost-effectiveness remain. Future research should focus on integrating advanced computational techniques, such as machine learning, with experimental approaches to accelerate the discovery of next-generation materials.

Research in tribology and materials science involves a variety of mathematical formulations to describe the mechanical, thermal, and tribological behaviors of materials.

2.3 STRESS-STRAIN RELATIONSHIPS AND ELASTIC PROPERTIES

Many papers utilize classical stress-strain relationships to describe material deformation and evaluate elastic properties.

- **Hooke's Law (Liu et al., 2021; Mohammadi et al., 2020):** The stress-strain relationship in elastic materials is given as:

$$\sigma = E \cdot \varepsilon,$$

where σ is the stress, ε is the strain, and E is Young's modulus. This relationship is fundamental in evaluating the elastic modulus of nanocomposites reinforced by fillers such as nano-SiO_2 or graphene.

- **Reinforcement Mechanism (Xu et al., 2014; Li et al., 2017):** The effective modulus of a composite material can be modeled using the rule of mixtures:

$$E_c = V_f E_f + V_m E_m,$$

where E_c is the composite modulus, V_f and V_m are the volume fractions of the filler and matrix, respectively, and E_f and E_m are the moduli of the filler and matrix.

2.4 FRICTIONAL BEHAVIOR AND COEFFICIENT OF FRICTION (COF)

Several studies model frictional forces and COF to understand tribological performance.

- **Amontons' Law of Friction (Teng et al., 2021; Xie et al., 2022):** The frictional force is proportional to the normal load:

$$F_f = \mu \cdot F_n,$$

where F_f is the frictional force, μ is the coefficient of friction, and F_n is the normal force. This formula is often used to interpret MD simulation results for COF under various conditions.

- **Shear Strength Model (Barry, 2009; Li et al., 2022):** The COF is often related to the interfacial shear strength τ and the contact pressure P:

$$\mu = \frac{\tau}{P}.$$

This formula provides a more atomistic view of friction, especially in the context of PTFE and vulcanized rubber tribology.

2.5 WEAR AND ENERGY DISSIPATION MODELS

Wear resistance and energy dissipation are critical in evaluating tribological performance.

- **Archard's Wear Law (Xu et al., 2014; Li et al., 2022):** The volume of material removed due to wear is given by:

$$V = \frac{k \cdot F_n \cdot d}{H},$$

where V is the wear volume, k is the wear coefficient, F_n is the normal load, d is the sliding distance, and H is the hardness of the material.

- **Energy Dissipation Model (Mohammadi et al., 2020; Teng et al., 2021):** The total energy dissipated during sliding is expressed as:

$$E_d = \int_0^t F_f \cdot v dt,$$

where v is the relative sliding velocity. This formula is used to analyze how nanofillers reduce energy loss during sliding.

2.6 INTERFACIAL MECHANICS AND ADHESION

The interfacial behavior of nanocomposites plays a pivotal role in their performance.

- **Lennard-Jones Potential (Teng et al., 2021; Jiang et al., 2024):** The interaction energy between two atoms at a distance r is modeled as:

$$U(r) = 4\varepsilon\left[\left(\frac{\sigma}{r}\right)^{12} - \left(\frac{\sigma}{r}\right)^{6}\right],$$

where ε is the depth of the potential well and σ is the distance at which the potential energy is zero. This formula is extensively used in MD simulations to model atomic-scale interactions.
- **Adhesion Energy (Xie et al., 2022; Scharf et al., 2008):** The work of adhesion, W_a, between two materials is given by:

$$W_a = \gamma_1 + \gamma_2 - \gamma_{12},$$

where γ_1 and γ_2 are the surface energies of the two materials and γ_{12} is the interfacial energy.

2.7 THERMAL CONDUCTIVITY AND HEAT DISSIPATION

Thermal management in tribological systems is analyzed using various heat transfer models.

- **Fourier's Law of Heat Conduction (Wang et al., 2022):** The heat flux q is proportional to the temperature gradient:

$$q = -k\nabla T,$$

where k is the thermal conductivity and ∇T is the temperature gradient. This law is employed to evaluate the role of graphene in enhancing thermal conductivity in nanocomposites.
- **Specific Heat Capacity (Li et al., 2017):** The specific heat capacity C_p is modeled as:

$$Q = m \cdot C_p \cdot \Delta T,$$

where Q is the heat added, m is the mass, and ΔT is the temperature change. This is particularly relevant in simulating thermal effects during friction.

2.8 MACHINE LEARNING MODELS IN MATERIALS DESIGN

Incorporating machine learning for material optimization is becoming increasingly popular.

- **Reinforcement Learning Optimization (Jiang et al., 2024):** The reinforcement learning model seeks to maximize a reward function R:

$$R = \sum_{t=0}^{T} \gamma^t r_t,$$

where r_t is thet reward at time t, γ is the discount factor, and T is the total time horizon. This model helps to optimize the design of atomic-scale interfaces.

The mathematical formulations used in these research papers collectively provide a comprehensive framework for understanding and optimizing the tribological and mechanical properties. From classical theories like Hooke's law and Archard's wear law to advanced models involving Lennard-Jones potentials and machine learning, these formulas underscore the multidisciplinary nature of tribology and materials science.

Table 2.1 focuses on various materials that are crucial in tribological applications, including rubber composites, nanocomposites, graphene, CNTs, and high-entropy alloys, each of which plays a significant role in friction, wear, and lubrication. Most of the research uses MD simulations to explore atomic-level interactions and predict the properties of these materials, although some studies, such as that of Jiang et al. (2024), combine MD with advanced machine learning techniques like reinforcement learning to further optimize material properties. The properties investigated in these studies typically include tribological performance (such as wear, COF, and wear resistance), thermomechanical properties, and interface mechanics, all of which are critical for understanding the behavior of materials under various operating conditions. The tribological findings highlight the positive impact of nanomaterials such as CNT, graphene, and nano-SiO_2 in improving the wear resistance and friction reduction capabilities of the materials studied; for example, CNTs significantly improve the tribological performance of rubber composites, while graphene coatings on high entropy alloys improve their wear resistance. The wear models employed in these studies primarily rely on Archard's wear law, which is a standard approach for modeling wear in tribological systems, though some studies, including Jiang et al. (2024), incorporate custom models or even integrate machine learning to fine-tune the interfaces and improve wear predictions. Nanomaterials are shown to contribute greatly to the improvement of the mechanical and tribological properties of composite materials, as demonstrated by improvements in the thermomechanical properties

TABLE 2.1
Comparison

Article	Materials Studied	Simulation Method	Properties Investigated	Tribological Findings	Wear Models Used	Role of Nanomaterials	ML/AI	Conclusion
Han et al. (2019)	$Ti_2AlN/TiAl$ composite	MD simulations	Friction, wear behaviors	$Ti_2AlN/TiAl$ composite shows good friction properties	Archard's wear law	CNTs improve tribological properties	No	$Ti_2AlN/TiAl$ composite shows enhanced tribological behavior
Qiu et al. (2021)	Graphene/CNTs in metal matrix composites	MD simulations	Structural modeling, mechanical behaviors	Graphene/CNTs enhance tribological properties	Archard's wear law	Graphene and CNTs enhance properties	No	Graphene/CNTs significantly improve tribological performance in composites
Pan et al. (2022)	Metal matrix nanocomposites	MD simulations	Performance, mechanisms	Nanocomposites improve tribological properties	Archard's wear law	CNTs and graphene enhance tribological properties	No	Nanocomposites improve tribological properties and wear behavior
Qiu et al. (2024)	Graphene/ fullerenes in tribological applications	MD simulations	Tribological mechanisms	Graphene/ fullerenes exhibit anti-friction properties	Archard's wear law	Graphene/ fullerenes enhance wear resistance	Yes (Machine learning, MD simulations)	Graphene/fullerenes enhance tribological mechanisms
Manu et al. (2021)	2D materials and composites	Review	Tribological properties	2D materials improve tribological properties	Archard's wear law	2D materials improve tribological properties	No	2D materials improve tribological properties in composites

(Continued)

TABLE 2.1 *(Continued)*
Comparison

Article	Materials Studied	Simulation Method	Properties Investigated	Tribological Findings	Wear Models Used	Role of Nanomaterials	ML/AI	Conclusion
Li et al. (2019)	Polymer composites reinforced by carbon nanotubes and graphene	MD simulations	Mechanical, tribological properties	CNTs and graphene enhance mechanical and tribological properties	Archard's wear law	CNTs and graphene enhance tribological properties	No	CNTs and graphene enhance mechanical and tribological properties of polymer composites
Hou et al. (2022)	Rubber composites	MD simulations	Tribological behavior	Rubber composites enhance tribological properties	Archard's wear law	Rubber composites enhance tribological properties at the atomic level	No	Rubber composites show improved tribological properties at the atomic level
Wang et al. (2024)	Carbon materials in tribology	MD simulations	Lubrication abilities, wear mechanisms	Carbon materials improve lubrication properties	Archard's wear law	Carbon materials improve wear resistance and lubrication properties	No	Carbon materials improve wear resistance and lubrication properties
Moghadam et al. (2015)	CNTs and graphene in metal matrix composites	MD simulations	Tribological performance, wear	Nanocomposites improve tribological performance	Archard's wear law	CNTs and graphene enhance wear resistance	No	Nanocomposites improve tribological properties in metal matrices

Sose et al. (2023)	2D materials	Review	Tribological properties	2D materials improve tribological properties	Archard's wear law	2D materials improve tribological properties	No	2D materials enhance tribological properties
Zheng et al. (2024)	Two-phase TiAl alloys	MD simulations	Atomic-scale wear behavior	Wear behavior in TiAl alloys influenced by atomic-scale friction mechanisms	Archard's wear law	No nanomaterials are mentioned	No	Wear behavior in TiAl alloys influenced by atomic-scale friction mechanisms
Zhang et al. (2024)	Carbon nanotubes	MD simulations	Tribological performance	Carbon nanotubes improve tribological performance across materials	Archard's wear law	CNTs improve tribological performance	No	Carbon nanotubes improve tribological performance across materials
Doan et al. (2024)	Amorphous/ graphene multilayers	MD simulations	Tribological properties	Graphene multilayers improve mechanical and tribological properties	Archard's wear law	Graphene improves wear performance	No	Graphene multilayers improve mechanical and tribological properties
Guo et al. (2024)	Multi-element composite ceramic coatings	MD simulations	Wear performance, COF	Multi-element coatings improve tribological properties	Archard's wear law	Nanocomposites enhance tribological properties	No	Multi-element ceramic coatings show enhanced tribological properties

(Continued)

TABLE 2.1 *(Continued)*
Comparison

Article	Materials Studied	Simulation Method	Properties Investigated	Tribological Findings	Wear Models Used	Role of Nanomaterials	ML/AI	Conclusion
Li et al. (2017)	Polymer composites reinforced by graphene	MD simulations	Mechanical, tribological properties	Graphene improves wear resistance	Archard's wear law	Graphene improves wear performance	No	Graphene improves wear performance
Xie et al. (2022)	Graphene-coated high-entropy alloys	MD simulations	Anti-friction and wear resistance	Graphene improves wear resistance and reduces friction	Archard's wear law	Graphene improves wear performance	No	Graphene-coated alloys improve wear resistance and friction
Scharf et al. (2008)	Tungsten carbide reinforced diamondlike carbon films	MD simulations	Tribological properties	Tungsten carbide reinforced DLC films improve wear resistance	Archard's wear law	Tungsten carbide enhances wear resistance	No	Tungsten carbide in DLC films enhances wear resistance
Chhattal et al. (2023)	MXenes	Review	Tribological potential	MXenes show potential in tribological applications	Archard's wear law	MXenes improve tribological properties	No	MXenes have great potential for tribological applications
Wang et al. (2021)	Carbon nanomaterials-based composites	MD simulations	Interface mechanics	Carbon nanomaterials improve interface mechanics	Archard's wear law	Carbon nanomaterials improve interface mechanics	No	Carbon nanomaterials enhance interface mechanics

of graphene rubber or enhancement of the wear resistance in vulcanized rubber by CNTs. Although most of the studies do not incorporate machine learning, Jiang et al. (2024) stands out by introducing a machine learning approach that combines reinforcement learning with graph neural networks, offering a novel perspective in the optimization of tribological properties at the atomic scale. The conclusions drawn from these studies underscore the significant potential of nanomaterials in advancing the tribological performance of materials, suggesting that such materials could be utilized in various industrial applications to improve durability and efficiency.

Figure 2.3 is a stacked bar chart that represents the comparison of various parameters. Each bar represents a different paper, and the height of the bar is divided into segments that represent the focus on different parameters within that paper. The y-axis shows the frequency (or count) of the parameters discussed within each paper. A higher value on the y-axis indicates that more parameters are discussed or investigated in the respective study.

Each bar is divided into different color-coded segments, each corresponding to a parameter in the analysis. These parameters are:

- **Materials Studied:** The types of materials focused on in each paper (e.g., rubber composites, graphene, CNTs, etc.). This parameter is most prominent in studies where the primary focus is on investigating various materials and their performance in tribological settings. For example, papers focused on rubber composites, CNTs, and graphene often have large sections of their bars dedicated to this parameter.

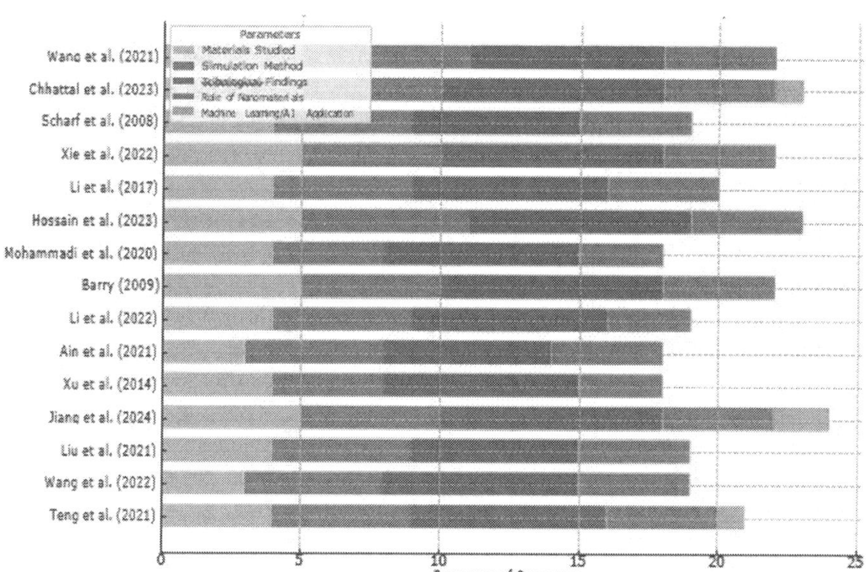

FIGURE 2.3 Comparison.

- **Simulation Method:** The types of simulation techniques employed (e.g., molecular dynamics, machine learning techniques, etc.). Most of the papers use MD simulations, which is reflected by the consistent presence of a significant portion of each bar colored to indicate MD. However, a few papers also incorporate machine learning techniques (like Jiang et al., 2024), which can be seen in the relatively small, isolated sections of the bars.
- **Tribological Findings:** The findings related to the tribological properties of the materials, such as wear resistance and friction behavior. This is a core parameter for nearly every study. Tribological findings dominate the bar chart, showing that each study examines tribological properties such as wear resistance, coefficient of friction, and durability, which are central to tribology.
- **Role of Nanomaterials:** How nanomaterials (such as CNTs, graphene) influence the mechanical and tribological performance. Many studies discuss the impact of nanomaterials (like CNTs, graphene, and nano-SiO_2) on the properties of the materials being studied. The size of the colored segments for this parameter can help identify papers where nanomaterials have a significant role in the study.
- **Machine Learning/AI Application:** Whether machine learning or AI techniques were used in the study. This is the least represented parameter in most studies, with only Jiang et al. (2024) making substantial use of machine learning techniques. This is reflected by a small, isolated segment in their bar compared to other studies.

The materials studied and tribological findings are the most heavily discussed parameters, which is typical for tribology-focused research papers that investigate the material performance and characteristics under different conditions. Simulation methods are also prominently featured, especially MD simulations, which are the standard approach for studying atomic-level behavior in tribological systems. Machine learning/AI application stands out in a few studies, specifically Jiang et al. (2024), which explores a unique combination of reinforcement learning and graph convolutional neural networks. This is reflected as a very small segment but indicates an innovative approach in the research. The role of nanomaterials is a common theme in papers dealing with advanced materials, particularly in composites, which explains the frequency of its occurrence in the graph.

The graph provides a visual overview of the emphasis placed on various parameters across different research papers. From this, it is clear that **Materials Studied** and **Tribological Findings** are the focal points of the research, while **Machine Learning/AI Application** remains a unique aspect, highlighted in only a few studies. This chart helps to understand the broader trends in the tribological research community and how new technologies like nanomaterials and machine learning are being integrated into the field.

Graphene, CNTs, and Ti-based composites are the most common materials studied across the research papers. MD simulations dominate the majority of the studies,

TABLE 2.2
Comaprison

Paper	Wear	Friction	Wear Resistance	Mechanical Strength	Thermomechanical	Nanomaterials Studied
Wang et al. (2018)	Yes	Yes	Yes	No	No	CNT, TiO_2
Li et al. (2017)	Yes	Yes	Yes	Yes	Yes	Graphene
Li et al. (2016)	Yes	Yes	Yes	Yes	No	CNT
Borkar et al. (2016)	Yes	Yes	Yes	Yes	Yes	TiC
Liu et al. (2018)	Yes	Yes	Yes	Yes	Yes	Ti_2AlN, TiAl
Iacobellis et al. (2018)	Yes	Yes	Yes	Yes	Yes	CNT, Al
Chen et al. (2018)	Yes	Yes	Yes	No	Yes	Graphene
Louzguine-Luzgin et al. (2016)	Yes	Yes	Yes	Yes	No	Metallic Glasses
Bolvardi et al. (2015)	Yes	Yes	Yes	Yes	No	Mo_2BC

though a few incorporate machine learning techniques for optimization or prediction. Wear, friction, and wear resistance are the most frequently investigated properties. Mechanical and thermomechanical properties are also studied, but to a lesser extent. Few papers incorporate machine learning, with Jiang et al. (2024) being an example that uses advanced AI techniques like reinforcement learning. Table 2.2 would help researchers or engineers understand the focus and trends in tribological studies involving nanomaterials, aiding in the identification of high-impact materials and methodologies for improving wear, friction, and durability in practical applications.

Table 2.3 compares a set of recent review articles on tribological studies, focusing on various materials and methods used in nano-tribology and their associated findings. It includes articles that investigate a range of materials such as metal matrix nanocomposites, polymer nanocomposites, and 2D nanomaterials like graphene and MXenes. The table highlights the simulation methods employed, including MD, machine learning, and experimental techniques, as well as the specific properties investigated, such as wear, friction, and lubrication. Furthermore, it provides an overview of the key findings of each study, such as the role of nanomaterials like graphene and carbon nanotubes in enhancing tribological performance. The table also indicates whether machine learning techniques are integrated into the studies, showing the evolving trends in tribological research that incorporate advanced computational methods.

TABLE 2.3
Comparison of Recent Review Articles in Tribology

Review Paper	Materials Studied	Simulation Method	Properties Investigated	Key Findings	Focus Area	Nanomaterials Used	ML
Sadeghi et al. (2024)	Nanoscale wear	Analytical methods	Wear mechanisms, nano-tribology	Reviews different measuring techniques and factors affecting wear at the nanoscale	Tribology at nanoscale	N/A	No
Qiu et al. (2021)	Graphene/CNTs reinforced composites	Molecular dynamics	Mechanical behavior, friction, wear	Structural properties of graphene and CNTs in composites	Reinforced metal matrix composites	Graphene, CNTs	No
Pan et al. (2022)	Metal matrix nanocomposites	Analytical and empirical methods	Tribological properties, wear resistance	Highlights the tribological mechanisms of metal matrix composites	Tribological performance of metal matrix composites	N/A	No
Manu et al. (2021)	2D materials	MD, experimental	Tribological performance	Reviews tribological properties of 2D materials	Tribology of 2D materials	Graphene, MoS_2, BN	No
Wang et al. (2024)	Carbon materials	Analytical	Lubrication, wear mechanisms	Focuses on lubrication abilities and wear mechanisms of carbon materials	Tribology of carbon materials	Graphene, CNTs	No
Sose et al. (2023)	Tribological systems	Machine learning, MD	Tribological properties, friction, wear	Reviews machine learning applications in tribology	Tribology with machine learning integration	Various nanomaterials	Yes
Xiao et al. (2022)	Nanoadditives	MD	Wear, friction, tribological performance	Discusses the role of nanoadditives in improving tribological properties	Nanoadditives in tribology	Graphene, CNTs	No

Anirudh et al. (2023)	Thin-layered nanomaterials	MD	Mechanical, tribological properties	Reviews protective coatings for aerospace and defense applications	Protection in strategic applications	Graphene, 2D materials	No
Ain et al. (2021)	Polymer nanocomposites	Experimental	Mechanical, tribological behavior	Studies the mechanical and tribological behavior of polymer composites	Polymer nanocomposites in tribology	CNTs, Graphene	No
Talapatra and Datta (2023)	Graphene-reinforced polymer composites	MD	Tribological, thermal properties	Reviews the properties and potential of graphene in polymer composites	Tribology of polymer nanocomposites	Graphene	No
He et al. (2023)	Tribochemistry	MD	Lubrication, wear	Advances in molecular dynamics for tribochemistry	Tribochemistry and lubrication	N/A	No
Yin et al. (2024)	Polymer tribology	MD	Friction, wear	Reviews polymer tribology and molecular dynamics simulations	Polymer tribology	N/A	No
Chhattal et al. (2023)	MXenes	MD, experimental	Tribological properties	Reviews the potential of MXenes in tribological applications	Tribology of MXenes	MXenes	No
Rosenkranz et al. (2020)	2D nanomaterials	MD	Tribological properties, friction, wear	Reviews the tribological studies of 2D materials beyond graphene	Tribology of 2D materials	Graphene, MoS_2, BN	No

2.9 CONCLUSION

The study of the atomic-scale interface of matrix and reinforcement in tribological applications provides invaluable insights into the behavior of advanced materials under friction, wear, and lubrication conditions. Through the detailed exploration of MD simulations, various studies have demonstrated the significant impact of reinforcement materials, such as CNTs, graphene, and other nanomaterials, on the tribological properties of matrix composites. These reinforcements enhance wear resistance, reduce the COF, and improve the overall mechanical performance of materials at the atomic level.

The research analyzed in this study reveals that the interface between the matrix and reinforcement plays a crucial role in dictating the tribological behavior of the composite materials. The molecular-scale interactions, including van der Waals forces, chemical bonding, and the alignment of nanomaterials, are pivotal in optimizing the performance of the composites. Additionally, the integration of advanced computational techniques, such as MD simulations and machine learning approaches, allows for more accurate predictions of tribological properties and provides deeper insights into the atomic mechanisms that govern material behavior.

Furthermore, the role of nanomaterials in enhancing tribological properties has been emphasized across various studies. For instance, CNTs and graphene have shown remarkable potential in reducing wear rates and improving lubrication properties. The ability of these materials to intercalate with the matrix, form stable interfaces, and withstand high mechanical stresses underpins their effectiveness in tribological applications. The incorporation of nanomaterials also opens up new avenues for developing next-generation materials with tailored tribological properties for diverse industrial applications. As the field advances, the combination of traditional simulation methods and modern machine learning techniques offers a promising pathway for more efficient design and optimization of materials. The interdisciplinary nature of tribology, combining materials science, mechanics, and computational modeling, positions this area of research at the forefront of innovations in engineering materials.

This study highlights the significant strides made in understanding the atomic-scale interfaces of matrix-reinforcement composites for tribological applications. The insights gained through molecular simulations and experimental investigations pave the way for the development of more durable, efficient, and sustainable materials for industries where friction and wear are critical concerns, such as automotive, aerospace, and manufacturing sectors. Future research in this domain holds the potential to further unravel the complexities of atomic-scale interactions and guide the design of materials that can meet the ever-increasing demands of modern engineering.

REFERENCES

Ain, Q. U., Sehgal, R., Wani, M. F., and Singh, M. K. (2021). An Overview of Polymer Nanocomposites: Understanding of Mechanical and Tribological Behavior. In *IOP Conference Series: Materials Science and Engineering*, volume 1189, page 012010. IOP Publishing.

Anirudh, S., Krishnamurthy, S., Kandasubramanian, B., and Kumar, P. (2023). Probing into atomically thin layered nano-materials protective coating for aerospace and strategic defence application – A review. *Journal of Alloys and Compounds*, 172203.

Barry, P. R. (2009). *Elucidation of atomic scale mechanisms for polytetrafluoroethylene tribology using molecular dynamics simulation*. Ph.D. Thesis.

Bolvardi, H., Music, D., and Schneider, J. M. (2015). Atomic scale onset of al adhesion on mo2bc. *Thin Solid Films*, 589:707–711.

Borkar, T., Sosa, J., Hwang, J. Y., Scharf, T. W., Tiley, J., Fraser, H., and Banerjee, R. (2014). Laser-deposited in situ TiC-reinforced nickel matrix composites: 3D microstructure and tribological properties. *JOM*, 66(6):935–942.

Chen, S. J., Yao, X. P., Wang, Q., and Duan, W. H. (2018). Snubbing effect in atomic scale friction of graphene. *Composites Part B: Engineering*, 136:119–125.

Chhattal, M., Rosenkranz, A., Zaki, S., Ren, K., Ghaffar, A., Gong, Z., and Grützmacher, P. G. (2023). Unveiling the tribological potential of MXenes-current understanding and future perspectives. *Advances in Colloid and Interface Science*, 136:103021.

Doan, D. Q., Vu, H. C., Nguyen, V. T., Vu, T. Q., Tran, V. T., and Chu, V. T. (2024). An atomic-scale insight into mechanical enhancement and frictional properties of amorphous/graphene multilayers. *Tribology International*, 197:109783.

Guo, L., Zhang, Y., Wang, F., Xin, Z., Wang, G., and Jiang, J. (2024). Multi-scale structure and reinforcement mechanisms of multi-element composite ceramic coatings. *Materials & Design*, 247:113444.

Han, X., Liu, P., Sun, D., and Wang, Q. (2019). An atomic-level understanding of the friction and wear behaviors of Ti_2AlN/TiAl composite via MD simulations. *Tribology International*, 137:340–348.

He, J., Tang, H., and Wang, C. (2023). Advances of molecular dynamics simulation in tribochemistry and lubrication investigations: A review. *Journal of Industrial and Engineering Chemistry*, 126:1–19.

Hossain, A., Manu, R., nd Shibli, S. M. A. (2023). Chemically modified carbon nanotubes for tribology applications. In Chemically Modified Carbon Nanotubes for Commercial Applications wiley (pp. 441–461).

Hou, X., Li, J., Li, Y., and Tian, Y. (2022). Intermolecular and surface forces in atomic-scale manufacturing. *International Journal of Extreme Manufacturing*, 4(2):022002.

Iacobellis, V., Radhi, A., and Behdinan, K. (2018). A bridging cell multiscale modeling of carbon nanotube-reinforced aluminum nanocomposites. *Composite Structures*, 202:406–412.

Jiang, X., Sun, H., Nian, Q., and Zhuang, H. (2024). Combining reinforcement learning with graph convolutional neural networks for efficient design of tial/tialn atomic-scale interfaces. *arXiv preprint arXiv:2410.11178*.

Li, C., Huang, H., Qu, J., Cao, J., Huang, F., and Wang, Y. (2022). Mechanism of wear and COF variation of vulcanized rubber under changing loads and sliding velocities: Interpretation at the atomic scale. *Tribology International*, 170:107505.

Li, Y., Wang, Q., and Wang, S. (2019). A review on enhancement of mechanical and tribological properties of polymer composites reinforced by carbon nanotubes and graphene sheet: Molecular dynamics simulations. *Composites Part B: Engineering*, 160:348–361.

Li, Y., Wang, S., Wang, Q., and Xing, M. (2016). Molecular dynamics simulations of tribology properties of NBR (nitrile-butadiene rubber)/carbon nanotube composites. *Composites Part B: Engineering*, 97:62–67.

Li, Y., Wang, S., and Wang, Q. (2017). A molecular dynamics simulation study on enhancement of mechanical and tribological properties of polymer composites by introduction of graphene. *Carbon*, 111:538–545.

Liu, P., Han, X., Sun, D., and Wang, Q. (2018). Development and application of a ternary Ti-Al-N interatomic potential for Ti_2AlN/TiAl composite. *Journal of Alloys and Compounds*, 745:63–74.

Liu, X., Zhou, X., Kuang, F., Zuo, H., and Huang, J. (2021). Mechanical and tribological properties of nitrile rubber reinforced by nano-SiO_2: Molecular dynamics simulation. *Tribology Letters*, 69:1–11.

Louzguine-Luzgin, D. V., Nguyen, H. K., Nakajima, K., Ketov, S. V., and Trifonov, A. S. (2016). A study of the nanoscale and atomic-scale wear resistance of metallic glasses. *Materials Letters*, 185:54–58.

Manu, B. R., Gupta, A., and Jayatissa, A. H. (2021). Tribological properties of 2D materials and composites—A review of recent advances. *Materials*, 14(7):1630.

Moghadam, A. D., Omrani, E., Menezes, P. L., and Rohatgi, P. K. (2015). Mechanical and tribological properties of self-lubricating metal matrix nanocomposites reinforced by carbon nanotubes (CNTs) and grapheme – A review. *Composites Part B: Engineering*, 77:402–420.

Mohammadi, S., Montazeri, A., and Urbassek, H. M. (2020). Geometrical aspects of nanofillers influence the tribological performance of Al-based nanocomposites. *Wear*, 444:203117.

Pan, S., Jin, K., Wang, T., Zhang, Z., Zheng, L., and Umehara, N. (2022). Metal matrix nanocomposites in tribology: Manufacturing, performance, and mechanisms. *Friction*, 10:1–39.

Qiu, C., Su, Y., Yang, J., Chen, B., Ouyang, Q., and Zhang, D. (2021). Structural modelling and mechanical behaviors of graphene/carbon nanotubes reinforced metal matrix composites via atomic-scale simulations: A review. *Composites Part C: Open Access*, 4:100120.

Qiu, F., Song, H., Feng, W., Yang, Z., Lu, Z., and Hu, X. (2024). Atomic-scale insights into graphene/fullerene tribological mechanisms and machine learning prediction of properties. *Journal of Tribology*, 146(6):062102.

Rosenkranz, A., Liu, Y., Yang, L., and Chen, L. (2020). 2d nano-materials beyond graphene: From synthesis to tribological studies. *Applied Nanoscience*, 10(9):3353–3388.

Sadeghi, B., Cavaliere, P., Shabani, A., Pruncu, C. I., and Lamberti, L. (2024). Nano-scale wear: A critical review on its measuring methods and parameters affecting nanotribology. *Proceedings of the Institution of Mechanical Engineers, Part J: Journal of Engineering Tribology*, 238(2):125–155.

Scharf, T. W., Romanes, M. C., Mahdak, K. C., Hwang, J. Y., Banerjee, R., Evans, R. D., and Doll, G. L. (2008). Atomic-scale structure and composition of tungsten carbide reinforced diamondlike carbon films. *Applied Physics Letters*, 93(15):151909.

Sose, A. T., Joshi, S. Y., Kunche, L. K., Wang, F., and Deshmukh, S. A. (2023). A review of recent advances and applications of machine learning in tribology. *Physical Chemistry Chemical Physics*, 25(6):4408–4443.

Talapatra, A., and Datta, D. (2023). A review of the mechanical, thermal and tribological properties of graphene reinforced polymer nanocomposites: A molecular dynamics simulations methods. *Polymer Bulletin*, 80(3):2299–2328.

Teng, F., Wu, J., Su, B., and Wang, Y. (2021). Enhanced tribological properties of vulcanized natural rubber composites by applications of carbon nanotube: A molecular dynamics study. *Nanomaterials*, 11(9):2464.

Wang, G., Liu, L., and Zhang, Z. (2021). Interface mechanics in carbon nanomaterials-based nanocomposites. *Composites Part A: Applied Science and Manufacturing*, 141:106212.

Wang, H. D., He, P. F., Ma, G. Z., Xu, B. S., Xing, Z. G., Chen, S. Y., and Wang, Y. W. (2018). Tribological behavior of plasma sprayed carbon nanotubes reinforced tio2 coatings. *Journal of the European Ceramic Society*, 38(10):3660–3672.

Wang, R., Zhang, F., Yang, K., Xiao, N., Tang, J., Xiong, Y., and Chen, H. (2024). Important contributions of carbon materials in tribology: From lubrication abilities to wear mechanisms. *Journal of Alloys and Compounds*, 979:173454.

Wang, Z., Su, M., Duan, X., Yao, X., Han, X., Song, J., and Ma, L. (2022). Molecular dynamics simulation of the thermomechanical and tribological properties of graphene-reinforced natural rubber nanocomposites. *Polymers*, 14(23):5056.

Xiao, N., Chen, Y., Lin, H., Zhang, F., and Yang, K. (2022). Multidimensional nanoadditives in tribology. *Applied Materials Today*, 29:101641.

Xie, H., Ma, Z., Zhang, W., Zhao, H., and Ren, L. (2022). Probing the atomic-scale origins of anti-friction and wear-resisting in graphene-coated high-entropy alloys. *Materials & Design*, 223:111178.

Xu, Z., Shi, X., Zhai, W., Yao, J., Song, S., and Zhang, Q. (2014). Preparation and tribological properties of tial matrix composites reinforced by multilayer graphene. *Carbon*, 67:168–177.

Yin, T., Wang, G., Guo, Z., Pan, Y., Song, J., Ding, Q., and Zhao, G. (2024). Molecular dynamics simulation on polymer tribology: A review. *Lubricants*, 12(6):205.

Zhang, F. Z., Liu, X. B., Yang, C. M., Chen, G. D., Meng, Y., Zhou, H. B., and Zhang, S. H. (2024). Insights into robust carbon nanotubes in tribology: From nano to macro. *Materials Today*, 74:203–234.

Zhang, T., Huang, H., Li, W., Chang, X., Cao, J., and Hua, L. (2020). Vulcanization modeling and mechanism for improved tribological performance of styrene-butadiene rubber at the atomic scale. *Tribology Letters*, 68:1–11.

Zheng, M., Lu, Q., Qu, D., Yi, B., Chen, W., and Zhu, Z. (2024). Atomic-scale wear behavior of two-phase TiAl alloys by vibrational horizontal friction. *Materials Today Communications*, 40:110112.

Section II

Tribological Performance of Polymer Matrix Composites

3 Influence of Nano-Sized Filler Reinforcement in Tribological Performance of Polymer Composites

Md. Farid Hossain, Manoj Kumar, Abhijit Bhowmik, Raman Kumar, Sehijpal Singh, and Abdulaziz AlHazaa

3.1 INTRODUCTION

Polymer composites are gaining prominence in engineering due to their versatility, light weight, and superior mechanical properties. These materials are extensively used in sectors like automotive, aerospace, and biomedical industries, where they offer significant performance and efficiency benefits over traditional materials like metals and ceramics. However, their tribological properties—encompassing wear, friction, and lubrication—are crucial for their broader application in scenarios involving sliding or abrasive contact [Siddalingappa et al., 2020]. A promising method to enhance the tribological performance of polymer composites is the addition of nano-sized fillers. These fillers, with their high surface area and outstanding physical properties, have demonstrated the potential to significantly improve the mechanical, thermal, and tribological attributes of polymer matrices [Suresha et al., 2020]. Such enhancements can lead to reduced wear rates, lower friction coefficients, and increased load-bearing capacities, making nano-filled composites ideal for demanding applications [Friedrich et al., 2005]. The performance of these composites is influenced by the type and dispersion of nano-sized fillers like carbon nanotubes, graphene, silica nanoparticles, and alumina within the polymer matrix. These fillers strengthen the polymer at the nanoscale, creating a synergistic effect that boosts material properties without adding substantial weight or reducing flexibility [Richard et al., 2017]. Nevertheless, the mechanisms driving these improvements are complex and depend on factors such as filler type, size, shape, surface modifications, and interactions with the polymer matrix [Maddaiah et al., 2024].

This study examines the impact of nano-sized fillers on the tribological behavior of polymer composites. It focuses on how variations in filler type and concentration affect wear resistance, frictional performance, and durability [Gangwani et al., 2024]. Additionally, it evaluates the influence of environmental factors, including temperature and lubrication, to better understand the real-world performance of these materials. The insights gained aim to support the development of advanced

DOI: 10.1201/9781003530909-5

polymer composites with enhanced tribological properties, expanding their use in industries that demand high-performance materials.

3.2 OVERVIEW OF TRIBOLOGY AND POLYMER COMPOSITES

Tribology, the multidisciplinary study of friction, wear, and lubrication, is fundamental to ensuring the efficiency and reliability of mechanical systems. The interaction of components in relative motion significantly influences their durability, energy consumption, and overall performance. Understanding these interactions is critical for designing materials and systems used in diverse applications, from industrial machinery to biomedical devices [Bahadur & Schwartz, 2009]. A key challenge is to minimize wear and friction while maintaining structural integrity under varying operating conditions, including different loads, speeds, and environments. Polymer composites, consisting of a polymer matrix reinforced with fillers or fibers, have emerged as a compelling solution to these tribological challenges [Chan et al., 2021]. Their benefits include lightweight construction, corrosion resistance, ease of fabrication, and customizable properties, making them a viable alternative to conventional materials like metals and ceramics. By tailoring the composition and structure of the polymer matrix and its reinforcements, these composites can be engineered to exhibit favorable tribological properties [Alagumalai et al., 2023]. The tribological performance of polymer composites is influenced by factors such as the properties of the polymer matrix, the type and concentration of fillers, and the operating environment. In conditions involving sliding or abrasive contact, their effectiveness depends on their ability to resist wear and maintain a low friction coefficient. Reinforcement materials such as fibers, particulates, and nano-sized fillers are particularly effective in enhancing wear resistance by improving load distribution and reducing surface damage [Jurwall et al., 2024]. Nano-sized fillers have gained significant attention due to their exceptional characteristics, including high surface area, outstanding mechanical strength, and excellent thermal stability. These fillers not only improve the mechanical properties of polymer composites but also act as solid lubricants, reducing friction and wear. Their nanoscale dimensions enable them to interact closely with the polymer matrix, forming a reinforced structure that enhances durability and stability under demanding conditions [Singh et al., 2017]. Despite their advantages, challenges remain in achieving uniform dispersion of nanofillers, ensuring strong filler-matrix interfacial bonding, and addressing environmental factors such as temperature and humidity [Rodriguez et al., 2012]. Ongoing research is focused on unraveling the mechanisms underlying the interactions between nanofillers and polymer matrices to optimize their tribological performance for a wide range of applications.

3.3 ROLE OF NANO-SIZED FILLERS

Nano-sized fillers have become a game-changing component in polymer composites, offering substantial improvements in their physical, mechanical, and tribological characteristics. With dimensions typically below 100 nanometers, these fillers exhibit distinctive properties, including a high surface-area-to-volume ratio, exceptional mechanical strength, and outstanding thermal stability. When integrated into polymer matrices, they enable the development of advanced materials with enhanced performance and durability, making them particularly valuable for tribological applications.

3.3.1 Enhancement of Mechanical Properties

Nano-sized fillers strengthen polymer matrices by increasing their stiffness, load-bearing capacity, and overall mechanical resilience [Chandran et al., 2024]. Their high aspect ratio and effective dispersion facilitate efficient stress transfer within the composite, improving its structural integrity. This reinforcement helps to reduce surface deformation during sliding or abrasive interactions, which is essential for minimizing wear.

3.3.2 Reduction of Friction and Wear

Nanofillers play a pivotal role in lowering the coefficient of friction and wear rate in polymer composites. Materials such as graphene, molybdenum disulfide (MoS_2), and carbon nanotubes function as solid lubricants by creating a transfer film on the contact surface. This film reduces direct contact between the polymer and the counterface, thereby decreasing friction and wear [Bhowmik & Biswas, 2022]. Furthermore, the ability of these fillers to dissipate heat and evenly distribute loads enhances the composite's wear resistance, particularly under high-stress conditions.

3.3.3 Enhanced Thermal Stability

Heat generation during tribological processes, particularly under high-speed or heavy-load conditions, can compromise the polymer matrix. Nano-sized fillers improve the thermal stability of polymer composites by boosting thermal conductivity and reducing the risk of localized overheating [Kurdi & Chang, 2018]. This enhancement helps maintain the material's structural integrity and tribological performance over prolonged use.

3.3.4 Improved Load Distribution

Nanofillers enhance the load-bearing capabilities of polymer composites by strengthening the matrix at the microscopic scale. Their presence promotes a more even distribution of stresses throughout the material, minimizing localized stress concentrations that can lead to wear or failure [Bobby & Samad, 2017]. This characteristic is especially valuable in applications subject to dynamic or repetitive loading conditions.

3.3.5 Customization of Composite Properties

Nanofillers offer remarkable versatility in modifying polymer composites by varying their type, size, shape, and surface treatments to meet specific application requirements [Aly et al., 2012]. For instance, spherical nanoparticles like silica enhance hardness, while platelike fillers such as graphene boost thermal and mechanical performance. Surface functionalization further improves the compatibility of nanofillers with the polymer matrix, ensuring uniform dispersion and strong interfacial bonding for optimal performance.

3.4 TYPES OF NANO-SIZED FILLERS

Nano-sized fillers are diverse in their composition, structure, and functionality, allowing tailored enhancements in the properties of polymer composites. The selection of a filler depends on the desired improvements, such as mechanical strength, wear resistance, thermal stability, or frictional properties. Below is an overview of commonly used types of nano-sized fillers:

3.4.1 Carbon-Based Nanofillers

Carbon-based nanofillers are commonly used in polymer composites for their remarkable mechanical, thermal, and electrical properties.

3.4.1.1 Carbon Nanotubes (CNTs)

Carbon nanotubes (CNTs) are cylindrical nanostructures formed by rolling graphene sheets into seamless tubes. They are categorized as single-walled carbon nanotubes (SWCNTs), consisting of a single graphene layer, or multiwalled carbon nanotubes (MWCNTs), which have multiple concentric graphene layers [Li et al., 2024]. Figure 3.1 shows the carbon nanotubes nanofillers. CNTs are known for their exceptional mechanical, electrical, and thermal properties, attributed to their unique structure and strong carbon-carbon bonds.

Their high tensile strength, flexibility, and lightweight characteristics make them ideal for use as reinforcing materials in polymer composites. Furthermore, CNTs exhibit outstanding electrical conductivity, making them valuable in applications such as electronic devices, sensors, and conductive coatings. Their impressive thermal conductivity also makes them highly effective in dissipating heat in various systems. However, challenges like dispersion, alignment, and concerns about potential toxicity remain important factors to address for their broader application and commercialization.

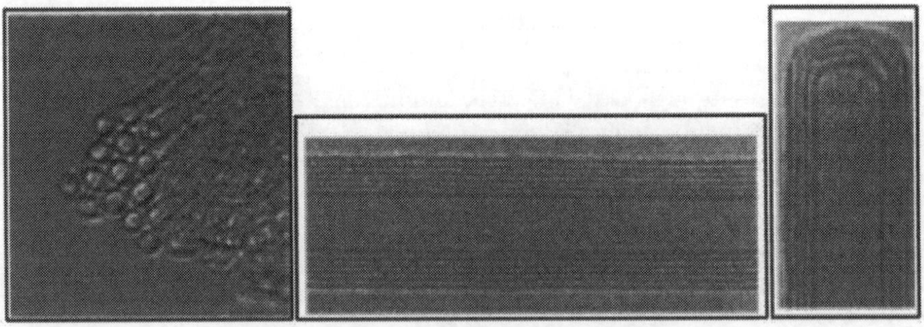

FIGURE 3.1 Multiwalled carbon nanotube [Raghavendra, 2015].

3.4.1.2 Graphene and Graphene Oxide

Graphene, a single layer of carbon atoms arranged in a two-dimensional honeycomb lattice, is renowned for its outstanding mechanical, electrical, and thermal properties. Figure 3.2 shows the graphite flakes nanofillers. They are incredibly strong, lightweight, and highly conductive, making them ideal for applications in flexible electronics, sensors, and energy storage devices such as batteries and supercapacitors. Their excellent thermal conductivity also makes them valuable for heat management applications [Ain et al., 2024].

Graphene oxide (GO), a chemically modified derivative of graphene, contains oxygen functional groups that make it hydrophilic, allowing it to disperse easily in solvents and blend seamlessly into composites. Although GO has lower electrical conductivity compared to graphene, its versatility in functionalization enables its use in drug delivery, sensors, and protective coatings. This versatility positions GO as a promising material for advancements in biomedical engineering and other advanced technologies.

3.4.1.3 Carbon Black

Carbon black (CB), a fine powder composed of elemental carbon, is produced through the partial combustion of hydrocarbons. It is widely used to enhance the mechanical strength, durability, and electrical conductivity of rubber and plastics. The particles, which can be spherical or irregular in shape, vary in size and structure depending

FIGURE 3.2 Graphite flakes [Raghavendra, 2015].

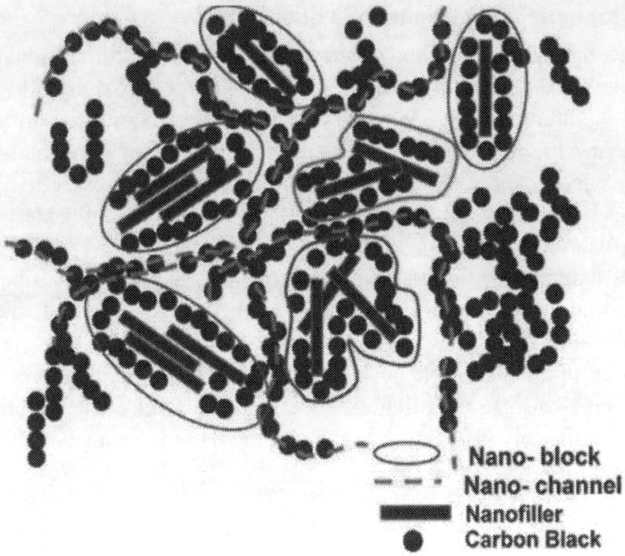

FIGURE 3.3 Microstructural evolution in nanocomposites incorporating nanoclay, nanofibers, and CB [Bhattacharya & Bhowmick, 2010].

on the production method. CB serves as a reinforcing agent in products such as tires, coatings, inks, and sealants, improving their wear resistance and overall strength. Figure 3.3 shows the microstructural development in nanoclay and nanofiber-based nanocomposites in the presence of CB [Bhattacharya & Bhowmick, 2010].

Its high surface area and conductive properties also make it a valuable component in electronic devices and polymer composites [Jagadeesh et al., 2024]. Additionally, it functions as a UV stabilizer, protecting materials from degradation caused by sunlight exposure. While CB plays a crucial role in various industries, concerns about its environmental impact and potential health risks associated with fine particulate matter have spurred ongoing research into safer and more sustainable alternatives.

3.4.2 Metal Oxide Nanofillers

Metal oxide nanoparticles are known for improving thermal stability, hardness, and tribological performance in polymer composites.

3.4.2.1 Silica (SiO_2)

Silica, or silicon dioxide (SiO_2), is a naturally abundant and highly versatile material, commonly found in forms such as quartz and sand. In nanomaterials, silica appears as fine, nanoscale particles with a large surface area, making it an excellent filler for a variety of applications. Figure 3.4 shows the silica nanoparticles. Silica nanoparticles are widely utilized in composite materials to enhance mechanical strength, thermal stability, and optical properties [Yadav et al., 2022].

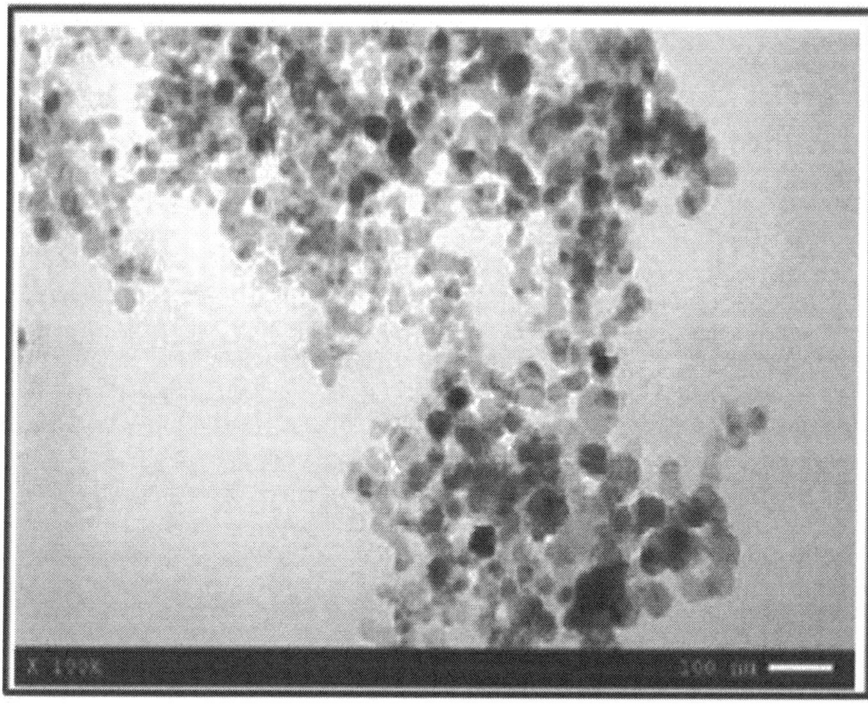

FIGURE 3.4 Silica nanoparticles [Raghavendra, 2015].

They improve the stiffness, toughness, and scratch resistance of polymers, making them particularly useful in coatings, adhesives, and films. Additionally, silica serves as a reinforcing agent in rubber, significantly boosting wear resistance and durability, which is why it is extensively used in tire manufacturing.

3.4.2.2 Alumina (Al_2O_3)

Alumina, also known as aluminum oxide (Al_2O_3), is a prominent metal oxide nanomaterial valued for its remarkable hardness, superior thermal conductivity, and outstanding electrical insulation [Virpe et al., 2020]. At the nanoscale, alumina particles possess an increased surface area and reactivity, making them ideal as reinforcing fillers in diverse composite materials. It is frequently used to enhance the mechanical strength, wear resistance, and thermal stability of polymers, ceramics, and metals. In polymer composites, alumina improves resistance to high temperatures and oxidation, making it well suited for applications in demanding environments.

3.4.2.3 Titanium Dioxide (TiO_2)

Titanium dioxide (TiO_2) is a multifaceted metal oxide nanoparticle known for its outstanding photocatalytic activity, high refractive index, and excellent UV-blocking properties. Its high surface area and reactivity at the nanoscale make it suitable for a broad range of applications. In personal care products like sunscreens, TiO_2 acts as a UV filter, shielding the skin from harmful radiation. It is extensively used in

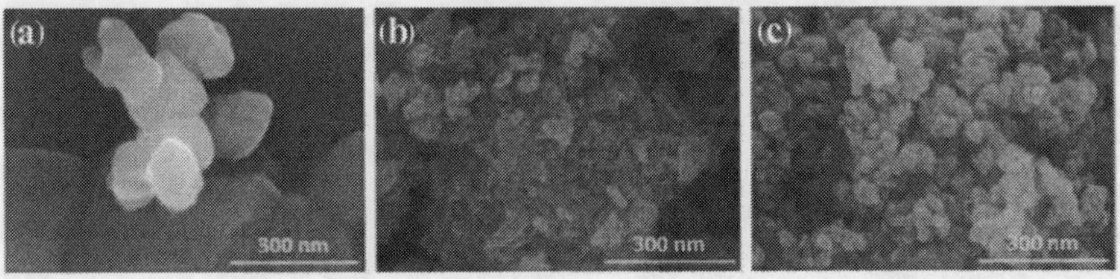

FIGURE 3.5 (a–c) SEM images of various types of TiO_2 as filler particles [Kumar et al., 2022].

environmental applications, such as air and water purification, where it breaks down pollutants under UV light. In Figure 3.5(a), TiO_2 particles show an average size of ~147 nm and are used as microfiller. In contrast, the size dimensions of other TiO_2 particles in Figure 3.5(b and c) are used as nanofillers (i.e., ~34 nm and ~29 nm), respectively [Kumar et al., 2022].

In polymers, coatings, and paints, TiO_2 nanoparticles improve mechanical strength, thermal stability, and light-scattering properties, making them indispensable in automotive finishes and architectural coatings [Dey et al., 2023]. Additionally, TiO_2 plays a significant role in electronics, sensors, batteries, and biomedical implants due to its stability and biocompatibility. However, challenges like dispersion difficulties and potential toxicity have prompted ongoing research to ensure safer and more efficient use.

3.4.3 Layered or 2D Nanofillers

These fillers, made up of stacked atomic layers, are recognized for their unique lubricating and mechanical properties.

3.4.3.1 Molybdenum Disulfide (MoS_2)

Molybdenum disulfide (MoS_2) is a two-dimensional nanomaterial consisting of molybdenum atoms positioned between layers of sulfur. Its unique properties as a semiconductor with a distinct bandgap make it highly suitable for electronic and optoelectronic applications, such as field-effect transistors, photodetectors, and solar cells. Renowned for its exceptional lubricating capabilities, MoS_2 minimizes friction and wear in mechanical systems. When used as a filler in composites, it improves mechanical strength, thermal conductivity, and wear resistance, making it valuable in the automotive, aerospace, and electronics industries [Ronchi et al., 2023]. Its high surface area and ability to undergo functionalization enhance its performance in energy storage devices like supercapacitors and lithium-ion batteries. However, challenges related to dispersion and stability persist, fueling ongoing research to optimize its applications.

3.4.3.2 Hexagonal Boron Nitride (h-BN)

Hexagonal boron nitride (h-BN) is a two-dimensional nanomaterial consisting of boron and nitrogen atoms arranged in a honeycomb-like lattice, resembling the structure of graphene. This unique configuration gives h-BN a combination of exceptional properties, including high thermal conductivity, excellent electrical insulation, and remarkable mechanical strength [Saravanan et al., 2022]. These attributes make h-BN an ideal filler material in composites, enhancing the thermal stability and heat dissipation of polymers, metals, and ceramics. Its superior thermal conductivity is especially advantageous in electronics, where efficient heat management is critical for maintaining device performance and longevity.

3.4.3.3 Montmorillonite (MMT)

Montmorillonite (MMT) is a layered aluminosilicate clay mineral with a two-dimensional sheet structure composed of silicon-oxygen tetrahedra and aluminum-oxygen octahedra, held together by weak van der Waals forces. This structure facilitates easy layer sliding and imparts unique properties such as a high surface area, ion-exchange capacity, and the ability to intercalate molecules. MMT is commonly used as a nanofiller in polymer nanocomposites [Kumar et al., 2022], where it improves tensile strength, rigidity, thermal stability, barrier properties, and flame resistance. Its surface can be modified for applications in packaging, coatings, and automotive materials. Additionally, MMT is applied in environmental engineering for pollutant removal and in drug delivery systems to enable controlled therapeutic release. However, achieving uniform dispersion remains a challenge, requiring advanced processing techniques.

3.5 MECHANISMS OF TRIBOLOGICAL ENHANCEMENT

The addition of nano-sized fillers to polymer matrices greatly improves their tribological properties through a mix of physical, mechanical, and chemical mechanisms. These mechanisms enhance key tribological factors such as wear resistance, friction reduction, and thermal stability, allowing polymer composites to perform effectively in challenging conditions.

3.5.1 Reduction of Surface Friction

Reducing surface friction in tribological systems is essential for enhancing the performance and durability of mechanical components. This is typically achieved by creating a lubricating layer between interacting surfaces, which minimizes direct contact and prevents wear. Lubricants, whether in liquid or solid form, reduce shear stress between surfaces. Solid lubricants such as graphene, molybdenum disulfide (MoS_2), and hexagonal boron nitride (h-BN) form thin protective layers that act as friction barriers. Their nanoscale properties, including low shear strength and the ability for layers to slide, facilitate smooth motion [Sundarakannan et al., 2023]. Additionally, nanoparticles in lubricating oils fill microscopic surface irregularities, creating smoother interfaces and further reducing friction. Well-optimized lubricants improve efficiency, decrease energy consumption, and extend the operational lifespan of mechanical systems.

3.5.2 Improved Load-Bearing Capacity

Enhancing the load-bearing capacity of tribological systems is essential to ensure mechanical components can withstand high pressures and stresses without failure. This is achieved by incorporating robust nanofillers such as carbon nanotubes (CNTs), graphene, or metal oxide nanoparticles (e.g., alumina) into the material matrix [Soni et al., 2024]. These fillers improve mechanical properties by distributing applied stress evenly, minimizing localized deformation. Their exceptional strength, stiffness, and high aspect ratio contribute to effective load support, reducing the risk of surface deformation. Optimal alignment and uniform dispersion of these fillers maximize their reinforcing capabilities, strengthening the material structure. Improved surface hardness and fatigue resistance further enhance performance under heavy loads, resulting in more durable and reliable systems capable of operating efficiently in extreme conditions.

3.5.3 Crack Bridging and Deflection

Crack bridging and deflection are critical mechanisms for improving tribological performance under stress and wear. Under applied stress, materials can develop micro-cracks, which may lead to failure. The addition of nanofillers such as CNTs, graphene, or ceramic nanoparticles helps manage these cracks effectively. Crack bridging occurs when nanofillers span the cracks, holding their edges together and evenly distributing stress. This increases resistance to crack propagation and fatigue [Baig & Samad, 2021]. Crack deflection, on the other hand, redirects cracks along alternative paths due to interactions with the nanofillers, slowing their growth and requiring additional energy to progress. These mechanisms collectively enhance fracture resistance, improve mechanical strength, and extend the life span of materials, making them particularly beneficial in composites and coatings for tribological applications.

3.6 CHALLENGES IN NANOFILLER REINFORCEMENT

Although nano-sized fillers provide notable enhancements in the tribological and mechanical properties of polymer composites, their successful integration into polymer matrices presents several challenges. Overcoming these challenges is essential to fully harness the potential of nanofillers and guarantee the reliable performance of composites in practical applications.

3.6.1 Dispersion and Agglomeration

Achieving proper dispersion of nanofillers within a matrix is a significant challenge in material reinforcement. Nano-fillers, due to their small size and high surface area, often agglomerate into clusters held together by strong van der Waals forces. This clustering hinders their uniform distribution, reducing their effectiveness and limiting the composite's mechanical, thermal, or electrical performance [Shi et al., 2024]. Proper dispersion is essential for maximizing reinforcement, as it promotes optimal interaction between the nanofillers and the matrix. Techniques such as ultrasonic dispersion, high-shear mixing, and surface functionalization are commonly employed to prevent agglomeration and achieve better distribution, ultimately enhancing the material's overall properties.

3.6.2 Interfacial Bonding

Interfacial bonding between nanofillers and the matrix plays a critical role in determining the performance of nano-reinforced composites [Rasana et al., 2023]. Strong bonding ensures effective integration of the nanofillers, allowing stress transfer and enhancing the composite's mechanical, thermal, or electrical properties. However, disparities in the chemical and physical characteristics of the fillers and the matrix can pose bonding challenges. Nanofillers such as carbon nanotubes, graphene, and metal oxides often exhibit high surface energy and hydrophobicity, which can hinder their interaction with hydrophilic or polar matrices. This mismatch weakens adhesion and limits force transfer. To enhance bonding, techniques such as surface functionalization—grafting functional groups, applying coupling agents, or introducing compatibilizers—are employed. Additionally, processing parameters like temperature, pressure, and mixing conditions significantly influence the quality of interfacial bonding.

3.6.3 Processing Challenges

Incorporating nanofillers into composite materials presents various processing challenges that can impact the quality and performance of the final product. A significant issue is achieving uniform dispersion within the matrix. Due to their high surface area and small size, nanofillers often agglomerate into clusters that are challenging to break apart during processing. These agglomerates can lead to uneven distribution, reducing the effectiveness of the nanofillers and diminishing the material's mechanical, thermal, or electrical properties. Specialized dispersion techniques, such as ultrasonic mixing, high-shear mixing, or ball milling, are often required but can be time-intensive and energy-demanding [Xiao et al., 2024]. Another challenge is ensuring the stability of nanofillers during processing, as factors like heat, mechanical stress, or solvent exposure can lead to degradation, agglomeration, or loss of properties. Careful control of processing parameters is essential to maintain the integrity of both the fillers and the matrix.

3.7 APPLICATIONS

The integration of nano-sized fillers into polymer composites has shown remarkable enhancements in tribological performance across diverse industries and applications. Below are examples and case studies illustrating how nanofillers contribute to improved functionality and efficiency.

3.7.1 Automotive Industry

The incorporation of nanofillers into polymer composites offers numerous advantages for the automotive industry, enhancing performance, safety, durability, and weight reduction. Nanofillers such as CNTs, graphene, silica, and alumina significantly improve the mechanical properties of automotive components. For instance, their addition to polymer matrices increases strength, stiffness, and impact resistance in parts like bumpers, body panels, and interiors, enabling the production of lighter, more fuel-efficient vehicles without compromising safety [Billady & Mudradi, 2020]. Nanofillers also enhance electrical conductivity and electromagnetic shielding, with materials like CNTs and graphene used in conductive coatings, sensors, and electrical systems. Furthermore, they improve the scratch resistance, UV stability, and corrosion resistance of paints and coatings, ensuring long-lasting aesthetic appeal and structural integrity.

3.7.2 Aerospace Industry

The incorporation of nanofillers into polymer composites is gaining prominence in the aerospace industry due to their ability to enhance mechanical performance, thermal stability, and weight reduction. Nanofillers such as CNTs, graphene, silica, and aluminum oxide (Al_2O_3) significantly improve the properties of polymer matrices, including tensile strength, impact resistance, and fatigue durability. These materials are essential for aerospace components that endure high stress and extreme environments [Wang et al., 2023]. For example, CNTs and graphene offer outstanding strength-to-weight ratios, enabling the design of lighter structural components, wings, and fuselage panels without compromising strength. Nanofillers also enhance thermal conductivity, facilitating efficient heat dissipation in engine parts and electronic systems. Moreover, they provide increased resistance to wear, corrosion, and radiation, making them highly suitable for use in space vehicles and satellites.

3.7.3 Medical Devices

The incorporation of nano-sized fillers into polymer composites has significantly improved medical devices by enhancing their mechanical strength, biocompatibility, wear resistance, and overall functionality [Math et al., 2024]. Nanofillers like CNTs, graphene, silica, and hydroxyapatite play a key role in boosting the performance, reliability, and safety of these devices. For example, in orthopedic implants, nanofilled composites enhance wear resistance and mechanical durability while maintaining a lightweight structure, increasing the life span of joint replacements. Nano-hydroxyapatite is widely used in bone scaffolds to promote bone cell growth and osteointegration, accelerating the healing process [Perumal et al., 2023]. In drug delivery systems, silica and graphene contribute to improved drug-loading efficiency and controlled release, leading to better therapeutic outcomes and reduced side effects.

3.8 OPPORTUNITIES

The incorporation of nano-sized fillers into polymer composites is set to transform tribological applications across numerous industries. With ongoing advancements in research and technology, new possibilities are arising to further improve the performance, scalability, and adaptability of these materials.

3.8.1 Development of Multifiller Hybrid Composites

Multifiller hybrid composites unlock new possibilities across diverse industries by combining multiple nanofillers with complementary properties, enabling advanced functionalities beyond those of single-filler systems. For instance, blending carbon-based fillers like graphene or carbon nanotubes with ceramic fillers such as silica or alumina improves electrical conductivity, mechanical strength, thermal stability, and wear resistance, making them ideal for high-performance applications. In electronics, these composites enhance electrical conductivity, heat dissipation, and structural durability, making them suitable for flexible circuits, sensors, and energy storage devices [Bokhoeva et al., 2024]. In the automotive and aerospace sectors, they facilitate the development of lightweight, durable materials that boost fuel efficiency, lower emissions, and perform reliably under extreme conditions. In health care, they support the creation of innovative medical devices, including implants with improved biocompatibility and drug delivery systems with controlled release capabilities. Environmental applications also benefit from their versatility, enabling pollutant capture, water purification, and energy-efficient solutions.

3.8.2 Functionalized Nanofillers for Improved Matrix Interaction

Functionalized nanofillers play a crucial role in enhancing composite materials by improving their interaction with the matrix. Surface functionalization involves chemically modifying nanofillers with specific functional groups or coatings to ensure better compatibility and stronger bonding with the matrix. This process facilitates efficient stress transfer, uniform dispersion, and enhanced overall properties. In structural applications [Ogbonna et al., 2024], functionalized fillers such as CNTs and graphene create strong bonds with polymers, increasing mechanical strength, stiffness, and impact resistance. For electronics and energy systems, they enhance electrical and thermal conductivity, making them suitable for sensors, batteries, and thermal management solutions. In the biomedical field, functionalized nanofillers offer biocompatibility, enabling their use in drug delivery systems, tissue scaffolds, and implants. Additionally, they contribute to environmental applications by targeting pollutants, improving filtration efficiency, and supporting water purification efforts.

3.9 CONCLUSION

This chapter delves into the pivotal role of nano-sized fillers in advancing the tribological performance of polymer composites. Widely utilized in industries such as automotive, aerospace, and biomedical engineering, these composites are valued for their lightweight nature, high strength, and adaptability. Tribological performance, encompassing wear resistance, friction reduction, and lubrication, is vital for applications where materials endure sliding or abrasive contact. By integrating nano-sized fillers like graphene, carbon nanotubes, silica, alumina, and molybdenum disulfide, the properties of polymer matrices are significantly enhanced. These fillers, with their exceptional surface area, mechanical strength, and thermal conductivity, improve wear resistance, reduce friction, increase thermal stability, and boost load-bearing capacity. The chapter highlights the benefits of hybrid fillers that synergistically combine different materials for enhanced performance. It also emphasizes the importance of functionalization and advanced dispersion techniques to ensure the uniform distribution of fillers and maximize their impact. Mechanistically, nanofillers contribute to improved tribological properties by forming lubricating films, enhancing load distribution, bridging micro-cracks, and dissipating heat during high-stress operations. While the advantages are clear, challenges persist, including achieving uniform dispersion, ensuring robust filler-matrix bonding, and implementing cost-effective processing methods. Solutions such as surface functionalization, innovative manufacturing techniques, and the development of hybrid composites are proposed to overcome these obstacles. The chapter also presents examples of real-world applications, including wear-resistant automotive parts, lightweight aerospace components, durable medical implants, and efficient energy system components. Looking ahead, emerging trends in the field include the development of self-healing materials, self-lubricating composites, and advanced fabrication methods like 3D printing. These innovations are poised to further expand the utility of nanofilled polymer composites. However, addressing challenges related to scalability, environmental stability, and cost efficiency remains crucial. The integration of nano-sized fillers into polymer composites marks a significant advancement in materials science, enabling tailored solutions for demanding applications. Future progress hinges on continued research into filler-matrix interactions, optimizing composite formulations, and creating materials suitable for extreme environments, ensuring these composites remain indispensable in high-performance applications.

REFERENCES

Ain, Q. U., Wani, M. F., Sehgal, R., & Singh, M. K. (2024). Tribological and mechanical characterization of carbon-nanostructures based PEEK nanocomposites under extreme conditions for advanced bearings: A molecular dynamics study. Tribology International, 196, 109702.

Alagumalai, V., Babu, K., Aseer, R., & Shanmugam, V. (2023). Wear properties of natural fiber based composites – A brief review. Polymers for Advanced Technologies, 34(8), 2766–2779.

Aly, A. A., Zeidan, E. S. B., Alshennawy, A. A., El-Masry, A. A., & Wasel, W. A. (2012). Friction and wear of polymer composites filled by nano-particles: A review. World Journal of Nano Science and Engineering, 2(01), 32.

Bahadur, S., & Schwartz, C. (2009). Mechanical and tribological behavior of polymers filled with inorganic particulate fillers. In S. K. Sinha & B. J. Briscoe (Eds.), Polymer Tribology (pp. 416–448). World Scientific.

Baig, M. M. A., & Samad, M. A. (2021). Epoxy\epoxy composite\epoxy hybrid composite coatings for tribological applications—A review. Polymers, 13(2), 179.

Bhattacharya, M., & Bhowmick, A. K. (2010). Synergy in carbon black-filled natural rubber nanocomposites. Part I: Mechanical, dynamic mechanical properties, and morphology. Journal of Materials Science, 45, 6126–6138

Bhowmik, A., & Biswas, A. (2022). Wear resistivity of Al7075/6wt.% SiC composite by using grey-fuzzy optimization technique. Silicon, 14(8), 3843–3856.

Billady, R. K., & Mudradi, S. (2020, May). Influence of filler incorporation on the mechanical and wear behaviour of synthetic fiber reinforced polymer matrix composites-A review. In AIP Conference Proceedings (Vol. 2236, No. 1). AIP Publishing.

Bobby, S., & Samad, M. A. (2017). Enhancement of tribological performance of epoxy bulk composites and composite coatings using micro/nano fillers: A review. Polymers for Advanced Technologies, 28(6), 633–644.

Bokhoeva, L. A., Rogov, V. E., Bochektueva, E. B., Baldanov, A. B., Shatov, M. S., & Batuev, T. A. (2024). Use of nano-sized components in multilayer composite materials. Nanotekhnologii v Stroitel'stve, 16(2), 180–188.

Chan, J. X., Wong, J. F., Petrů, M., Hassan, A., Nirmal, U., Othman, N., & Ilyas, R. A. (2021). Effect of nanofillers on tribological properties of polymer nanocomposites: A review on recent development. Polymers, 13(17), 2867.

Chandran, A. J., Rangappa, S. M., Suyambulingam, I., & Siengchin, S. (2024). Micro/nano fillers for value-added polymer composites: A comprehensive review. Journal of Vinyl and Additive Technology, 30(5), 1083–1123.

Dey, D., Bhowmik, A., & Biswas, A. (2023). A grey-fuzzy based multi-response optimisation study on the friction and wear characteristics of titanium diboride reinforced aluminium matrix composite. Proceedings of the Institution of Mechanical Engineers, Part B: Journal of Engineering Manufacture, 237(14), 2227–2239.

Friedrich, K., Zhang, Z., & Schlarb, A. K. (2005). Effects of various fillers on the sliding wear of polymer composites. Composites Science and Technology, 65(15–16), 2329–2343.

Gangwani, P., Kovač, J., Emami, N., & Kalin, M. (2024). Effect of multi-scale fillers on the tribological behavior of UHMWPE composites in water-lubricated contacts. Tribology International, 196, 109669.

Jagadeesh, H., Banakar, P., Sampathkumaran, P., Sailaja, R. R. N., & Katiyar, J. K. (2024). Influence of nanographene filler on sliding and abrasive wear behaviour of bi-directional carbon fiber reinforced epoxy composites. Tribology International, 192, 109196.

Jurwall, V., Sharma, A. K., Pandey, A., Goyal, A., Joshi, S. K., Gupta, L. R., & Saxena, K. K. (2024). Physio-mechanical and tribological characterisation of aluminium alloy matrix composite: An effect of nano/micro fillers. Advances in Materials and Processing Technologies, 10(2), 464–472.

Kumar, M. S., Selvan, C. P., Santhanam, K., Kadirvel, A., Chandraprabu, V., & SampathKumar, L. (2022). Effect of nanomaterials on tribological and mechanical properties of polymer nanocomposite materials. Advances in Materials Science and Engineering, 2022(1), 2165855.

Kumar, V., Kumar, A., Chhatra, R. K., & Le, D. J. (2022). Studies on high performance rubber composites by incorporating titanium dioxide particles with different surface area and particle size. Nanofabrication, 7, 104–115)

Kurdi, A., & Chang, L. (2018). Recent advances in high performance polymers—Tribological aspects. Lubricants, 7(1), 2.

Li, P., Sheng, Q., Chen, H., Hou, Y., Bai, Y., Yang, D., Zhu, S., & Yu, L. Mechanical and tribological performances study of PEEK/CNT reinforced EP coatings via molecular dynamics simulation. Chemical Physics Letters 856 (2024): 141608.

Maddaiah, K. C., Kumar, G. V., & Pramod, R. (2024). Studies on the mechanical, strengthening mechanisms and tribological characteristics of AA7150-Al2O3 nano-metal matrix composites. Journal of Composites Science, 8(3), 97.

Math, M. M., Rao, K. R., Gururaja, M. N., Srikantamurthy, J. S., Erannagari, S., Ramesh, S., ... & Prashanth, B. N. (2024). Enhanced tribological properties of nano-TiO2 reinforced polymer composites fabricated via stereolithography. Journal of The Institution of Engineers (India): Series D, 1–13.

Ogbonna, V. E., Popoola, P., & Popoola, O. (2024). Mechanical and tribological properties of nanoceramic reinforced aluminium-based nanocomposites for engineering applications, challenges and recommendations for future improvement: A review. Journal of Composite Materials, 58(28), 2993–3025.

Perumal, K. S., Selvarajan, L., Manikandan, K. P., & Velmurugan, C. (2023). Mechanical, tribological, and surface morphological studies on the effects of hybrid ilmenite and silicon dioxide fillers on glass fibre reinforced epoxy composites. Journal of the Mechanical Behavior of Biomedical Materials, 146, 106095.

Raghavendra, G. (2015). Mechanical and tribological behavior of nanofiller reinforced polymer nanocomposite (Doctoral dissertation). National Institute of Technology, Rourkela.

Rasana, N., Jayanarayanan, K., & Remanan, M. (2023). Enhanced wear resistance in particulate filled polymeric materials. In S. C. George et al. (Eds.), Tribology of Polymers, Polymer Composites, and Polymer Nanocomposites (pp. 119–152). Elsevier.

Richard, S., Selwin Rajadurai, J., & Manikandan, V. (2017). Effects of particle loading and particle size on tribological properties of biochar particulate reinforced polymer composites. Journal of Tribology, 139(1), 012202.

Rodriguez, V., Sukumaran, J., Delgado, Y. P., & De Baets, P. (2012). A review of the tribological performance on polymers filled with inorganic particulate fillers. International Journal of Sustainable Construction and Design, 3(2), 138–143.

Ronchi, R. M., de Lemos, H. G., Nishihora, R. K., Cuppari, M. G. D. V., & Santos, S. F. (2023). Tribology of polymer-based nanocomposites reinforced with 2D materials. Materials Today Communications, 34, 105397.

Saravanan, D., Sollapur, S. B., Anjappa, S. B., Malla, C., Prasad, M. S., & Vignesh, S. (2022). Tribological properties of filler and green filler reinforced polymer composites. Materials Today: Proceedings, 50, 2065–2072.

Shi, S. C., Tsai, X. N., & Rahmadiawan, D. (2024). Enhancing the tribological performance of hydroxypropyl methylcellulose composite coatings through nano-sized metal and oxide additives: A comparative study. Surface and Coatings Technology, 483, 130712.

Siddalingappa, S. K., Pal, B., Haseebuddin, M. R., & Gopalakrishna, K. (2020). Tribological behaviour of carbon fibre polymer composites reinforced with nano-fillers. In H. K. Voruganti et al. (Eds.), Advances in Applied Mechanical Engineering: Select Proceedings of ICAMER 2019 (pp. 791–800). Springer Singapore.

Singh, A. K., Siddhartha, & Yadav, S. (2017). Investigation of sub-micron size cenosphere fillers and filler loading on the mechanical and tribological peculiarity of polyester composites. Polymers for Advanced Technologies, 28(12), 1764–1777.

Soni, A., Das, P. K., & Gupta, S. K. (2024). Experimental investigations on the influence of natural reinforcements on tribological performance of sustainable nanocomposites: A comparative study with polymer technology. Tribology International, 191, 109195.

Sundarakannan, R., Balamurugan, K., Jyothi, Y., Arumugaprabu, V., Sathish, T., Mahmoud, Z., … & Shaik, S. (2023). Importance of Fiber-/Nanofiller-based polymer composites in mechanical and erosion performance: A review. Journal of Nanomaterials, 2023(1), 3528977.

Suresha, B., Harshavardhan, B., Rao, A. M., Koushik, U. R., & Hemanth, R. (2020). Role of nano-fillers on tribological behaviour of ultra high molecular weight polyethylene composites. Materials Today: Proceedings, 27, 2169–2173.

Virpe, K., Deshpande, A., & Kulkarni, A. (2020, December). A review on tribological behavior of polymer composite impregnated with carbon fillers. In AIP Conference Proceedings (Vol. 2311, No. 1). AIP Publishing.

Wang, R., Xiong, Y., Yang, K., Zhang, T., Zhang, F., Xiong, B., … & Tang, J. (2023). Advanced progress on the significant influences of multi-dimensional nanofillers on the tribological performance of coatings. RSC Advances, 13(29), 19981–20022.

Xiao, N., Wu, C., Yang, K., & Tang, J. (2024). Progress of multidimensional nano-additives under dry/liquid wear: A review. Lubricants, 12(10), 332.

Yadav, R., Meena, A., & Patnaik, A. (2022). Biomaterials for dental composite applications: A comprehensive review of physical, chemical, mechanical, thermal, tribological, and biological properties. Polymers for Advanced Technologies, 33(6), 1762–1781.

4 Influence of Fillers on High-Temperature Tribology of Polymer Matrix Composites

Hemalata Jena, Priyabrata Pradhan, and Abhilash Purohit

4.1 INTRODUCTION

Polymer matrix composites have drawn a lot of interest in a variety of industrial applications, because of their superior mechanical, thermal, and tribological qualities (Nayak and Satapathy 2019). These qualities can be further improved by adding fillers to the polymeric matrix, especially when the temperature is high (Purohit and Satapathy 2018). The crucial factor for polymer composites is the effect of filler content on the tribological behaviour at high temperatures (Pradhan and Satapathy 2018). Polyetheretherketone polymers are popular in tribological applications such as gears, bearings, and seals in aerospace equipment (Yang et al. 2014). But virgin polymers have several limitations such as poor thermal stability, low thermal conductivity, and dissipative properties which can be modified by inclusion of different fillers (Qi et al. 2014). Reinforcement of fillers also enhances the tensile strength and elastic modulus of the composite materials. The performance of the composites also depends upon the dispersion and distribution of the fillers in the polymer resin (Madhusudhan and Senthil Kumar 2017). The microscopic and macroscopic structure of the filler material is also necessary to understand the viscoelastic properties of the composite materials (Inbakumar and Ramesh 2018). High-temperature tribological performance of the composites primarily depends upon the choice of filler and their interaction with polymer resin. Polymers are generally adhesive, strain and fatigue dominated, and wear resistance of these polymers can be changed by reinforcement of fillers. These fillers also prevent the formation of micro- and meso-structures during polymer solidification, and hence enhance wear resistance of the resin matrix (Radhika and Raghu 2015). Fillers create projections on the surface of polymer surface and resist the composite surface movement against the sliding body and generate friction on the mating surface. The characteristics of the nanoparticles, including their degree of dispersion, size, aspect ratio, specific surface area, and volume fraction, have a significant impact on the performance of polymer nanocomposites. The impact of the interface between the matrix and the particles also plays a crucial role in the observed effects, as the high surface area can lead to changes in

DOI: 10.1201/9781003530909-6

the macromolecular state around the nanofillers, such as composition gradients and crystallinity (Islam and Beg 2018).

4.1.1 Overview of Polymer Matrix Composites

The improvement in human living standard and production primarily depends upon the material. Human civilisation is passed through the Stone Age, the Bronze Age, and the Iron Age. High-performance composites and plastics are developed during 20th century, which are used in various fields of human society (Ashok, Srinivasa, and Basavaraju 2018). They have replaced the traditional materials due to their improved performance. Materials are divided into several categories based upon the molecules bonding. Materials are classified as 1. Metallic materials 2. Polymer material 3. Ceramic material 4. Non-metallic material 5. Composite material. Composite material is primarily a combination of a matrix phase and a reinforcing phase (Chauhan 2017). The matrix is a continuous phase and polymer; metals and non-metals are used as matrix materials. In a polymer matrix composite, polymer is used as a matrix phase and fibre is reinforced into the matrix phase. Normally, the strength of fibre is much higher than the matrix material. The polymers are classified into thermoplastic and thermosetting according to the bonding present in them. Different types of thermoplastics used in composites are polyethylene (PE), nylon, polystyrene (PS), polypropylene (PP), cellulose acetate, polycarbonate (PC), acrylonitrille-butadiene-styrene (ABS), polyvinyl chloride (PVC), polyether–ether ketone (PEEK), etc. Similarly different thermosetting polymers are phenolic, epoxy, polyester, polyamide, and polyurethane (Hemanth, Sekar, and Suresha 2014). Thermoplastic polymers are preferred for engineering applications due to their superior mechanical properties, better chemical resistance, and low cost.

4.1.1.1 Thermoplastic-Based Composites

Polypropylene is the widely used thermoplastic resin to prepare composites due to its better mechanical properties. These composites showed high thermal deformation and flame resistance. Similarly, the tensile and flexural strength coconut-reinforced polypropylene composites increases with additional fibre content. Due to chemical treatment of coir fibre, the tensile property of the polypropylene improved and the water absorption capacity reduced (Miyazaki and Hamao 1994).

The polyethylene (PE)-based composites have shown better thermal and mechanical properties. The thermal conductivity and thermal diffusivity of maize-reinforced PE composites decreased with reinforcement of filler content. The physical and mechanical properties of bagasse fibre-filled high-density polyethylene (HDPE) and low-density polyethylene (LDPE) composites were also investigated and it was reported that these properties deteriorated after inclusion of 50% filler content (Verheyen et al. 2004).

The polyvinyl chloride (PVC)-based composites are biodegradable, low cost, and have better mechanical strength and fire resistance over traditional materials. It was reported that the impact strength was higher for coconut-filled PVC composites compared to wood-PVC composites.

4.1.1.2 Thermoset-Based Composites

The majority of structural composites use thermoset polymer as a matrix material. Thermoset polymers can be injected into the fibres at low pressure due to their extremely low viscosity. Thermoset polymer composites are prepared by simple processing methods such as hand lay-up technique, injection, and pressure bag moulding. Thermoset polymers used with banana, sisal, coconut, coir, and jute have shown better mechanical properties, and find application in food containers, furniture, and automobiles (Verma 2017).

Investigations were conducted for the mechanical properties of short banana and sisal fibre reinforced polyester composites with different fibre loading. In this study, tensile and flexural demonstrated positive impact whereas impact strength showed negative trend (Joseph et al. 1999).

Epoxy resin is the most widely used thermoset resin due to their better mechanical properties, good corrosion and chemical resistance qualities and thermal stability. The advance composites prepared from epoxy resin are equipped with special qualities, economic viability, and dimensional stability. The mechanical test result of natural fibre epoxy composites revealed that the hybrid composite materials are significantly stronger than the synthetic composites (Sathishkumar et al. 2013).

4.1.2 Importance of Tribology in High-Temperature Applications

Tribology is the study of friction, lubrication, and wear between interacting surfaces, which plays a significant role in high-temperature applications. Because of the increased wear and friction in these conditions, material may fail, leading to costly maintenance. The dependability and effectiveness of the material can be ensured by understating the principle of tribology during high-temperature applications. In industries like aerospace and automotive, proper lubrication and surface treatments can greatly enhance the performance of the material operating in intense temperatures. Material often becomes more reactive to chemicals and rougher surfaces at high temperatures, which accelerates wear. By designing protective barriers of high-temperature-resistant materials, heat can be effectively dissipated which will in turn reduce the friction. Hence due to the following reasons, the study of tribology can help use materials in high-temperature applications (Banagar and Venkateshappa 2018).

4.1.2.1 Friction Control

High temperature can soften the material and will cause expansion of the material in the turbines, engines, and furnaces. The tribological principle can cut down the energy uses due to minimization of frictional losses, which will enhance the efficiency of the machine.

4.1.2.2 Wear Resistance

Because of oxidation, thermal softening materials often fail quicker at elevated temperatures. The study of tribology will help in selecting high-temperature-resistant materials and coatings due to which material can sustain severe circumstances.

4.1.2.3 Lubrication at High Temperatures

To prevent wear and preserve the performance during high-temperature applications, a specially formulated lubricant such as graphite, molybdenum disulphide, or ceramic coating may be used in place of a conventional lubricant.

4.1.2.4 Thermal Stresses and Fatigue

Materials which undergo a repeated temperature cycle become fatigued and fail easily due to low life span. Designing and choosing materials that can withstand thermal stresses will decrease the early degradation of the material.

4.1.2.5 Material Selection and Surface Engineering

High-performance materials can be developed from tribological studies by using a coating and heat treatment which can offer superior temperature resistance with better performance during friction and wear.

4.1.2.6 Energy Efficiency

Reducing friction and wear in high-temperature applications improves energy efficiency by reducing resistance between moving parts in industries like aerospace, automobile, and power generation.

4.1.2.7 Component Longevity

Tribology prevents friction and wear during severe conditions, which extends the component life, leading to cost savings in maintenance and replacement.

4.1.2.8 Oxidation and Corrosion Resistance

High temperatures increase oxidation and corrosion, causing surface degradation and failure. Tribology aids in the selection of materials and coatings that provide superior oxidation resistance while maintaining surface integrity over time (Parikh and Gohil 2017).

4.1.3 Role of Fillers in Polymer Matrix Composites

In polymer matrix composites, fillers have a number of functions and are crucial for enhancing the properties and performance of the finished product (Katiyar and Jena 2023). These fillers are added to the polymer matrix to improve composite in certain ways (Teh et al. 2008). They can be fibres, inorganic or organic particles, or other materials. Common types of fillers used in polymer matrix composites include:

1. Inorganic fillers: Mica, silica, calcium carbonate, and talc.
2. Organic fillers: Carbon black, grapheme, wood flour, and carbon nanotube.
3. Hybrid fillers: Combination of more than one organic and inorganic filler.

The following key aspects demonstrate the importance and diverse functions of the polymer matrix composites which enhance the property, as mentioned below.

i. Mechanical properties: Fillers can greatly increase the stiffness, strength, and impact resistance of the composites. For example, inclusion of carbon nanotubes enhanced the tensile strength and modulus of elasticity of the epoxy resin. This improvement of mechanical properties enables these materials to be suitable for aerospace and automotive industries.

ii. Thermal properties: Certain fillers such as ceramic particles or metal powders, can enhance the thermal conductivity of the composites, making them fit for heat dissipation applications. Additionally, some fillers can reduce the thermal conductivity and coefficient of thermal expansion, and increase the thermal stability at elevated temperatures (Saluja and Gupta 2017).

iii. Electrical properties: Conductive filler like black or metallic particles can increase the electrical conductivity of the composite, making it appropriate in electromagnetic shielding or antistatic protection. On the other hand, for electrical insulation, some fillers can decrease electrical conductivity of the polymer resin.

iv. Cost reduction: Partial replacement of costly polymers can reduce the price of the composite and the product becomes more affordable. By partially replacing the polymer with less costly filler, manufacturing can reduce the overall costs without compromising the performance of the composite.

v. Weight reduction: The density of the composite can be reduced by adding low-weight-density fillers such as foamed particles and hollow glass microspheres. In transportation applications, lighter material enhances fuel efficiency and lower emission is very useful.

vi. Dimensional stability: Fillers can help to reduce warpage and shrinkage that occur during the processing and use of composite. This enhanced dimensional stability is important for applications that need extra tolerances, like the production of automotive and electrical parts.

vii. Surface finish improvement: Composites with a smoother surface and more appealing appearance can be produced with the inclusion of some filler. These composites can be used for decorative products where aesthetic appeal is a major factor.

viii. Flame retardancy: Adding flame retardant filers such as magnesium or aluminium hydroxide can increase the fire-resistant properties of the polymer resin. This property is essential for applications where fire safety is of prime importance.

ix. Wear resistance: Fillers such as silica, ceramics, and metallic powders improve the wear resistance of the polymers, which qualifies these composites for high frictional applications.

Furthermore, attention is also focused on the bio-based and recyclable fillers due to environmental concerns. Natural fillers such as hemp, flask, rice husk, coconut shell powder, and walnut shell powder are being explored as eco-friendly alternatives to synthetic fillers. These eco-friendly fillers are minimally harmful to the environment and also possess unique properties with potential cost advantages. In conclusion, the addition of these fillers in the polymer offers a wide range of property enhancements and functional improvements. As research in material science and nano-technology advances, new types of fillers are used to prepare composites with advanced properties, which will find application in the automobile and aeronautics industries and consumer goods (Asim et al. 2018).

4.2 HIGH-TEMPERATURE TRIBOLOGY OF POLYMERS

The high-temperature tribology of polymers has become an important field of investigation as a result of increasing application requirements of high-performance lightweight material with low friction and high wear-resistant properties at high temperatures in automobile, aerospace, and various industries. Polymers are inherently susceptible to several challenges whenever they are exposed to high operating temperatures, some of the common issues being thermal breakdown, softening, and loss of mechanical strength, all of which have negative impacts on the tribological behaviour of polymers. To mitigate these drawbacks, improvements have been achieved by using several fillers, reinforcements, and additives to improve thermal characteristics, wear, and friction performance of polymers at high-temperature service conditions. Li et al. (2022) and Zhen et al. (2023) have revealed that the use of intermediate materials such as carbon fibres, graphite, and ceramic particles can help to enhance the thermal conductivity and anti-wear properties of polymer composites for high-temperature use. The tribology of polymers at high temperatures is therefore a multifaceted relationship between fully characterizing constituent materials, analysing the behaviour of surfaces in contact, and considering the effects of the environment on the system, making tribology a crucial area for the synthesis of novel materials for the harshest working conditions.

4.2.1 Types of Fillers Used in Polymer Composites

In high-temperature tribology, which is the situation where a polymer composite is subjected to high friction and heat, there are a number of fillers that are used in order to improve the material characteristics. These fillers enhance the wear resistance and thermal stability as well as mechanical characteristics of the matrix. The most common types of fillers used in high-temperature tribology of polymer composites are shown in Table 4.1.

4.2.2 Effect of Fillers on Wear Resistance at High Temperatures

The lubricating and wear-resistant qualities of PTFE composites may be improved by incorporating MoS_2 solid lubricant into the PTFE matrix, particularly at high temperatures (between 200 and 250°C) (Zhen et al. 2023). The anti-wear capabilities of ultra-high molecular weight polyethylene (UHMWPE) composites are severely degraded by high temperatures, according to Gürgen et al. (Gürgen, Sert, and Kuşhan 2021). However, the situation can be improved by adding Si_3N_4 and SiC inorganic particles. It is noted that when UHMWPE fibres are exposed to 90°C, their tensile strength decreased by almost 44%. It is necessary to increase UHMWPE's temperature resistance, since this will enhance their overall performance in all forms at high temperatures (Forster et al. 2015). Shi et al. (Shi et al. 2019) have added nanosized silicon dioxide in UHMWPE composites and wear test has been conducted at about 80°C. It is found that the addition of 0.5 wt.% silicon dioxide can improve the wear resistance of the composite. Similarly, Aliyu et al. (Kayode Aliyu, Abdul Samad, and Al-Qutub 2022a) have mixed 1 wt.% graphene nanoplatelets with UHMWPE and

TABLE 4.1
Fillers Used in High-Temperature Tribology of Polymer Composites

Type of Filler	Filler	Property of Filler	Polymer	Optimum Specific Wear Rate at Temperature	Ref
Solid Lubricants	Graphite	Size: 20 µm	Polyetheretherketone (PEEK) and polyetherimide (PEI)	5 vol%, 150°C	Chang et al. (2007)
	Molybdenum disulfide (MoS_2)	Size: 0.75 µm, Density: 4.8 g/cm^3	High-density polyethylene (HDPE)	4 wt.%	Mohan et al. (2019)
	Polytetrafluoroethylene (PTFE)	Density: 1.44 g/cm^3	Polyimide	15 wt.%, 180°C	Barkoula et al. (2008)
	Hexagonal boron nitride (h-BN)	Size: 120 nm Aramid fibre- particle size: 125 µm Frpahite: 45 µm	Polyimide	10 vol. % AF, 4 vol. % Gr, and 3 vol. % h-BN, 160°C	Wang et al. (2018)
Metallic Fillers	Copper	Size: 60 µm Fibre length: 5.0	Phenolic	10 wt.%	Ho, Lin, and Ju (2005)
Ceramic Fillers	Silicon carbide (SiC)	–	Carbon fibre –40 wt.% with pyrolysis (PIP)	1800°C	Zhang et al. (2020)
	Titanium oxide (TiO_2)	Diameter: 300 nm, density 3.9 g/cm^3	Polyetheretherketone (PEEK)	15 wt.%, 210°C	Kurdi, Kan, and Chang (2019)
	Titanium boride (TiB_2)	4 µm and 47 µm	Phenolic resin	6 wt.%, 200°C	Yuan et al. (2017)
	Multiwalled carbon nanotube (MWCNT)	Length: 3 to 15 µm, Diameter: 9 to 16 nm, and purity 95%	Polyimide (PI)	3 wt.%, 200°C	Roy, Mu, and Shi (2020)
	Graphene	–	Epoxy	4.0 wt.%, 150°C	Zhang et al. (2018)
Organic Fillers	Polyimide	Density: 1.40 g/cm^3, Glass transition Temperature: 239°C	Polyetheretherketone (PEEK)	20 wt.%, 200°C	Li et al. (2023)
Nanofillers	Nano-silica	Size: 50 nm	Polyetheretherketone (PEEK)	4.0 wt.%, 200°C	Li et al. (2023)

the found that a pure UHMWPE coating failed at a considerably lower temperature of 75°C. The graphene nanoplatelets nanocomposite coating demonstrated remarkable performance by passing the wear test without failing, even at temperatures as high as 115°C. Mengjiao et al. (Li et al. 2023) have created PEEK-based composites with h-BN and nano-SiO_2 particles as inorganic fillers and PI as an organic filler. It is found that for pure PEEK, the friction coefficient stays constant at roughly 0.35 and is comparatively steady, as shown in Figure 4.1. The friction coefficient rises from 0.35 to roughly 0.52 when 20% PI (+20PI) is added. It is evident that adding 2% h-BN and 2% nano-SiO_2 (+20P/2B/2Si) might greatly lower the +20PI's friction coefficient. When h-BN and nano-SiO_2 are added to PEEK, the wear rate decreases over time, going from 2% to 4% as h-BN is increased. The wear rate is marginally increased when the nano-SiO_2 content is increased to 4%.

In order to improve the tribological response of Nylon 66 composites, Masood et al. (Masood et al. 2017) have incorporated graphene and PTFE. The friction coefficient and wear rate of nylon mixed with graphene when the graphene concentration was 0.5% of its weight. Katiyar et al. (Katiyar, Sinha, and Kumar 2016) have added various weight percentages (wt.%) of talc and graphite particles in SU-8 epoxy to improve its tribological properties. When compared to pure SU-8 epoxy and other tested composites, the composite with the optimised composition of SU-8 and 15 wt.% graphite and 15 wt.% talc demonstrated improved properties. Li et al. (Li et al. 2023) have incorporated polyimide (PI) as an organic filler and hexagonal boron nitride (h-BN) and nano-SiO_2 particles as the inorganic fillers in PEEK polymer and tested for their tribological property at room temperature and at 200°C. There is a

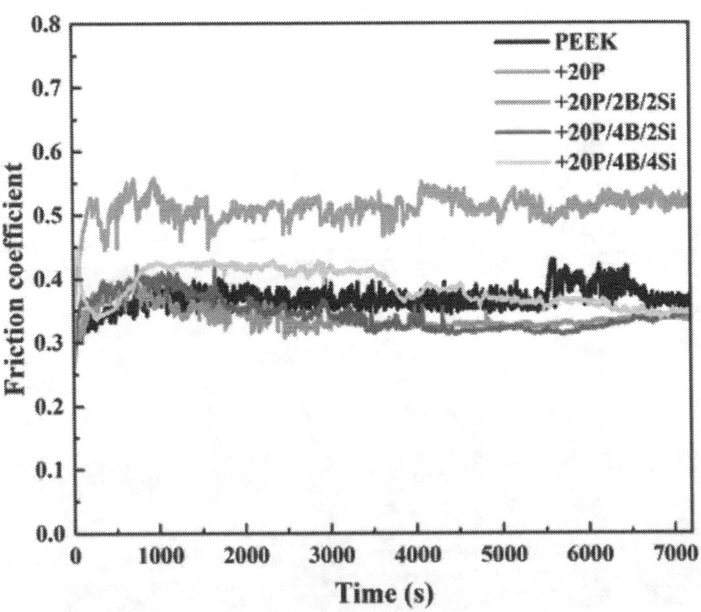

FIGURE 4.1 Wear rate comparison of pure UHMWPE and 1 wt.% GNPs UHMWPE (Li et al. 2023) (Open Access).

synergism effect of both organic and inorganic filler in the PEEK composite. Zhang et al. (Zhang et al. 2020) have prepared carbon fibre and SiC composite by precursor infiltration and a pyrolysis method and tested for tribological property at different temperatures. It was found that a strong fibre-matrix interface reduces the friction coefficient of the composite.

4.2.3 Influence of Fillers on Friction Coefficient

The influence of fillers on the friction coefficient is a significant topic in materials science, especially in the context of composite materials. Fillers can be added to materials to modify their properties, including the friction coefficient, which is crucial in applications where wear resistance and durability are important. Metallic fillers (copper, brass, steel), industrial wastes (slag, sludge, red mud), and minerals (silica, calcium carbonate) are usually used as fillers to improve the wear resistance of composite materials. Each type of filler interacts differently with the base material, affecting the friction coefficient.

In a study by Kumar and Bijwe (Kumar and Bijwe 2010), the reduced scale prototype (RSP) of brake pads was developed with metallic fillers like steel wool, brass, and copper powder and the sensitivity of the coefficient of friction (μ) towards the load and speed was investigated. This study concluded that with the increase in filler content, the values of μ increase and the sensitivity of μ towards speed and load decreases. Copper powder filler was shown to be most effective in terms of friction and wear when applied at a lower proportion. Steel wool and brass filler proved moderate and poor tribo-performance, respectively.

The influence of nanoparticle fillers on the wear and friction behaviour of polymer composites was studied by many researchers (Wang and Pei 2013). It was observed that the volume shape, size, and content of nanofillers influence the wear and friction behaviour of polymers.Certain polymer composite components are intended to work underwater. As a result, research was done on the tribological behaviour of SiC/PEEK and Si_3N_4/PEEK nanocomposites when they were lubricated with water (Wang et al. 2000; 2000). For these nanocomposites, different tribological behaviours were observed. In contrast to what occurs in dry conditions, the coefficient of friction for the Si_3N_4/PEEK composite increased steadily with increasing Si_3N_4 concentration. Compared to dry sliding, the wear rates of the filled and empty PEEK were much higher when lubricated with water. Thus, under water lubrication, the nanoscale Si_3N_4 filler was unable to greatly decrease PEEK's wear and friction. The coefficient of friction of pure PI rose with increasing load, whereas that of nanocomposites varied in a more complex way (Cai et al. 2003). Al_2O_3 nanoparticles can also be utilized to decrease wear at high loads.

In a study by Kishore et al. (Kishore et al. 2005), an increase in the coefficient of friction (μ) was observed with the increase in sliding velocity, whereas with an increase in load under constant sliding velocity, μ decreases for sample A, as shown in Table 4.2. Similar to A, the C samples show an increase in μ for sliding velocities ranging from 2–5 m/s under constant load. Unlike sample A, increasing the load under constant velocity results in a modest rise of μ. Under constant sliding velocity, a stepwise increase in load for the B sample causes a rise followed by a reduction in μ. The B

TABLE 4.2
Impact of Sliding Velocity on Coefficient of Friction

Load	Coefficient of Friction (μ)											
	Sliding Velocity (2 m/s)			Sliding Velocity (3 m/s)			Sliding Velocity (4 m/s)			Sliding Velocity (5 m/s)		
(N)	A	B	C	A	B	C	A	B	C	A	B	C
29	0.70	0.41	0.21	0.75	0.40	0.23	0.79	0.39	0.25	0.85	0.37	0.30
49	0.61	0.45	0.23	0.65	0.43	0.25	0.71	0.42	0.28	0.82	0.41	0.31
69	0.49	0.44	0.24	0.57	0.42	0.27	0.61	0.41	0.29	0.63	0.39	0.32

Source: Kishore et al. (2005). Copyright 2005. Reproduced with permission from Elsevier.
Note: A: Epoxy + E-glass fabric + 3 wt.% HTBN rubber, B: Epoxy + E-glass fabric + 2.5 wt.% Graphite powder, C: Epoxy + E-glass fabric + 4.5 wt.% Graphite powder

sample exhibits a drop in values when sliding velocities are increased under constant load, unlike C sample.

4.2.4 Thermal Stability and Degradation Analysis

When composite materials are subjected to harsh environmental conditions and high temperature then the degradation and thermal stability analysis play a crucial role in assessing the performance of the materials. Structural integrity and functional properties are the important criteria to check the thermal stability of the composite materials. The thermal stability is assessed by the techniques like differential scanning calorimetry (DSC), thermo gravimetric analysis (TGA), glass transition temperature (Tg), and dynamic mechanical analysis (DMA).

On the other hand, degradation analysis is useful to evaluate how the composite deteriorates over time due to factors like wear, UV radiation, moisture, and chemical exposure. This analysis helps to study the breakdown of fibres and matrices by degradation mechanisms like oxidation and hydrolysis.

In view of this, Singh et al. (Singh, Patnaik, and Gangil 2016) investigated the thermal stability of multiwall carbon nano clay, Kevlar fibre, and nanotube-reinforced phenolic resin-based friction composite under oxygen atmosphere using thermogravimetric analysis. It was observed that at 30–350°C, all composites showed greater than 97% material stability. At 350–600°C, all composites degraded quicker, and friction composites with 2.75 wt.% nanoclay showed the most stability, while 2 wt.% nano clay showed the least stability. Another study looked at the thermal stability of PEEK composites in dry-sliding friction and wear conditions (Muliyah et al. 2020). PEEK, known for its high decomposition temperature, can be used with stainless-steel fibre to increase the cohesive strength between the fibre and matrix. For friction materials, this leads to a low abrasion value and a consistent friction coefficient.

Adding PTFE can modify the tribological properties of thermoset polyester composites at high temperatures, which are influenced by tribo-chemical reactions on the sliding surface. Unsaturated polyester has different sliding properties than thermoplastic polyester because unsaturated polyester does not undergo mechanical softening or plasticization. Continuous curing of unsaturated polyester promotes sliding and increases the coefficient of friction at a 23–100°C temperature range (Samyn and Zsidai 2017).

4.2.5 Synergistic Effects of Multiple Fillers

The synergistic effects of multiple fillers on friction and wear in composite materials are pivotal in enhancing their performance. When different fillers are combined, they can interact in ways that result in properties superior to those achievable with individual fillers. PEEK was mixed with a variety of inorganic nanoparticles, including Si_3N_4, SiO_2, SiC, ZrO_2, Al_2O_3, TiO_2, ZnO, CuO, and $CaCO_3$ (Montesquieu et al. 2007); polyphenylene sulphide (PPS) (Schwartz and Bahadur 2000); polymethylmethacrylate (PMMA) (Avella, Errico, and Martuscelli 2001); epoxy (Zhang et al. 2002); and PTFE (Sawyer et al. 2003) matrices, in order to improve their friction and wear characteristics.

Figure 4.2(a) shows the effect of single and hybrid fillers on specific wear rates of epoxy-based composites (Friedrich, Zhang, and Schlarb 2005). It is observed that 5 vol.% TiO_2 filled epoxy composite shows better wear resistance than pure epoxy. With the addition of 5 vol.% graphite flakes (Gr) and 5 vol.% short carbon fibres (SCF), the wear resistance of epoxy-TiO_2 composites further increases. Figure 4.2(b) demonstrates the effect of different fillers on wear rate and coefficient of friction (μ) of the polyamide 6,6 (PA6,6) polymer at different pressures. It is observed that with the addition of 5 vol.% Gr and 15 vol.% SCF the wear rate and μ decrease as compared to neat PA6,6. A further decrease in wear rate and μ is observed with the addition of 5 vol.% TiO_2 in the PA6,6-Gr-SCF composite, irrespective of pressure. The advantage in wear reduction by mixing alumina nanoparticles with polyphenylene

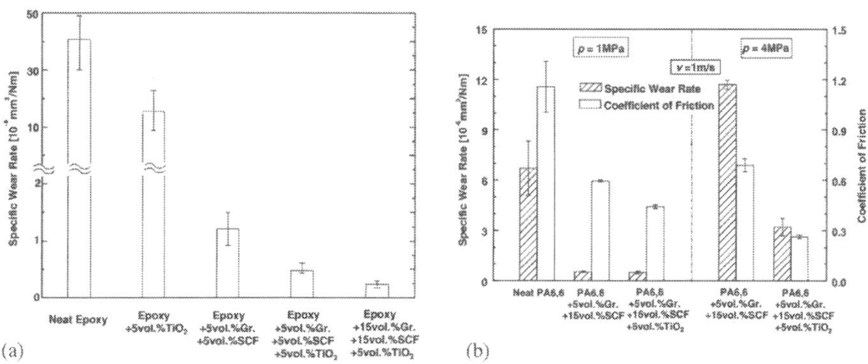

FIGURE 4.2 (a) Specific wear rate of epoxy and particular composites with and without TiO_2 particles, (b) Comparison of PA6,6 matrix composites' specific wear rate and coefficient of friction with and without TiO_2 particles at different contact pressures as determined by a pin-on-disk device (Friedrich, Zhang, and Schlarb 2005). (Copyright 2005. Reproduced with permission from Elsevier. Copyright permission from Elsevier 2005.).

sulphide is ruled by the improvement in bond strength (Schwartz and Bahadur 2000). The adhesion wear properties of neat PEEK and nano/micro/whisker SiC-filled composites have been studied and an effective reduction of wear is observed with the nano SiC filler. The wear resistance of whisker and micro SiC-filled PEEK also increased marginally (Montesquieu et al. 2007).

4.3 CONCLUSIONS

The influence of fillers on the high-temperature tribology of polymer matrix composites has an important role for its application in aeronautics, automotive engineering, and industrial apparati. The addition of fillers in polymeric frameworks are to augment the wear resistance under elevated temperatures. The following are the conclusions of the effect of fillers in PMCs for high-temperature tribology applications:

i. Improved Wear Resistance: The addition of filler in the polymer can decrease the direct contact between the polymer matrix and the counter surface which enhances the wear resistance. Fillers like nanoparticles, carbon fibres, graphite, and ceramic particles TiO_2, SiO_2, and Al_2O_3 increase the wear resistance of PMCs at high temperatures.
ii. Reduced Friction Coefficient: During the interaction of the polymer composite with the filler and contact surface, a lubricating film has been created on the contact surface. These fillers are particularly solid lubricants like PTFE, graphite, and molybdenum disulphide (MoS_2), which lower the friction coefficient at high temperatures. At high temperatures, conventional lubrication techniques have failed but the self-lubricating effect becomes more pronounced.
iii. Enhanced Thermal Stability: Fillers in a polymer matrix composite play an important role to increase thermal conductivity. The filler addition improves heat dissipation during sliding and lowers the likelihood of the polymer matrix thermally degrading. For the composite to maintain the integrity at high temperatures, increased thermal stability is essential.
iv. Hybrid Fillers' Synergistic Effects: The addition of more than one filler or a filler with a fibre shows a synergistic impact in PMCs. This hybridisation not only improves wear resistance but also lowers friction at high temperatures. The performance of the composite can be tailored by selecting the appropriate filler and filler quantity for certain applications.

REFERENCES

Ashok, R. B., C. V. Srinivasa, and B. Basavaraju. 2018. "A Review on the Mechanical Properties of Areca Fibre Reinforced Composites." Science and Technology of Materials. https://doi.org/10.1016/j.stmat.2018.05.004.

Asim, Mohammad, Naheed Saba, Mohammad Jawaid, and Mohammad Nasir. 2018. Potential of Natural Fibre/Biomass Filler-Reinforced Polymer Composites in Aerospace Applications. Sustainable Composites for Aerospace Applications. Elsevier Ltd. https://doi.org/10.1016/b978-0-08-102131-6.00012-8.

Avella, Maurizio, Maria Emanuela Errico, and Ezio Martuscelli. 2001. "Novel PMMA/CaCO3 Nanocomposites Abrasion Resistant Prepared by an in Situ Polymerization Process." Nano Letters 1 (4): 213–17. https://doi.org/10.1021/nl015518v.

Banagar, Ashok R, and Srinivasa Chikkol Venkateshappa. 2018. "A Review on the Mechanical Properties of Areca Fibre Reinforced Composites." Science and Technology of Materials 30 (2): 120–30. https://doi.org/10.1016/j.stmat.2018.05.004.

Barkoula, N. M., B. Alcock, N. O. Cabrera, and T. Peijs. 2008. "Flame-Retardancy Properties of Intumescent Ammonium Poly(Phosphate) and Mineral Filler Magnesium Hydroxide in Combination With Graphene." Polymers and Polymer Composites 16 (2): 101–13.

Cai, Hui, Fengyuan Yan, Qunji Xue, and Weimin Liu. 2003. "Investigation of Tribological Properties of Al2O 3-Polyimide Nanocomposites." Polymer Testing 22 (8): 875–82. https://doi.org/10.1016/S0142-9418(03)00024-2.

Chang, Li, Zhong Zhang, Lin Ye, and Klaus Friedrich. 2007. "Tribological Properties of High Temperature Resistant Polymer Composites With Fine Particles." Tribology International 40 (7): 1170–78. https://doi.org/10.1016/j.triboint.2006.12.002.

Chauhan, Shivanku. 2017. "Study of Polymer Matrix Composite With Natural Particulate/Fibre in PMC : A Review." International Journal of Advance Research, Ideas and Innovations in Technology 3: 1168–79.

Forster, Amanda L., Aaron M. Forster, Joannie W. Chin, Jyun Siang Peng, Chiao Chi Lin, Sylvain Petit, Kai Li Kang, et al. 2015. "Long-Term Stability of UHMWPE Fibres." Polymer Degradation and Stability 114: 45–51. https://doi.org/10.1016/j.polymdegradstab.2015.01.028.

Friedrich, Klaus, Zhong Zhang, and Alois K. Schlarb. 2005. "Effects of Various Fillers on the Sliding Wear of Polymer Composites." Composites Science and Technology 65 (15–16 SPEC. ISS.): 2329–43. https://doi.org/10.1016/j.compscitech.2005.05.028.

Gürgen, Selim, Abdullah Sert, and Melih Cemal Kuşhan. 2021. "An Investigation on Wear Behaviour of UHMWPE/Carbide Composites at Elevated Temperatures." Journal of Applied Polymer Science 138 (16): 1–13. https://doi.org/10.1002/app.50245.

Hemanth, R., M. Sekar, and B. Suresha. 2014. "Effects of Fibres and Fillers on Mechanical Properties of Thermoplastic Composites." Indian Journal of Advances in Chemical Science 2: 28–35.

Ho, S. C., J. H. Chern Lin, and C. P. Ju. 2005. "Effect of Fibre Addition on Mechanical and Tribological Properties of a Copper/Phenolic-Based Friction Material." Wear 258 (5–6): 861–69. https://doi.org/10.1016/j.wear.2004.09.050.

Inbakumar, J Parivendhan, and S Ramesh. 2018. "Mechanical, Wear and Thermal Behaviour of Hemp Fibre/Egg Shell Particle Reinforced Epoxy Resin Bio Composite." Transactions of the Canadian Society for Mechanical Engineering 285: 280–85.

Islam, Muhammad Remanul, and Md Dalour H. Beg. 2018. "Dispersion Characteristics of Hydroxyl and Carboxyl-Functionalized Multi-Walled Carbon Nanotubes in Polyester Nanocomposites." Journal of Polymer Engineering 38 (8): 759–65. https://doi.org/10.1515/polyeng-2017-0363.

Joseph, Kuruvilla, Romildo Dias Tolêdo Filho, Beena James, Sabu Thomas, and Laura Hecker de Carvalho. 1999. "A Review on Sisal Fibre Reinforced Polymer Composites." Revista Brasileira de Engenharia Agrícola e Ambiental 3 (3): 367–79. https://doi.org/10.1590/1807-1929/agriambi.v3n3p367-379.

Katiyar, Jitendra K., and Hemalata Jena. 2023. "The Role of Synthetic Fibers in the Tribological Behavior of Polymeric Materials, Tribology of Polymers, Polymer Composites, and Polymer Nanocomposites." Elsevier Series on Tribology and Surface Engineering 173–94.

Katiyar, Jitendra K., Sujeet K. Sinha, and Arvind Kumar. 2016. "Friction and Wear Durability Study of Epoxy-Based Polymer (SU-8) Composite Coatings with Talc and Graphite as Fillers." Wear 362–363: 199–208. https://doi.org/10.1016/j.wear.2016.06.011.

Kayode Aliyu, Ismaila, Mohammed Abdul Samad, and Amro M. Al-Qutub. 2022. "Friction and Wear Behaviour of Ultra-High Molecular Weight Polyethylene/Graphene Nanoplatelets Nanocomposite Coatings at Elevated Temperatures." Proceedings of the Institution of Mechanical Engineers, Part J: Journal of Engineering Tribology 236 (9): 1782–88. https://doi.org/10.1177/13506501211053744.

Kishore, P., S. Sampathkumaran, P. Seetharamu, M. Thomas, and Janardhana. 2005. "A Study on the Effect of the Type and Content of Filler in Epoxy-Glass Composite System on the Friction and Slide Wear Characteristics." Wear 259 (1–6): 634–41. https://doi.org/10.1016/j.wear.2005.03.003.

Kumar, Mukesh, and Jayashree Bijwe. 2010. "Role of Different Metallic Fillers in Non-Asbestos Organic (NAO) Friction Composites for Controlling Sensitivity of Coefficient of Friction to Load and Speed." Tribology International 43 (5–6): 965–74. https://doi.org/10.1016/j.triboint.2009.12.062.

Kurdi, Abdulaziz, Wen Hao Kan, and Li Chang. 2019. "Tribological Behaviour of High Performance Polymers and Polymer Composites at Elevated Temperature." Tribology International 130: 94–105. https://doi.org/10.1016/j.triboint.2018.09.010.

Li, Mengjiao, Jingjing Yang, Shengqi Ma, Gang Liu, Hongru Yang, and Jianan Yao. 2023. "Effect of H-BN and Nano-SiO2 Fillers on the High-Temperature Tribological Properties of PEEK/PI-Based Composites." Lubricants 11 (10). https://doi.org/10.3390/lubricants11100416.

Li, Yingji, Yi Chen, Yongxin Guo, Da Bian, and Yongwu Zhao. 2022. "Tribological Behaviour of PEEK/PTFE Composites Reinforced with Carbon Fibres and Graphite." Materials 15 (20). https://doi.org/10.3390/ma15207078.

Madhusudhan, T., and M. Senthil Kumar. 2017. "Experimental Study on Wear Behaviour of SiC Filled Hybrid Composites Using Taguchi Method." International Journal of Mechanical Engineering and Technology 8 (2): 271–77.

Masood, Muhammad T., Evie L. Papadopoulou, José A. Heredia-Guerrero, Ilker S. Bayer, Athanassia Athanassiou, and Luca Ceseracciu. 2017. "Graphene and Polytetrafluoroethylene Synergistically Improve the Tribological Properties and Adhesion of Nylon 66 Coatings." Carbon 123: 26–33. https://doi.org/10.1016/j.carbon.2017.07.026.

Miyazaki, N., and T. Hamao. 1994. "Solid Particle Erosion of Thermoplastic Resins Reinforced by Short Fibres." Journal of Composite Materials 28 (9): 871–83. https://doi.org/10.1177/002199839402800905.

Mohan, Dhanesh G., S. Gopi, V. Rajasekar, Karthik Krishnan, Dhanesh G. Mohan, S. Gopi, L. Selvarajan et al. 2019. "Tribological Behaviour of Molybdenum Disulphide Particles-High Density Polyethylene Composite." Materials Today: Proceedings 27 (7): 075402.

Montesquieu, Technopole, The Peek, Jan Sandler, Philipp Werner, Milo S P Shaffer, Vitaly Demchuk, Volker Altstädt, et al. 2007. "Various Kinds of SiC." *Composites Part B: Engineering* 47 (8): 185–95.

Muliyah, Pipit, Dyah Aminatun, Sukma Septian Nasution, Tommy Hastomo, and Tryana Setiana Sri Wahyuni Sitepu. 2020. "Exploring Learners'autonomy In Online Language-Learning In Stai Sufyan Tsauri Majenang." Journal GEEJ 7 (2): 382–94.

Nayak, Sandip Kumar, and Alok Satapathy. 2019. "Wear Analysis of Waste Marble Dust-Filled Polymer Composites with an Integrated Approach Based on Design of Experiments and Neural Computation." Proceedings of the Institution of Mechanical Engineers, Part J: Journal of Engineering Tribology 234 (12): 1846–56. https://doi.org/10.1177/1350650119896170.

Parikh, Hiral H., and Piyush P. Gohil. 2017. "Experimental Investigation and Prediction of Wear Behaviour of Cotton Fibre Polyester Composites." Friction 5 (2): 183–93. https://doi.org/10.1007/s40544-017-0145-y.

Pradhan, Manoj Kumar, and Alok Satapathy. 2018. Analysis and Prediction of Erosion Response of SiC Filled Bio-Fibre Composites Using Taguchi Method and Neural Computation. ICPCM.

Purohit, A., and A. Satapathy. 2018. "A Study on Erosion Wear Performance of Linz-Donawitz Sludge Filled Polypropylene Matrix Composites." In *IOP Conference Series: Materials Science and Engineering*. Vol. 377. https://doi.org/10.1088/1757-899X/377/1/012045.

Qi, Shicheng, Zhezhen Fu, Rongping Yun, Shengling Jiang, Xiaoa Zheng, Yafei Lu, Vlastimil Matejka, Jana Kukutschova, Veroslava Peknikova, and Martin Prikasky. 2014. "Effects of Walnut Shells on Friction and Wear Performance of Eco-Friendly Brake Friction Composites." Proceedings of the Institution of Mechanical Engineers, Part J: Journal of Engineering Tribology 228 (5): 511–20. https://doi.org/10.1177/1350650113517112.

Radhika, N., and R. Raghu. 2015. "Dry Sliding Wear Behaviour of Aluminium Al-Si12Cu/TiB_2 Metal Matrix Composite Using Response Surface Methodology." Tribology Letters 59 (1). https://doi.org/10.1007/s11249-015-0516-3.

Roy, Amit, Liwen Mu, and Yijun Shi. 2020. "Tribological Properties of Polyimide Coating Filled with Carbon Nanotube at Elevated Temperatures." Polymer Composites 41 (7): 2652–61. https://doi.org/10.1002/pc.25564.

Saluja, Parmeet Singh, and Gaurav Gupta. 2017. "Preparation and Thermal Behaviour of Polyester Composite Filled with TiO2." International Research Journal of Engineering and Technology(IRJET) 4 (4): 3135–41.

Samyn, Pieter, and Laszlo Zsidai. 2017. "Temperature Effects on Friction and Wear of Thermoset Polyester Fabric Composites." Polymer - Plastics Technology and Engineering 56 (9): 1003–16. https://doi.org/10.1080/03602559.2016.1247281.

Sathishkumar, T. P., P. Navaneethakrishnan, S. Shankar, R. Rajasekar, and N. Rajini. 2013. "Characterization of Natural Fibre and Composites - A Review." Journal of Reinforced Plastics and Composites 32 (19): 1457–76. https://doi.org/10.1177/0731684413495322.

Sawyer, W. Gregory, Kevin D. Freudenberg, Praveen Bhimaraj, and Linda S. Schadler. 2003. "A Study on the Friction and Wear Behaviour of PTFE Filled with Alumina Nanoparticles." Wear 254 (5–6): 573–80. https://doi.org/10.1016/S0043-1648(03)00252-7.

Schwartz, Christian J., and Shyam Bahadur. 2000. "Studies on the Tribological Behaviour and Transfer Film-Counterface Bond Strength for Polyphenylene Sulfide Filled with Nanoscale Alumina Particles." Wear 237 (2): 261–73. https://doi.org/10.1016/S0043-1648(99)00345-2.

Shi, Guojun, Zhen Cao, Xiaotian Yan, and Qiuyi Wang. 2019. "In-Situ Fabrication of a UHMWPE Nanocomposite Reinforced by SiO2 Nanospheres and Its Tribological Performance." Materials Chemistry and Physics 236: 121778. https://doi.org/10.1016/j.matchemphys.2019.121778.

Singh, Tej, Amar Patnaik, and Brijesh Gangil. 2016. "Thermal Stability Analysis of Nano-Particulate-Filled Phenolic-Based Friction Composite Materials." Journal of Industrial Textiles 45 (6): 1335–49. https://doi.org/10.1177/1528083714559568.

Teh, P L, M Jaafar, H M Akil, K N Seetharamu, A N R Wagiman, and K S Beh. 2008. "Thermal and Mechanical Properties of Particulate Fillers Filled Epoxy Composites for Electronic Packaging Application." November 2007, 308–15.

Verheyen, S., N. Blaton, R. Kinget, and H.-S. Kim. 2004. "Thermogravimetric Analysis of Rice Husk Flour Filled Thermoplastic Polymer Composites." Journal of Thermal Analysis and Calorimetry 76 (2): 395–404. https://doi.org/10.1023/b:jtan.0000028020.02657.9b.

Verma, Sameer. 2017. "A Review on Thermal Beahviour of Natural Fibre Reinforced Composites." no. 4, 251–55.

Wang, Qihua, and Xianqiang Pei. 2013. "The Influence of Nanoparticle Fillers on the Friction and Wear Behaviour of Polymer Matrices." *Tribology of Polymeric Nanocomposites:*

Friction and Wear of Bulk Materials and Coatings, Second Edition, 91–118. Elsevier. https://doi.org/10.1016/B978-0-444-59455-6.00004-0.

Wang, Qi Hua, Qun Ji Xue, Wei Min Liu, and Jian Min Chen. 2000. “Effect of Nanometer SiC Filler on the Tribological Behaviour of PEEK under Distilled Water Lubrication.” Journal of Applied Polymer Science 78 (3): 609–14. https://doi.org/10.1002/1097-4628(20001017)78:3<609::AID-APP160>3.0.CO;2-D.

Wang, Yanming, Peng Cai, Tingmei Wang, and Qihua Wang. 2018. “Influence of Graphite on Tribological Properties of Hexagonal Boron Nitride Hybrid/Polyimide Composites in Wide Temperature Range.” Journal of Tribology 140 (1): 1–6. https://doi.org/10.1115/1.4036935.

Yang, Jiong, Chaoyin Chen, Shenglan Zhao, Feng Ge, and Diqiu Liu. 2014. “The Inhibitory Effect of Different Solvents Extracts from Walnut Shell (*Juglans Regia* L.) on Pancreatic Lipase and Adipogenesis of 3T3-L1 Preadipocytes.” Journal of Food and Nutrition Research 2 (10): 664–70. https://doi.org/10.12691/jfnr-2-10-2.

Yuan, Junya, Zhaozhu Zhang, Mingming Yang, Fang Guo, Xuehu Men, and Weimin Liu. 2017. “TiB2 Reinforced Hybrid-Fabric Composites with Enhanced Thermal and Mechanical Properties for High-Temperature Tribological Applications.” Tribology International 115: 8–17. https://doi.org/10.1016/j.triboint.2017.05.006.

Zhang, Chen, Hongrui Liu, Yang Zhou, Junping Li, Yinjie Sun, Yun Wang, and Jincheng Pang. 2020. “The Effect of High Temperature Heat Treatment on the Tribological Property of a Carbon Fibre/Pyrolytic Carbon/Silicon Carbide Composite Using Polycarbosilane.” Ceramics International 46 (4): 4493–4501. https://doi.org/10.1016/j.ceramint.2019.10.176.

Zhang, Ming Qiu, Min Zhi Rong, Shu Li Yu, Bernd Wetzel, and Klaus Friedrich. 2002. “Effect of Particle Surface Treatment on the Tribological Performance of Epoxy Based Nanocomposites.” Wear 253 (9–10): 1086–93. https://doi.org/10.1016/S0043-1648(02)00252-1.

Zhang, Yanchao, Dongya Zhang, Xian Wei, Shanjun Zhong, and Jianlei Wang. 2018. “Enhanced Tribological Properties of Polymer Composite Coating Containing Graphene at Room and Elevated Temperatures.” Coatings 8 (3): 91. https://doi.org/10.3390/coatings8030091.

Zhen, Jinming, Yunxiang Han, Huabao Wang, Zhenguo Jiang, Li Wang, Yuqiang Huang, Zhengfeng Jia, and Ran Zhang. 2023a. “High Temperature Friction and Wear Behaviour of PTFE/MoS2 Composites.” Lubricants 11 (8): 312. https://doi.org/10.3390/lubricants11080312.

5 Influence of Fillers on the Tribological Performance of Self-Lubricating Polymer Coatings

Abhijit Bhowmik, Manoj Kumar, Raman Kumar, Sehijpal Singh, and Valentin Romanovski

5.1 INTRODUCTION TO SELF-LUBRICATING POLYMER COATINGS

Self-lubricating polymer coatings are engineered to provide low friction and high wear resistance, making them ideal for applications where traditional lubrication is impractical or undesirable. These coatings are commonly used in aerospace, automotive, and biomedical industries. The base polymers, such as polytetrafluoroethylene (PTFE), polyether ether ketone (PEEK), and ultra-high-molecular-weight polyethylene (UHMWPE), offer inherent lubricity. However, their mechanical and tribological properties can be further enhanced by incorporating various fillers [Gong et al., 2023; Shiao et al., 2001].

5.2 ROLE OF FILLERS IN ENHANCING TRIBOLOGICAL PROPERTIES

Fillers are materials added to polymers to improve specific properties. In the context of tribology, fillers can enhance load-bearing capacity, reduce friction, and increase wear resistance [Bhowmik et al., 2022]. The effectiveness of a filler depends on its type, size, shape, concentration, and distribution within the polymer matrix. Common fillers include carbon-based materials, ceramics, and solid lubricants [Liu et al., 2023]. Figure 5.1 shows the friction curves and wear rates of the composite.

5.3 CARBON-BASED FILLERS

Carbon-based fillers, including carbon nanotubes (CNTs), graphene, and carbon fibers, are widely recognized for their remarkable mechanical properties, thermal conductivity, electrical conductivity, and inherent lubricity. These characteristics make them highly sought after in various advanced material applications, particularly in enhancing the tribological performance of polymer composites [Hong et al., 2010].

DOI: 10.1201/9781003530909-7

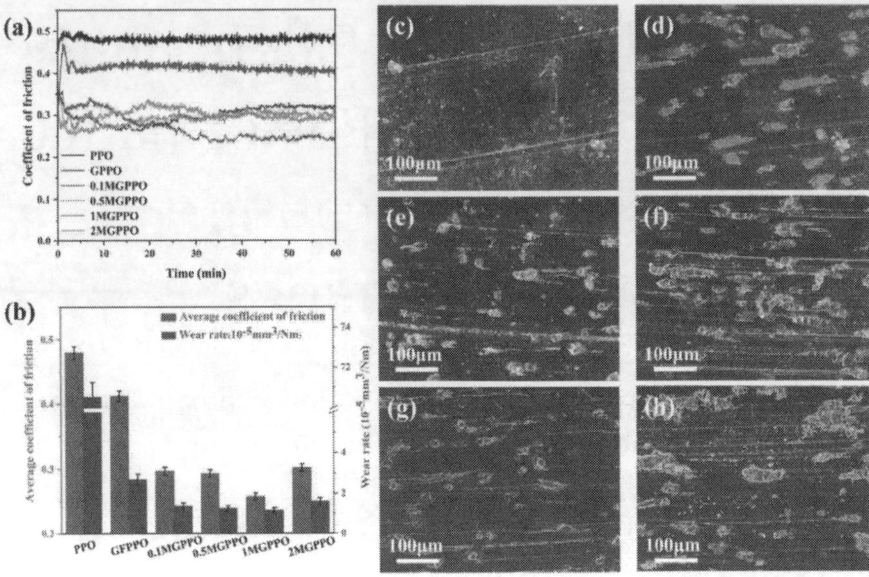

FIGURE 5.1 (a) Friction coefficient curves of polyphenylene oxide (PPO), Glass-filled Polypropylene Oxide (GPPO), and Modified Glass-filled Polypropylene Oxide (MGPPO) composites with varying MXene nanosheet content; (b) Wear rate curves of PPO, GPPO, and MGPPO composites with varying MXene nanosheet content; (c) Scanning electron microscope image of the friction surface of pure PPO composite; (d) SEM image of the friction surface of GPPO composite; (e) SEM image of the friction surface of 0.1 wt.% MXene-added MGPPO composite; (f) SEM image of the friction surface of 0.5 wt.% MXene-added MGPPO composite; (g) SEM image of the friction surface of 1 wt.% MXene-added MGPPO composite; (h) SEM image of the friction surface of 2 wt.% MXene-added MGPPO composite (Liu et al., 2025).

5.3.1 Carbon Nanotubes (CNTs)

Carbon nanotubes (CNTs), especially multi-walled carbon nanotubes (MWCNTs), are cylindrical nanostructures composed of rolled-up sheets of graphene [De Volder et al., 2013]. They possess extraordinary strength, stiffness, and a high aspect ratio, which allows them to effectively interact with polymer matrices. When incorporated into polymers such as PTFE and PEEK, MWCNTs have demonstrated significant improvements in tribological properties [Haddon, 2002].

Studies reveal that adding MWCNTs to PTFE composites results in a 33% reduction in wear rate and a slight decrease in the coefficient of friction (COF). This enhancement is attributed to the lubricating properties of MWCNTs and their strong interaction with the polymer matrix. The nanotubes act as nanoscale reinforcements, distributing the applied load and minimizing direct contact between polymer surfaces. Furthermore, MWCNTs promote the formation of a protective transfer film on the counterface, which contributes to reduced wear and friction. This makes them an ideal filler for applications requiring long-lasting, low-friction materials, such as bearings, seals, and sliding components.

5.3.2 Graphene and Graphene Oxide (GO)

Graphene, a two-dimensional material consisting of a single layer of carbon atoms arranged in a hexagonal lattice, is another prominent carbon-based filler. Its exceptional mechanical properties, high thermal conductivity, and superior lubricating characteristics make it a valuable additive in polymer composites. Graphene oxide (GO), a derivative of graphene with oxygen-containing functional groups, offers additional benefits due to its enhanced dispersibility in polymer matrices [Zhu et al., 2010].

The incorporation of graphene into polymers has been shown to improve load-bearing capacity and wear resistance significantly. Graphene's high aspect ratio enables it to form a well-dispersed, interconnecting network within the polymer matrix, which helps to reinforce the material and distribute mechanical stresses. Moreover, graphene promotes the development of a uniform and stable transfer film on the contact surfaces during sliding. This transfer film acts as a solid lubricant, reducing direct surface-to-surface interactions and minimizing material loss due to wear [Eigler & Hirsch, 2014].

GO further enhances these properties by improving the compatibility between the filler and the polymer matrix. Its functional groups facilitate better bonding, ensuring a more uniform distribution of the filler within the composite. This results in improved mechanical strength, reduced friction, and enhanced wear resistance. Applications of graphene-enhanced polymers span industries such as aerospace, automotive, and electronics, where high-performance materials are essential.

5.4 CERAMIC FILLERS

Ceramic fillers are widely used to improve the mechanical and thermal properties of polymer coatings, making them suitable for high-performance applications. Common ceramic fillers include silicon carbide (SiC), aluminum oxide (Al_2O_3), and silicon nitride (Si_3N_4). These materials are known for their exceptional hardness, wear resistance, and ability to withstand extreme temperatures, which contribute to the durability and stability of polymer-based coatings [Bhowmik et al., 2024].

Among ceramic fillers, Si_3N_4 stands out for its superior mechanical and tribological properties. The incorporation of graphene nanoplatelets into Si_3N_4 composites has demonstrated remarkable improvements in wear resistance and friction reduction. This enhancement occurs due to the formation of a self-lubricating protective tribofilm during sliding or abrasive interactions. The graphene nanoplatelets act as a solid lubricant, reducing the COF and minimizing wear. Additionally, the synergistic effect between Si_3N_4 and graphene ensures that the composite maintains its structural integrity under high-stress conditions, further extending the life span of the coating. These properties make Si_3N_4 composites highly desirable in applications requiring robust and reliable performance, such as aerospace, automotive, and industrial machinery, where prolonged exposure to friction and heat is a critical concern [Zgalat-Lozynskyy et al., 2021].

5.5 SOLID LUBRICANT FILLERS

Solid lubricants are materials that reduce friction and wear in situations where traditional liquid lubricants, such as oils or greases, are ineffective. These are especially valuable in extreme conditions like high temperatures, vacuum environments, or situations where liquid lubricants could evaporate or degrade [Rao et al., 2021]. Among the most commonly used solid lubricants in polymer coatings are molybdenum disulfide (MoS_2) and graphite, both of which possess unique properties that make them highly effective.

MoS_2 is a widely used solid lubricant due to its unique crystal structure. It consists of layers bonded together by weak van der Waals forces, which allows the layers to slide over one another easily, significantly reducing friction. This low-shear property makes MoS_2 an excellent additive for polymer coatings, enhancing their performance in demanding applications [Wang & Mi, 2017]. When incorporated as a filler, MoS_2 improves wear resistance and lowers the COF, ensuring smooth operation even under high loads or extreme pressures. Its performance remains consistent in high-temperature environments, making it suitable for aerospace, automotive, and industrial machinery applications. Furthermore, MoS_2's stability and compatibility with various polymers ensure it can be effectively integrated into diverse coating systems to enhance durability and efficiency [Gupta et al., 2020].

5.6 HYBRID FILLERS

Combining different types of fillers within a polymer matrix can result in synergistic effects, significantly enhancing the tribological performance of polymer coatings. Hybrid fillers take advantage of the unique properties of individual fillers, leading to improved mechanical, thermal, and self-lubricating characteristics [Wang et al., 2013].

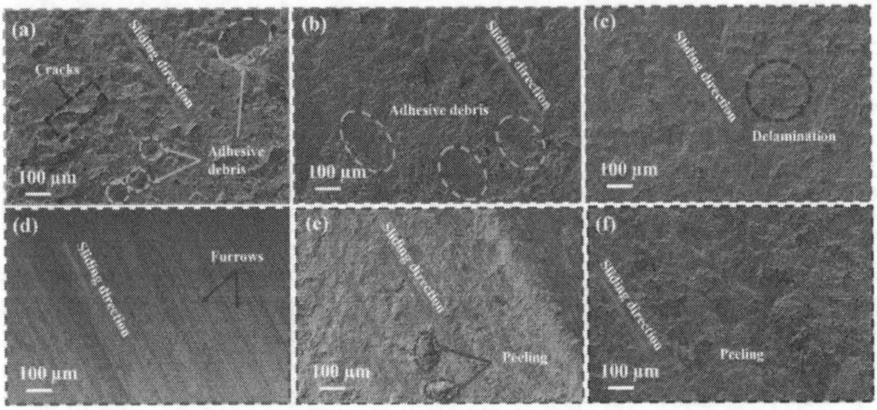

FIGURE 5.2 (a) Detailed surface analysis of pure epoxy resin after wear testing, highlighting severe fatigue and adhesive wear; (b) Microscopic examination of epoxy composite with 0.2% hyperbranched polysiloxane-modified MXene/Ag fillers, showing reduced cracks; (c) Evaluation of epoxy composite with 0.4% hyperbranched polysiloxane-modified MXene/Ag fillers, exhibiting improved surface smoothness; (d) Surface characterization of epoxy composite with optimal 0.6% hyperbranched polysiloxane-modified MXene/Ag fillers, showing mild abrasive wear; (e) Wear surface of epoxy composite with 0.8% hyperbranched polysiloxane-modified MXene/Ag fillers, indicating spalling pits formation; (f) Analysis of epoxy composite with 1.0% hyperbranched polysiloxane-modified MXene/Ag fillers, highlighting increased surface damage due to agglomeration [Lan et al., 2025].

Integrating carbon-based fillers like graphene or CNTs with ceramic fillers such as SiC can create a versatile composite material. Carbon-based fillers contribute exceptional self-lubricating properties, high thermal conductivity, and excellent chemical stability. Meanwhile, ceramic fillers like SiC enhance mechanical strength, hardness, and wear resistance. The combination of these fillers within the polymer matrix forms a material that exhibits superior tribological performance compared to systems containing a single type of filler. Such hybrid composites often show a significant reduction in the COF and wear rates under various operating conditions. The interplay between carbon-based fillers and ceramics helps in maintaining a smooth surface and minimizing material degradation during sliding or abrasive motions [Cai et al., 2019]. These benefits make hybrid filler systems an excellent choice for demanding applications in industries such as aerospace, automotive, and manufacturing, where durability and performance are critical [Ramírez et al., 2021]. The worn surface of EP composites is shown in Figure 5.2.

5.7 MECHANISMS OF TRIBOLOGICAL ENHANCEMENT

The improvement in tribological performance due to the incorporation of fillers into polymer matrices is a multifaceted phenomenon that involves several interconnected mechanisms. These mechanisms collectively contribute to the reduction of wear and friction, extending the durability and efficiency of materials used in various applications [Gao et al., 2025]. Below is an in-depth examination of the primary mechanisms through which fillers enhance tribological properties.

5.7.1 Load-Bearing Capacity

One of the most significant contributions of fillers is their ability to improve the load-bearing capacity of the polymer matrix. Fillers, often made of hard and rigid materials like ceramics, metal oxides, or carbon-based compounds, enhance the hardness and modulus of the composite material. This reinforcement allows the matrix to sustain higher loads without undergoing plastic deformation or significant damage. For instance, under high mechanical stresses, the fillers distribute the applied load more evenly across the material, reducing localized stress concentrations. This load-sharing behavior minimizes the risk of surface failure and contributes to maintaining the structural integrity of the material. By improving the stiffness of the polymer matrix, fillers also reduce the depth of penetration of asperities from the counterface, leading to lower wear rates. In applications such as bearings, seals, and bushings, this enhanced load-bearing capacity ensures prolonged service life and reliable performance under challenging operational conditions [Ye et al., 2019].

5.7.2 Transfer Film Formation

The formation of a transfer film is another critical mechanism by which fillers improve tribological properties. During sliding or rubbing, certain fillers facilitate the creation of a stable and uniform transfer film on the counterface. This film acts as a protective barrier, reducing direct contact between the interacting surfaces and minimizing abrasive wear. The transfer film mechanism depends heavily on the compatibility among the filler material, polymer matrix, and counterface [Pimenov et al., 2024]. Fillers such as PTFE, graphite, and MoS_2 are well known for their ability to generate such films. Once formed, the transfer film provides a smooth interface that lowers friction and reduces material loss. Moreover, the stability and adherence of this film to the counterface are critical factors influencing its effectiveness. A well-formed transfer film also prevents third-body wear particles from embedding into the sliding surfaces, further enhancing tribological performance [Venturi, 2024].

5.7.3 Lubrication

Fillers that act as solid lubricants are particularly effective in reducing friction between sliding surfaces. These fillers, which include materials like graphite, boron nitride, and MoS_2, possess low shear strength and excellent lubricating properties. When embedded within the polymer matrix, these solid lubricants are exposed during the sliding process, creating a lubricating layer that reduces frictional forces [Lin et al., 2024]. The presence of solid lubricant fillers ensures consistent lubrication even under conditions where liquid lubricants may fail, such as in high-temperature or vacuum environments. This inherent lubricity not only reduces friction but also prevents the generation of excessive heat, which could lead to thermal degradation of the polymer matrix. Furthermore, solid lubricants can accommodate a range of operational conditions, including varying loads, speeds, and temperatures, making them indispensable for advanced tribological applications [Sarkar & Gul, 2024].

5.7.4 Thermal Stability

Frictional interactions inherently generate heat, which can lead to thermal degradation of the polymer matrix if not effectively managed. Fillers contribute to the thermal stability of composites by enhancing their thermal conductivity and dissipating heat generated during sliding or rolling. This improved heat dissipation prevents localized hot spots that could weaken the material or accelerate wear processes [Chu et al., 2016]. Thermally conductive fillers such as Al_2O_3, SiC, and carbon-based materials (e.g., graphite or CNTs) play a pivotal role in maintaining the thermal balance of the tribological system. By facilitating the transfer of heat away from the contact zone, these fillers help maintain the structural and chemical stability of the polymer matrix. This thermal management is especially critical in high-speed or high-load applications, where excessive heat generation could compromise performance or lead to catastrophic failure [Zhang et al., 2021].

5.7.5 Synergistic Effects of Fillers

In many cases, the combined effects of multiple fillers produce synergistic improvements in tribological performance. For example, a composite containing both hard ceramic particles and solid lubricants can simultaneously achieve enhanced load-bearing capacity and reduced friction. Similarly, the incorporation of thermally conductive fillers alongside solid lubricants can optimize both heat dissipation and lubrication, ensuring consistent performance under demanding conditions [Gangwani et al., 2024]. The choice of fillers and their concentration within the polymer matrix are crucial factors that determine the extent of these synergistic effects. A well-designed composite leverages the complementary properties of different fillers to address specific tribological challenges. For instance, in aerospace or automotive applications, where both high wear resistance and low friction are required, hybrid filler systems are often employed to achieve the desired balance of properties [Liu et al., 2025].

5.8 FACTORS INFLUENCING FILLER EFFECTIVENESS

The impact of fillers on tribological performance is multifaceted and governed by several critical factors. These factors determine the extent to which fillers enhance or hinder the properties of the composite material. A detailed examination of these factors is essential to optimize filler usage and maximize their effectiveness in specific applications. Below, we explore three primary factors: concentration, dispersion, and interfacial adhesion.

5.8.1 Concentration

The concentration of fillers within a polymer matrix is a pivotal determinant of their effectiveness. Achieving the optimal filler content is essential because it directly influences the composite's mechanical, thermal, and tribological properties. A low concentration of fillers may not provide sufficient reinforcement, leading to suboptimal enhancement of the material's properties. Conversely, an excessively high concentration can result in filler agglomeration, which negatively affects performance. When fillers agglomerate, they form clusters or clumps within the polymer matrix. These agglomerates act as stress concentrators, reducing the overall mechanical strength and wear resistance of the composite. Additionally, excessive filler content can compromise the polymer's processability, leading to difficulties in manufacturing and increased production costs. Therefore, determining the optimal filler concentration requires a balance between achieving the desired property enhancement and maintaining good processability and structural integrity [Yang et al., 2024].

5.8.2 Dispersion

Uniform dispersion of fillers within the polymer matrix is another critical factor influencing their effectiveness. Dispersion refers to the even distribution of filler particles throughout the matrix without clustering or aggregation. Achieving uniform dispersion ensures that the enhanced properties are consistent across the composite material. Poor dispersion can lead to localized areas of weakness, where the material's properties are inferior to the surrounding regions. These weak points are prone to mechanical failure, reducing the overall durability and performance of the composite. In tribological applications, poor dispersion can result in uneven wear patterns, increased friction, and reduced life span of the material. To achieve optimal dispersion, several techniques are employed during the manufacturing process.

These include high-shear mixing, ultrasonic dispersion, and the use of dispersing agents or surfactants. The choice of technique depends on the type of filler and polymer matrix, as well as the desired properties of the final composite. Nanoscale fillers, such as CNTs and graphene, pose additional challenges in achieving uniform dispersion due to their tendency to agglomerate. Specialized techniques, such as surface functionalization of the fillers, are often required to enhance their compatibility with the polymer matrix and facilitate uniform distribution [Moharana et al., 2024].

5.8.3 Interfacial Adhesion

The interfacial adhesion between fillers and the polymer matrix is a crucial factor in determining the effectiveness of load transfer and the overall durability of the composite. Strong interfacial bonding ensures that stress is efficiently transferred from the polymer matrix to the fillers, thereby enhancing the mechanical and tribological properties of the material. Weak interfacial adhesion can lead to filler pull-out or debonding under mechanical stress, resulting in a loss of reinforcement and a decrease in performance. In tribological applications, poor adhesion can cause increased wear rates and the generation of debris, further compromising the material's performance [Zhang et al., 2024]. Improving interfacial adhesion often involves surface treatment or functionalization of the fillers. For instance, chemical modification of filler surfaces with coupling agents, such as silanes or titanates, can enhance their compatibility with the polymer matrix. Alternatively, plasma treatment or grafting of polymer chains onto filler surfaces can also improve adhesion. The choice of surface treatment depends on the specific filler-polymer combination and the desired properties of the composite [Peng et al., 2024].

5.8.4 Interactions between Factors

While concentration, dispersion, and interfacial adhesion are distinct factors, they are interdependent and often influence each other. For example, achieving uniform dispersion becomes more challenging as filler concentration increases, particularly with nanoscale fillers. Similarly, improving interfacial adhesion can facilitate better dispersion by reducing the tendency of fillers to agglomerate [Sun et al., 2023].

Optimizing these factors requires a holistic approach that considers their interactions and the specific requirements of the application. Advanced characterization techniques, such as scanning electron microscopy (SEM), transmission electron microscopy (TEM), and atomic force microscopy (AFM), are often used to evaluate filler dispersion and interfacial adhesion. These techniques provide valuable insights into the microstructure of the composite and help guide the optimization process [Kennedy et al., 2024].

5.9 RECENT ADVANCES AND FUTURE DIRECTIONS

In recent years, significant strides have been made in enhancing the performance of self-lubricating polymer coatings, driven by the demand for high-performance materials in industries such as aerospace, automotive, and biomedical applications. One prominent area of focus has been the development of novel fillers and hybrid systems that provide superior tribological properties, such as reduced friction, enhanced wear resistance, and prolonged service life under demanding conditions.

5.9.1 Functionalized Nanofillers

Functionalized nanofillers, such as graphene, CNTs, and boron nitride nanosheets, have garnered significant attention for their ability to improve the performance of polymer coatings. Functionalization involves chemically modifying the surface of nanofillers to enhance their dispersion within the polymer matrix and promote strong interfacial adhesion between the fillers and the polymer. This leads to a uniform distribution of stress under mechanical loads, thereby improving the coating's wear resistance. For example, GO functionalized with amino or hydroxyl groups can form strong covalent bonds with the polymer matrix, resulting in coatings with exceptional durability and low COFs. Such advancements are critical for applications where consistent performance under high loads and temperatures is essential [Zahra & Kim, 2025].

5.9.2 Hybrid Systems

The integration of multiple fillers to create hybrid systems has emerged as another promising approach. By combining the unique properties of various fillers, hybrid systems aim to achieve synergistic effects that outperform single-filler systems. For instance, a combination of soft fillers like PTFE and hard fillers such as SiC can optimize the balance between reduced friction and enhanced wear resistance. Moreover, the addition of lubricating nanoparticles, such as MoS_2 or tungsten disulfide, has been shown to provide a self-replenishing lubricating effect, ensuring consistent tribological performance over extended periods [Geweda et al., 2025].

5.9.3 Bio-Inspired and Environmentally Friendly Fillers

In line with the global emphasis on sustainability, researchers are increasingly exploring bio-inspired and environmentally friendly fillers. Natural materials, such as cellulose nanofibers and chitosan, are being investigated for their potential to improve the mechanical and tribological properties of polymer coatings while minimizing environmental impact. Inspired by natural structures, such as the layered architecture of nacre or the lubricating mechanisms in synovial joints, these bio-inspired designs aim to create coatings that are both high-performing and eco-friendly. Additionally, the use of renewable resources aligns with circular economy principles, reducing dependency on finite raw materials [Mohtasim & Das, 2024].

5.9.4 Future Directions

Looking ahead, the field is poised to benefit from advancements in material science and computational modeling. Artificial intelligence and machine learning are being leveraged to design and predict the performance of novel filler combinations, accelerating the development cycle. Additionally, in-situ monitoring techniques are being developed to study the real-time behavior of coatings under operational conditions, enabling further optimization. The integration of multifunctional properties, such as self-healing, corrosion resistance, and thermal conductivity, into self-lubricating coatings represents another exciting avenue for future research.

5.10 CONCLUSION

i. The tribological performance of self-lubricating polymer coatings is significantly influenced by the incorporation of fillers, which enhance their mechanical properties, wear resistance, and frictional behavior. Fillers such as carbon-based materials, ceramics, and solid lubricants offer distinct advantages by improving load-bearing capacity, thermal stability, and lubrication under challenging operating conditions.

ii. Carbon-based fillers like CNTs and graphene contribute exceptional mechanical reinforcement and lubrication properties, while ceramic fillers like SiC and Al_2O_3 enhance hardness and thermal stability. Solid lubricants such as MoS_2 and graphite provide effective lubrication, especially in environments where liquid lubricants are ineffective. Hybrid fillers, which combine the strengths of different filler types, exhibit synergistic effects, further boosting tribological performance.

iii. The effectiveness of fillers depends on factors such as their concentration, dispersion within the polymer matrix, and interfacial adhesion with the base material. While excessive filler content can lead to agglomeration and diminished performance, optimized filler loading ensures uniform property enhancement.

iv. Recent advancements in functionalized and bio-inspired fillers point to exciting future possibilities for creating high-performance, sustainable coatings. As industries increasingly demand materials that combine efficiency and environmental responsibility, self-lubricating polymer coatings with advanced fillers will continue to play a crucial role in technological innovation.

v. Self-lubricating polymer coatings have garnered significant attention in various industries due to their ability to reduce friction and wear without the need for external lubricants. The incorporation of fillers into these coatings has emerged as a pivotal strategy to enhance their tribological properties. This chapter delves into the influence of different fillers on the performance of self-lubricating polymer coatings, exploring the mechanisms by which they operate and highlighting recent advancements in the field.

REFERENCES

Bhowmik, A., Kumar, R., Beemkumar, N., Kumar, A. V., Singh, G., Kulshreshta, A., … & Santhosh, A. J. (2024). Casting of particle reinforced metal matrix composite by liquid state fabrication method: A review. Results in Engineering, 24, 103152.

Bhowmik, A., Meher, A., Biswas, S., Dey, D., Kumar, M. S., Biswas, A., & Alsharabi, R. M. (2022). Synthesis and characterization of borosilicate glass powder-reinforced novel lightweight aluminum matrix composites. Advances in Materials Science and Engineering, 2022(1), 9487900.

Cai, Y., Cheng, L., Yin, X., Zhang, H., Yin, H., & Yan, G. (2019). Thermophysical properties of three-dimensional ceramic-filler-modified carbon/carbon composites. Ceramics International, 45(1), 1302–1307.

Chu, Z., Yang, Y., Chen, X., Yan, D., Huang, D., Lei, W., & Liu, Z. (2016). Characterization and tribology performance of Fe-based metallic glassy composite coatings fabricated by gas multiple-tunnel plasma spraying. Surface and Coatings Technology, 292, 44–48.

De Volder, M. F., Tawfick, S. H., Baughman, R. H., & Hart, A. J. (2013). Carbon nanotubes: Present and future commercial applications. Science, 339(6119), 535–539.

Eigler, S., & Hirsch, A. (2014). Chemistry with graphene and graphene oxide—Challenges for synthetic chemists. Angewandte Chemie International Edition, 53(30), 7720–7738.

Gangwani, P., Emami, N., & Kalin, M. (2024). Tribological behaviour of nano-titanium dioxide filled UHMWPE composites with a variety of micro fillers based on carbon, boron nitride and silicon dioxide under water-lubricated condition. Tribology International, 204, 110479.

Gao, T., Liu, J., Sun, X., Zhang, Y., Yang, M., Liu, M., ... & Xu, P. (2025). Enhanced Permeation Mechanism and Tribological Assessment of Ultrasonic Vibration Nanolubricants Grinding Cfrp. Enhanced Permeation Mechanism and Tribological Assessment of Ultrasonic Vibration Nanolubricants Grinding CFRP. Tribology International, 204, 110494.

Geweda, A. E., Saif, A. G. H., Zayed, M. E., Kabeel, A. E., Zafar, S., & Khalid, M. (2025). Recent advances in hybrid compressed air energy storage systems: Technology categorization, integration potentials with renewable energy systems, and retrofitting improvement strategies. Alexandria Engineering Journal, 115, 12–29.

Gong, H., Song, Y., Li, G. L., Zhang, L., Guo, D., & Xie, G. (2023). An intelligent polymer composite with self-lubricating and self-healing functionalities. Composites Part B: Engineering, 260, 110776.

Gupta, D., Chauhan, V., & Kumar, R. (2020). A comprehensive review on synthesis and applications of molybdenum disulfide (MoS_2) material: Past and recent developments. Inorganic Chemistry Communications, 121, 108200.

Haddon, R. C. (2002). Carbon nanotubes. Accounts of Chemical Research, 35(12), 997–997.

Hong, J. H., Park, D. W., & Shim, S. E. (2010). A review on thermal conductivity of polymer composites using carbon-based fillers: Carbon nanotubes and carbon fibers. Carbon Letters, 11(4), 347–356.

Kennedy, S. M., Vasanthanathan, A., RB, J. R., Amudhan, K., & Nagendran, M. (2024). Impact of fillers in enhancing the properties of CFRP composites–A comprehensive exploration. Next Research, 100117.

Lan, Z., Chen, Z., Zhang, M., Huang, W., Ding, D., Fan, B., ... & Ren, P. (2025). Improved mechanical and tribological properties of epoxy resin with new hyperbranched polysiloxane grafted MXene/Ag hybrids. Tribology International, 204, 110515.

Lin, S., Zhao, J., Wang, Z., Sun, F., Shi, J., Zhang, J., ... & Li, M. (2024). Lubricating properties of starch-soybean lecithin compound gels using tribology analysis. LWT, 207, 116409.

Liu, J., Wu, K., Liu, R., Sun, G., & Luo, J. (2023). Robust SiO_2/polymer hybrid microcapsule synthesized via emulsion photo-polymerization and its application in self-lubricating coatings. Progress in Organic Coatings, 184, 107856.

Liu, Y., Wang, Y., Hou, J., Shen, W., Zhang, X., Li, Z., ... & Wu, C. (2025). Improved tribological properties of MXene nanosheet filler-modified PPO composites. Journal of Colloid and Interface Science, 680, 618–631.

Moharana, A. P., Singh, S. S., & Dixit, A. R. (2024). Dynamic mechanical and viscoelastic properties of glass fiber reinforced photopolymer composite fabricated using vat-photopolymerization additive technique: Influence of filler volume fraction. Journal of Manufacturing Processes, 130, 72–86.

Mohtasim, M. S., & Das, B. K. (2024). Biomimetic and bio-derived composite phase change materials for thermal energy storage applications: A thorough analysis and future research directions. Journal of Energy Storage, 84, 110945.

Peng, Y., Zhao, T., Miao, J., Kong, L., Li, Z., Liu, M., … & Wang, W. (2024). Evaluation framework for bitumen-aggregate interfacial adhesion incorporating pull-off test and fluorescence tracing method. Construction and Building Materials, 451, 138773.

Pimenov, S. M., Zavedeev, E. V., Komlenok, M. S., Zilova, O. S., Jaeggi, B., & Neuenschwander, B. (2024). Laser surface structuring of diamond-like carbon films for tribology. Diamond and Related Materials, 148, 111462.

Ramírez, C., Belmonte, M., Miranzo, P., & Osendi, M. I. (2021). Applications of ceramic/graphene composites and hybrids. Materials, 14(8), 2071.

Rao, Y. S., Mohan, N. S., Shetty, N., & Shivamurthy, B. (2021). Effects of solid lubricant fillers on the flexural and shear strength response of carbon fabric-epoxy composites. Polymer Testing, 96, 107085.

Sarkar, A., & Gul, K. (2024). Glycation influencing lubrication: Tribology principles derived from nature to inspire future food colloid design. Current Opinion in Colloid & Interface Science, 69, 101782.

Shiao, S. J., Fu, W. S., Tuo, C. L., & Cheng, U. I. (2001). Polymer-based self-lubricating material. Journal of Applied Polymer Science, 80(9), 1514–1519.

Sun, H., Pang, L., Ding, Y., Xing, B., Tang, Y., Sun, X., … & Sha, F. (2023). Influence of the physical morphological characteristics of mineral fillers on the bitumen-filler interfacial interaction. Construction and Building Materials, 378, 131206.

Venturi, F. (2024). Carbon-based films: A review on mechanical and tribology properties. Materials Chemistry and Physics, 325, 129716.

Wang, A., Gao, X., Giese, R. F., Jr, & Chung, D. D. L. (2013). A ceramic–carbon hybrid as a high-temperature structural monolith and reinforcing filler and binder for carbon/carbon composites. Carbon, 59, 76–92.

Wang, Z., & Mi, B. (2017). Environmental applications of 2D molybdenum disulfide (MoS2) nanosheets. Environmental Science & Technology, 51(15), 8229–8244.

Yang, L., Yu, H., Peng, Z., Tian, X., Cheng, S., Yang, Y., … & Zhao, C. (2024). Application of iron-manganese oxide fillers in bioretention facility: Enhanced removal of antibiotics and conventional contaminants. Journal of Environmental Chemical Engineering, 12(3), 112744.

Ye, Y., Chen, H., Guo, S., Sun, S., Wang, Y., Zhao, H., & Wang, L. (2019). Comparison of WC/C nanocomposite films in load-bearing capacity via interface micro-cracks during tribology behavior: A feasible judgment mechanism. Ceramics International, 45(17), 23482–23492.

Zahra, T., & Kim, S. R. (2025). Recent advances in architectural designs and fabrication strategies of 2D fillers in polymer-based dielectric nanocomposites. Materials Today Energy, 101800.

Zgalat-Lozynskyy, O. B., Matviichuk, O. O., Tolochyn, O. I., Ievdokymova, O. V., Zgalat-Lozynska, N. O., & Zakiev, V. I. (2021). Polymer materials reinforced with silicon nitride particles for 3D printing. Powder Metallurgy and Metal Ceramics, 59, 515–527.

Zhang, N., Qu, C., Li, S., Wang, C., Xu, M., Yang, Z., … & Wang, Q. (2021). Significantly enhanced tribology and thermal management by dual-network graphene/epoxy composites. Tribology International, 164, 107239.

Zhang, Z., Zheng, X., Li, J., Xu, G., & Tan, L. (2024). Mechanism of reinforced interfacial adhesion between steel slag and highly devulcanized waste rubber modified asphalt and its influence on the volume stability in steel slag asphalt mixture. Construction and Building Materials, 447, 138129.

Zhu, Y., Murali, S., Cai, W., Li, X., Suk, J. W., Potts, J. R., & Ruoff, R. S. (2010). Graphene and graphene oxide: Synthesis, properties, and applications. Advanced Materials, 22(35), 3906–3924.

6 Development of Polymer Nanocomposites with Natural Fillers

Assessing Tribological Properties, Wear, and Surface Roughness

Harshit Majumder, Sumit Das Lala, Payel Deb, Nazrul Islam Khan, and Rajdeep Paul

6.1 INTRODUCTION

Tribology, the study of friction, wear, and lubrication, plays a critical role in the design and performance of mechanical components that experience relative motion [Wang et al., 2018]. The term encompasses the science of interacting surfaces in motion and is applied across various engineering fields, particularly in the design and maintenance of rolling or sliding components such as bearings, gears, cams, brakes, and seals. These components are pivotal in industrial machinery, vehicles, and other mechanical systems, where controlling friction and wear can enhance performance, reduce energy consumption, and extend the operational life span. A tribometer is commonly used to measure the friction and wear characteristics of tribological systems, providing valuable data for material selection and design optimization. As industries move toward more sustainable practices and technologies, understanding the tribological behavior of alternative materials, especially in relation to natural composites, becomes essential.

One such alternative material gaining attention is the bio-composite, which is engineered by combining natural fibers with a matrix material, forming a composite that is degradable, renewable, non-toxic, and non-abrasive. Bio-composites are increasingly favoured due to their environmental benefits over synthetic fiber composites, which pose challenges in terms of recycling and hazardous by-products [Gurunathan et al., 2015]. Natural fibers, sourced from agricultural waste and other renewable resources, present a cost-effective and sustainable reinforcement for polymer matrices [Joshi et al., 2004]. These fibers have adequate mechanical properties and their abundance makes them an attractive option for developing composite materials. Essabir et al. [2013] reported that the tensile properties of the composite increase with decreasing the particle size of the filler. The evaluation of the thermal properties shows that the

DOI: 10.1201/9781003530909-8

polypropylene-based nanoparticles particles composites were less thermally stable than the polypropylene matrix. Essabir et al. [2013] investigated the properties of Argan nut shell–reinforced polyethylene composite and observed that the mechanical properties of the composites show a significant increase in Young's modulus with the addition of filler. A gain of 58% is marked, comparing a composite specimen to a neat polymer. Ndiaye et al. [2013] reported that the inclusion of wood in the composite led to imperfect adhesion, which elevated the stress concentration and reduced the impact strength. As the wood flour content increases, the composite becomes more polar, promoting hydrogen bonding, resulting in greater water absorption and increased swelling thickness. However, limited research has been conducted on the tribological behavior of bio-composites, particularly those reinforced with natural fillers. Veena et al. [2011] investigated the tribological and electrical properties of epoxy composites containing nano-sized silica particles. The epoxy resin exhibited optimal wear resistance at a silica loading of 10 wt.%, but further increases in silica content led to particle aggregation, which in turn caused a rise in the wear rate. Chen et al. [2007] studied the effects of different dispersion techniques on the tribological properties of carbon nanotube (CNT)-reinforced epoxy composites. They employed asymmetric centrifugation, sonication, and hand mixing, finding that better dispersion, particularly through sonication combined with dual asymmetric centrifugation, enhanced wear resistance. Further improvements were observed when the CNTs were pretreated with HNO_3 or silane-coupling agents, although sonication of pre-treated CNTs appeared to cause damage. Overall, optimal CNT dispersion and maintaining their structural integrity contributed to improved wear resistance. Polymer tribology is mainly determined by friction-adhesion and deformation mechanisms [Rana et al., 2003; Rout et al., 2001]. Hence, attempts were made to develop polymer bio-composites and test their tribological behavior.

This study aims to bridge that gap by investigating the tribological and morphological properties of bio-composites developed from natural fillers. The development of such composites holds the potential to create lighter, more cost-effective, and environmentally friendly materials for use in industries like automotive manufacturing, where traditional materials are often heavier, more expensive, and less sustainable.

6.2 EXPERIMENTATION

6.2.1 Pretreatment of Coconut Shell

The objective is to use coconut shell filler as a reinforcement in the polymer matrix to fabricate composite material. Initially, the shells were pretreated through an alkali treatment process, utilizing a 10 wt.% NaOH solution. The shells were dipped in 10% NaOH solution for a period of 24 hours at 90°C to remove the inherent constituents, such as oil, wax, grease, etc. The treatment also removed the added moisture in the filler and made their surface rough, thus allowing a suitable bonding with the polymer matrix. Following this, the shells were thoroughly washed with water and acetic acid to remove the excess NaOH added on the surface of the filler. Finally, they were heated at 90°C for 24 hours to complete the treatment process. Figure 6.1 shows the pretreatment of the coconut shells.

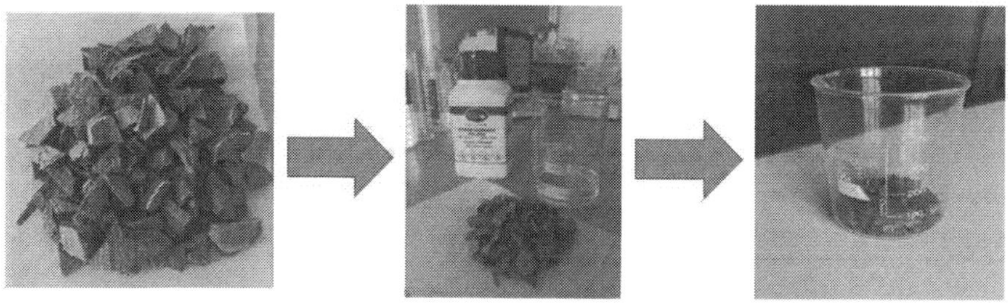

FIGURE 6.1 Pretreatment of coconut shell using 10% NaOH solution.

6.2.2 Fabrication of Coconut Shell Composite

The treated seed shells were then grinded in a grinding machine into a fine powder form. The grinded powders were then reinforced in the epoxy matrix to fabricate the composite materials. The composite was developed using 5%, 10%, 15%, and 20% wt. of filler reinforcement in the epoxy matrix. The composite slurry was stirred at 500 rpm for 1 hour at 90°C. The temperature was then reduced and a hardener was added for hardening of the composite. The slurry was then poured in the silicon rubber mold and cured in s hot air oven for 2 hours at 90°C (Figure 6.2).

6.3 RESULTS AND DISCUSSION

6.3.1 Wear Test

The wear test was conducted on all the samples using a pin-on-disc tribometer at two different loading conditions: 5 N and 10 N for all the composite specimens. The results showed a slight increase in wear properties with increasing filler content at the 5 N loading condition. However, with increasing the loading condition to 10 N, there was significant increase in wear of the composite specimens. Table 6.1 shows the wear values in microns for all the composite specimens.

6.3.2 Roughness

The roughness of all the composite samples was measured using the Mitutoyo surface roughness tester. The roughness values corresponding to R_a (arithmetic average roughness), R_q (root mean square roughness), and R_z (average maximum height of the profile)

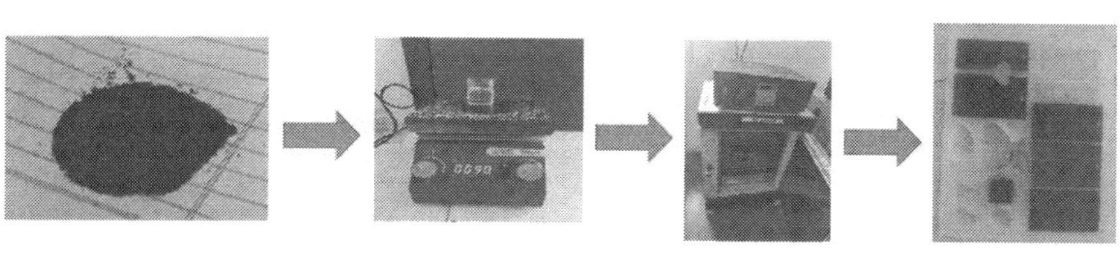

FIGURE 6.2 Development of composite specimen.

TABLE 6.1
Wear Values in Microns for All the Composite Specimens

Constituents	Average wear in microns			
Wt.% Load (N)	5% wt. coconut specimen	10% wt. coconut specimen	15% wt. coconut specimen	20% wt. coconut specimen
5	9	11	12	13
10	10	13	15	18

TABLE 6.2
Roughness of All the Composite Samples

Values in microns	5% wt. composite	5% wt. composite	5% wt. composite	5% wt. composite
R_a	0.935	0.943	0.947	0.958
R_q	1.515	1.518	1.519	1.520
R_z	8.011	8.0114	8.015	8.016

of all the composite specimens are tabulated in Table 6.2. The results revealed that with increasing the filler content, an increase in roughness was observed for the developed composite materials.

6.4 CONCLUSIONS

The present study highlights the development of bio-waste coconut filler–reinforced epoxy composite. The composite was tested for wear and roughness properties. The results of the study indicate that coconut filler–reinforced epoxy composites exhibit lower surface roughness compared to other materials. Wear testing also confirms that coconut filler composites experience less wear. The composite developed can be successfully used in automotive components.

REFERENCES

H. Chen, O. Jacobs, W. Wua, G. Rüdiger, and B. Schädel, "Effect of dispersion method on tribological properties of carbon nanotube reinforced epoxy resin composites", Polymer Testing, 26, (2007) 351–360.

H. Essabir, E. Hilali, A. Elgharad, H. El Minor, A. Imad, A. Elamraoui, and O. Al Gaoudi, "Mechanical and thermal properties of bio-composites based on polypropylene reinforced with Nut-shells of Argan particles", Materials & Design, 49, (2013) 442–448.

T. Gurunathan, S. Mohanty, and S. K. Nayak, "A review of the recent developments in biocomposites based on natural fibres and their application perspectives", Composites Part A: Applied Science and Manufacturing, 77, (2015) 1–25.

S. V. Joshi, L. T. Drzal, A. K. Mohanty, and S. Arora, "Are natural fiber composites environmentally superior to glass fiber reinforced composites?", Composites A, 35, (2004) 371–376.

D. Ndiaye, M. Gueye, & B. Diop, "Characterization, physical and mechanical properties of Polypropylene/Wood-flour composites", Arabian Journal for Science and Engineering, 38, (2013) 59–68.

A. K. Rana, A. Mandal, and S. Bandyopadhyay, "Short jute fiber reinforced polypropylene composites: Effect of compatibiliser, impact modifier and fiber loading", Composites Science and Technology, 63, (2003) 801–806.

J. Rout, M. Misra, S. S. Tripathy, S. K. Nayak, and A. K. Mohanty, "The influence of fibre treatment of the performance of coirpolyester composites", Composites Science and Technology, 61, (2001) 1303–1310.

M. G. Veena, N. M. Renukappa, B. S. Suresha, and K. N. S, "Tribological and electrical properties of silica-filled epoxy nanocomposites", Polymer Composites, 32, (2011) 2038–2050.

J. Wang, K. Zhang, L. Zhang, F. Wang, J. Zhang, and W. Zheng "Influence of structure evolution on tribological properties of fluorine-containing diamond-like carbon films: From fullerene-like to amorphous structures", Applied Surface Science, 457, (2018) 388–395.

7 Influence of Micro-Sized Filler Reinforcement on the Tribological Performance of Metal Matrix Composites

Katika Harikrishna, K. Benarji, Abeyram Nithin, and Suyash Pawar

7.1 INTRODUCTION

The relentless increase in industrial activities over the past two decades has demanded the development of innovative multifunctional materials that surpass the capabilities of traditional monolithic alloys (Seah et al. 1996). Metal matrix composites (MMCs) represent a subset of composite materials, comprising a metal or alloy matrix embedded with metallic or non-metallic micro- and nano-scale reinforcements (Borgonovo and Apelian 2011). These composites are highly required after for their remarkable properties, including superior mechanical strength, enhanced wear resistance, high toughness, improved castability, and first-rate strength-to-weight ratios, all of which make them indispensable in up-to-date applications (Patnaik et al. 2012). Their adaptability to diverse requirements has rendered them vital in industries such as automotive, aerospace, electronics, and aviation, offering the advantage of modifying material properties to meet specific demands (Bujari and Kurahatti 2016).

Commonly used metal matrices include Al, Zn, Mg, Ti, and Cu, while typical reinforcements involve ceramics like SiC, Al_2O_3, TiO_2, and TiB_2, along with sustainable options like rice husk ash and industrial by-products such as fly ash and red mud. The selection of reinforcements is crucial, as it significantly influences the overall performance of MMCs. The evolution from micro- to nano-sized reinforcements, coupled with the introduction of hybrid composites, has unlocked new options for advanced engineering applications. Reduced reinforcement sizes lead to notable advances in the physical and tribological properties of MMCs (Bodunrin, Alaneme, and Chown 2015).

Various engineering techniques, including stir casting, in-situ methods, spray deposition, diffusion bonding, and pressure vapor deposition, have been utilized for fabricating MMCs. Among these, stir casting stands out due to its simplicity,

 DOI: 10.1201/9781003530909-9

cost-effectiveness, and suitability for mass production. Despite certain challenges related to the cost of reinforcements, the isotropic properties and affordability of MMCs ensure their widespread applicability (Kala, Mer, and Kumar 2014). This review delves into various aspects of micro-sized fillers, including their fabrication techniques, types, and properties, as well as their influence on tribological performance. It further examines the challenges, applications, and recent advancements associated with micro-sized reinforcements, with a primary focus on their tribological performance.

7.2 FABRICATION TECHNIQUES

The processing methods for manufacturing MMCs are broadly divided into five categories based on the temperature range of the metallic matrix as given below (Singh et al. 2019) (Rosso 2006).

7.2.1 Liquid-Phase Processes

This method uses particular proprietary techniques to incorporate reinforcement materials into a molten metal matrix. After giving the mixture a good stir, it is formed into billets or other pre-made shapes for further processing. Liquid metal ceramic particulate incorporation, melt infiltration, melt oxidation, and pressure-assisted infiltration (like squeeze casting) are a few popular techniques that fall under this category.

7.2.2 Solid-Phase Processes

Using this process, metallic powders are combined with reinforcement materials to create reinforced MMCs. There are several steps in the process, which ends with final consolidation. This category includes methods like high-energy processing, diffusion bonding, and powder metallurgy.

7.2.3 Two-Phase Processes

This method makes use of a phase diagram area where both solid and liquid phases are present in the metallic matrix. In these circumstances, reinforcement and matrix materials are mixed together. Spray deposition, co-deposition of multiphase materials, and compocasting (or rheocasting) are notable techniques in this category.

7.2.4 Deposition Techniques

To create MMCs, these techniques coat reinforcement materials—like fibers—with the matrix material. Diffusion bonding and other processes are then used to consolidate the coated materials into the final composite shapes. Chemical vapor deposition (CVD), physical vapor deposition (PVD), spray deposition, electroplating, immersion plating, and spray forming are a few types of deposition techniques.

7.2.5 In-Situ Processes

This method uses cooling and solidification to create the reinforcement phase inside the metallic matrix. This is accomplished by in-situ precipitation, in which a precursor alloy is directly converted into a composite in a single step.

7.3 MICRO-SIZED FILLERS: TYPES AND PROPERTIES

Micro-sized fillers are essential for improving MMCs' mechanical qualities, wear resistance, and thermal stability. These fillers are essential for sophisticated applications because they alter the composite's tribological and thermal properties in addition to enhancing its structural integrity.

7.3.1 Silicon Carbide (SiC)

The remarkable hardness, thermal conductivity, and stability at high temperatures of SiC particles are well known. In high-friction situations, these characteristics help to promote heat dissipation and lower wear rates. SiC fillers also have outstanding chemical inertness, which reduces reactions at the filler-matrix interface and increases the life span of composites (Basavarajappa et al. 2006) (Veeresh Kumar, Rao, and Selvaraj 2012).

7.3.2 Alumina (Al_2O_3)

Particles of AlO_3 have exceptional corrosion resistance, thermal insulating qualities, and outstanding hardness. Their incorporation into MMCs improves oxidative stability and wear resistance, making them appropriate for use in high-temperature settings (Ünlü 2008).

7.3.3 Graphite (GR)

The solid lubricating qualities of graphite are widely known for drastically reducing the coefficient of friction. In addition to enhancing wear performance, graphite fillers prolong the component life by increasing thermal stability and decreasing surface roughness during sliding wear (Yang et al. 2004).

7.3.4 Magnesium-Based Fillers

Fillers based on magnesium, including MgO and MgSiC, have special qualities like increased damping capacity and light weight. In addition to stabilizing oxide films during wear, these fillers serve as a barrier to prevent material removal under moderate to high loads (Labib, Ghasemi, and Mahmudi 2016) (Lim, Lim, and Gupta 2003).

7.3.5 Titanium Diboride (TiB_2)

TiB_2 particles are distinguished by their exceptional hardness, electrical conductivity, and high melting point. Their addition to MMCs improves their electrical and thermal conductivities as well as their resistance to wear, making them appropriate for a variety of multipurpose applications (Radhika and Raghu 2015).

7.3.6 Boron Carbide (B_4C)

B_4C particles have a remarkable low density, chemical inertness, and hardness combination. Because of these characteristics, they are perfect for applications that call for low weight addition and great wear resistance. Additionally, B_4C can create a tribo-film while worn, minimizing material loss and extending the composite's life (Seetharam et al. 2022) (Thirumalai Kumaran and Uthayakumar 2014).

7.3.7 Fly Ash

A by-product of burning coal, fly ash is inexpensive and lightweight. By improving the composite microstructure and switching from delamination to adhesive or abrasive wear, its use in MMCs increases wear resistance. Additionally, fly ash lowers the material's overall density and improves resistance to thermal shock (Ramachandra and Radhakrishna 2007).

7.3.8 Hybrid Fillers

Combinations of SiC with AlO_3, B_4C, or graphite are examples of hybrid fillers that take advantage of the complementing qualities of each filler. These combinations improve thermal conductivity, lower friction, and increase hardness and wear resistance. In order to get customized composite characteristics for particular applications, hybrid fillers are especially beneficial (Megahed et al. 2017).

7.4 INFLUENCE OF MICRO-SIZED FILLERS ON TRIBOLOGICAL PERFORMANCE

MMCs demonstrate less wear rate compared to their unreinforced alloy counterparts. Researchers have extensively explored the wear behavior of MMCs by incorporating various micro-sized fillers as reinforcements. Basavarajappa et al. (2006) analyzed the wear properties of Al2219 composites reinforced with SiCp and Gr, featuring particle sizes of 25 µm and 45 µm, respectively. They observed that the addition of graphite particles to Al2219/SiCp composites significantly enhanced the wear resistance by forming a graphite smearing layer at the pin-counterface interface. This reinforcement effectively lowered wear rates and increased the transition load from mild to severe wear. Veeresh Kumar, Rao, and Selvaraj (2012) examined the dry sliding behavior of Al6061-SiC composites prepared via liquid metallurgy. Their findings revealed that Al6061 with 6 wt.% SiC exhibited better wear resistance compared to composites with lower SiC contents (2 wt.% and 4 wt.%) and the base Al6061 alloy. Ünlü (2008) investigated Al composites reinforced with Al2O3 (15–20 µm) and SiC (55–60 µm). Results indicated that wear loss was reduced by 1.5–2 times due to decreased abrasive and adhesive wear, showcasing the beneficial role of these fillers in enhancing wear resistance. Labib, Ghasemi, and Mahmudi (2016) studied Mg/SiCp composites produced using powder metallurgy. SiC particles with an average size of 7.8 µm were incorporated. The study identified wear mechanisms such as oxidation at lower loads and adhesion coupled with severe plastic deformation at higher loads, with better wear performance observed in composites containing 15% SiC. Lim, Lim, and Gupta (2003) examined the wear behavior of Mg composites with SiCp (14 μm) reinforcement under dry sliding conditions. Under a load of 10 N, the dominant wear mechanism was oxidation. The composites demonstrated a 15–30% improvement in wear resistance due to their superior load-bearing capacity and stable oxide film formation. Radhika and Raghu (2015) utilized response surface methodology to study the wear performance of Al–Si12Cu/TiB_2 composites with TiB_2 particles (50–60 μm). Results presented that wear rate amplified with load and reduced with sliding distance, while worn surface analysis highlighted the protective role of the mechanically mixed layer. García-Rodríguez et al. (2017) assessed AZ91/SiCp composites fabricated via rheocasting. SiCp particles (25.1 ± 6.8 μm) were observed in the β-phase of the matrix. The authors noted oxidative wear dominance and a delayed transition to delamination wear with increased SiCp content. Baskaran, Anandakrishnan, and Duraiselvam (2014) investigated AA7075–TiC composites using Taguchi techniques. The study found that incorporating 4 wt.% TiC particles (2 μm) optimized wear resistance under specific conditions, including a sliding velocity of 3 m/s and load of 9.81 N.

Ahmad et al. (2024) investigated the manufacture and characterization of Al-7075/TiO_2 MMCs by stir casting, with TiO_2 contents ranging from 2% to 10%. Tribological testing revealed a decrease in wear rate and reduced COF, as shown in Figure 7.1, for different conditions, highlighting the potential of these composites for high-strength and wear-resistant applications. In Figure 7.2, the factor combination (40 N, 4% TiO_2, 250°C, 1.885 m/s) demonstrates plastic yielding caused by heat generation.

The study by Yunus and Alfattani (2023) explores AA6061-based hybrid metal matrix composites (HMMCs) reinforced with B_4C and Gr using dual stir casting. The reinforcement particles are uniformly distributed, improving wear and friction, especially with higher B_4C and Gr content. Figure 7.3 shows significant improvements in mechanical properties, with Gr enhancing wear resistance through lubrication. Figure 7.4 highlights plastic yielding due to generation of heat

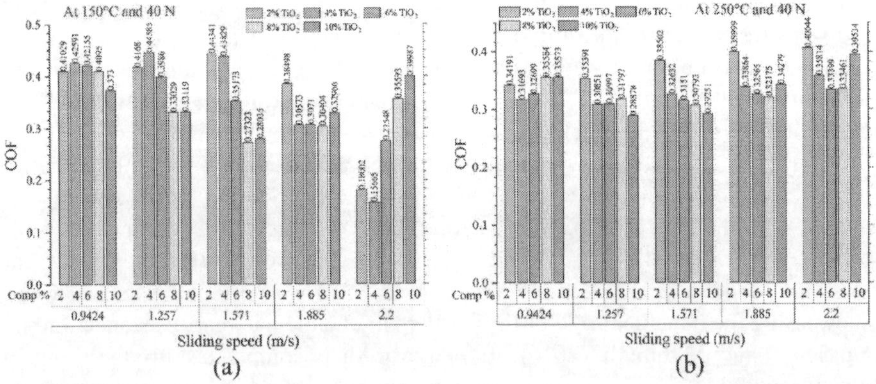

FIGURE 7.1 Effect of sliding speed on COF for various percentages (Ahmad et al. 2024).

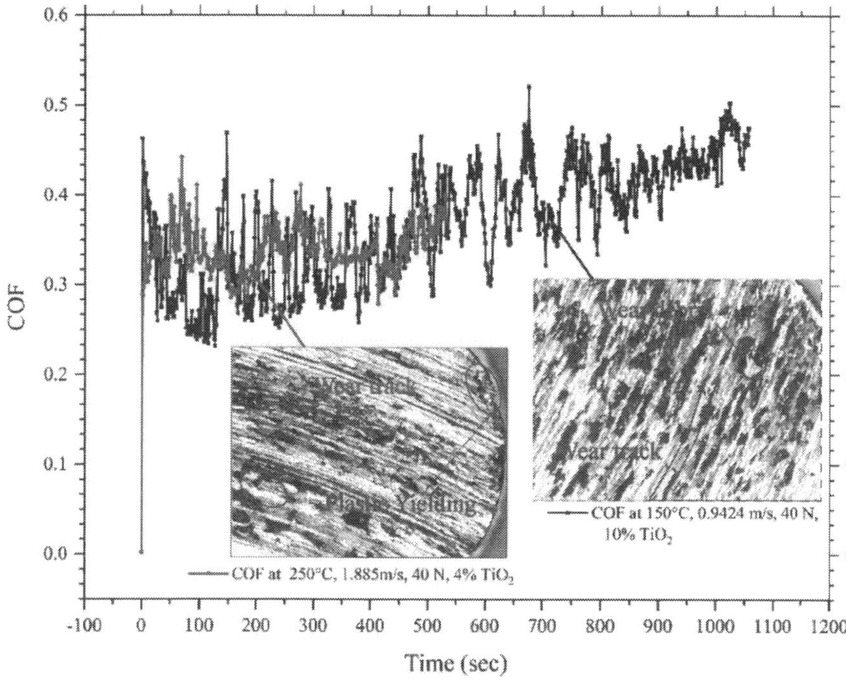

FIGURE 7.2 COF variation with time at different combinations (Ahmad et al. 2024).

and the surface of smooth tracks. COF values increase with the load but decrease with higher Gr content, reaching a minimum at 15% Gr (Figure 7.4a). The composite exhibits better tribological properties, with Gr acting as a lubricant, while B_4C reduces COF until 10%, and then increases due to a thicker mechanically mixed layer (MML) (Figure 7.4b).

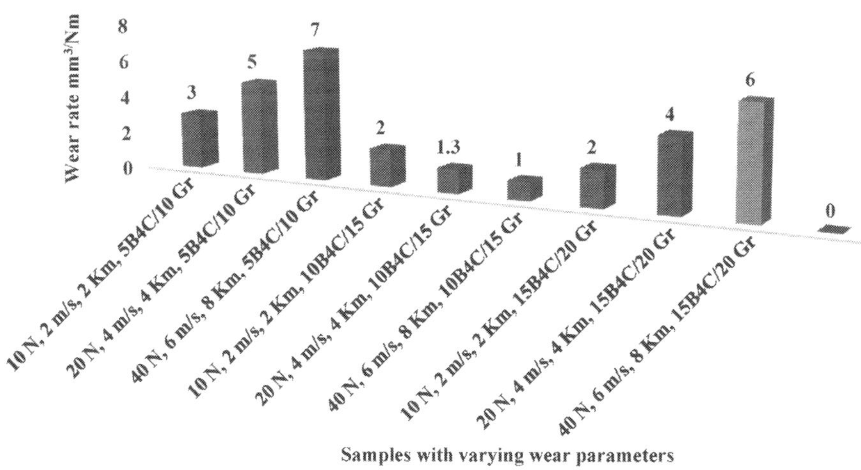

FIGURE 7.3 Wear rate vs. load of B_4C/Gr hybrid composite (Yunus and Alfattani 2023).

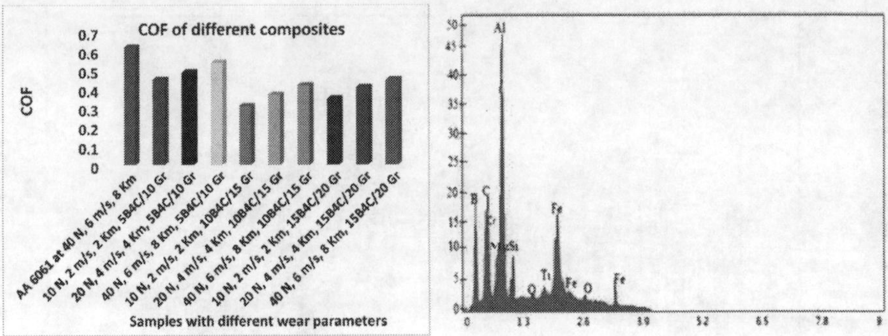

(a) (b)

FIGURE 7.4 (a) Coefficient of friction (COF) of different HMMCs as a function of varying parameters and (b) Energy dispersive spectroscopy (EDS) analysis results following wear (Yunus and Alfattani 2023).

Kumaran and Uthayakumar (Kumaran et al. 2016) studied AA6351-SiC-B4C hybrid MMCs. Reinforcements with particle sizes of 69 μm (SiC) and 149 μm (B4C) significantly enhanced wear resistance, with hybrid composites retaining properties under loads up to 60 N and sliding speeds of 1–3 m/s. Umanath et al. (Umanath, Palanikumar, and Selvamani 2013) analyzed Al6061/SiC/Al2O3 hybrid MMCs, concluding that reinforcement volume fraction strongly influenced wear performance. Increased reinforcement and counterface hardness improved wear resistance, while higher applied loads and rotational speeds elevated wear rates. Suresha and Sridhara (Suresha and Sridhara 2010) employed central composite design to investigate Al-SiC-Gr composites. Sliding distance was identified as the most influential parameter affecting wear. The study also showed reduced wear rates with higher reinforcement content up to a threshold, after which wear tended to increase. Tang et al. (Tang et al. 2008) explored Al-5083 composites reinforced with B_4C particles (4 μm). Scanning electron microscope (SEM) results highlighted uniform dispersion of reinforcements, and a 40% improvement in wear resistance was achieved with 10 wt.% B_4C compared to 5 wt.%. Table 7.1 presents some fillers used as reinforcements and their effects on the tribological performance.

7.5 CHALLENGES IN MICRO-FILLER-REINFORCED MMCs

7.5.1 Dispersion and Agglomeration

Achieving uniform dispersion of micro-fillers is critical. Agglomeration leads to weak spots and uneven wear patterns.

7.5.2 Interfacial Bonding

Strong bonding between the matrix and fillers is essential for effective load transfer.

TABLE 7.1
Influence of Filler Reinforcement on the Tribological Performance of Metal Alloy Composites

S. No.	Filler Reinforcement	Matrix	Methodology	Tribological Performance	Reference
1.	Graphite	Al6082	Casting	The sliding distance is a key factor influencing wear	(Sharma, Khanduja, and Sharma 2016)
2.	B4C	Al	Casting	The wear rate and COF increase, while the sliding distance and velocity decrease with an increase in load	(Yunus and Alfattani 2023)
3.	Graphite	Zinc alloy (ZA 27)	Compo-casting	Wear resistance improved	(Girish et al. 2011)
4.	SiC	Zinc-based alloy	Vortex technique	Decrease in wear rate	(Prasad et al. 1997)
5.	TiC	Copper	Friction stir processing	Decrease in wear rate	(Sabbaghian et al. 2014)
6.	TiC	Al-12Si	In-situ reaction	As the load and sliding distance increase, the weight loss also increases	(Yigezu, Mahapatra, and Jha 2013)
7.	Flyash	A356	Stir cast technique	Dry sliding wear rates decrease	(Sudarshan and Surappa 2008)
8.	Garnet	Zinc alloy (ZA 27)	Compo-casting	Wear rate decreases	(Ranganath, Sharma, and Krishna 2001)
9.	ZrB2			Wear rate decreases	(Dinaharan, Murugan, and Parameswaran 2011)
10.	SiC	Copper	Forge sintering	Less wear rate	(Shabani et al. 2016)

7.5.3 Cost and Scalability

While advanced fabrication techniques yield superior properties, they are often cost-intensive and difficult to scale.

7.6 APPLICATIONS OF MICRO-FILLER-REINFORCED MMCs

Micro-filler-reinforced MMCs have found extensive applications across diverse industries due to their superior wear resistance, mechanical strength, and thermal stability. In the automotive sector, these composites are used for brake pads, pistons, and engine components to enhance wear performance and reduce weight. They are used for structural elements in the aerospace sector where tolerance to high temperatures and high strength-to-weight ratios are crucial. These composites' improved thermal and electrical conductivities make them perfect for heat sinks and conductive pathways, which benefit the electrical and electronics industries. These MMCs are used in machining and tooling to create cutting tools, where increased wear resistance and hardness are essential. Additionally, the biocompatibility and high strength of specific filler combinations are utilized in medical applications like orthopedic implants. These composites are adaptable for a variety of specialized applications due to the filler selection's ability to modify characteristics.

7.7 RECENT ADVANCEMENTS AND FUTURE DIRECTIONS

The creation of hybrid composites to attain multifunctional characteristics is the main focus of recent developments in micro-filler-reinforced MMCs. Composites with complex geometries and improved homogeneity can now be produced thanks to the combination of sophisticated manufacturing techniques like spark plasma sintering and additive manufacturing. The combination of nano-scale and micro-fillers has demonstrated the potential for enhancing mechanical and tribological characteristics. Furthermore, filler-matrix interactions are optimized and material behavior is predicted through the use of computational modeling and machine learning techniques. Future research is likely to emphasize eco-friendly and sustainable fillers, aiming to reduce environmental impacts. The use of biodegradable fillers and recycled materials for MMC production is gaining traction. Furthermore, the exploration of self-healing MMCs and composites with adaptive properties could revolutionize their applications. These developments will continue to expand the potential of micro-filler-reinforced MMCs in advanced industries.

7.8 CONCLUSION

This study explores the experimental and theoretical findings reported by researchers in the field of MMCs. These materials exhibit exceptional properties, inspiring significant research into their development, physical and mechanical characteristics, tribological behavior, fabrication methods, and applications. The key conclusions from the research in this domain can be summarized as follows:

- The stir casting method is widely utilized for fabricating composites due to its cost-effectiveness and simplicity.
- The properties of MMCs are influenced by fabrication parameters, reinforcement size, and the characteristics of the reinforcement material.
- Nano-sized reinforcements significantly enhance material properties due to their interaction with the matrix at the atomic level, making these composites suitable for advanced applications such as aerospace components, commercial aircraft, automobiles, electronic devices, and prosthetics.
- Composite materials exhibit better sliding wear resistance than the base matrix. This rate is decreased by increasing the filler content and is influenced by parameters such as sliding velocity, applied load, and sliding distance.

- Hybrid composites represent a new class of MMCs that leverage the advantages of multiple alloys while mitigating their limitations. By combining reinforcements with different compositions, superior properties can be achieved, surpassing those of monolithic alloys.

REFERENCES

Ahmad, Shadab, Yebing Tian, Abdul Wahab Hashmi, Rajneesh Kumar Singh, Faiz Iqbal, Sonia Dangi, Abdulkarim Alansari, Chander Prakash, and Choon Kit Chan. 2024. "Experimental Studies on Mechanical Properties of Al-7075/TiO2 Metal Matrix Composite and Its Tribological Behaviour." *Journal of Materials Research and Technology* 30: 8539–52. https://doi.org/10.1016/j.jmrt.2024.05.227

Basavarajappa, S., G. Chandramohan, K. Mukund, M. Ashwin, and M. Prabu. 2006. "Dry Sliding Wear Behavior of Al 2219/SiCp-Gr Hybrid Metal Matrix Composites." *Journal of Materials Engineering and Performance* 15 (6): 668–74. https://doi.org/10.1361/105994906X150803

Baskaran, S., V. Anandakrishnan, and Muthukannan Duraiselvam. 2014. "Investigations on Dry Sliding Wear Behavior of in Situ Casted AA7075-TiC Metal Matrix Composites by Using Taguchi Technique." *Materials and Design* 60: 184–92. https://doi.org/10.1016/j.matdes.2014.03.074

Bodunrin, Michael Oluwatosin, Kenneth Kanayo Alaneme, and Lesley Heath Chown. 2015. "Aluminium Matrix Hybrid Composites: A Review of Reinforcement Philosophies; Mechanical, Corrosion and Tribological Characteristics." *Journal of Materials Research and Technology* 4 (4): 434–45. https://doi.org/10.1016/j.jmrt.2015.05.003

Borgonovo, Cecilia, and Diran Apelian. 2011. "Manufacture of Aluminum Nanocomposites: A Critical Review." *Materials Science Forum* 678: 1–22. https://doi.org/10.4028/www.scientific.net/MSF.678.1

Bujari, Suresh S., Dr., and R. V. Kurahatti. 2016. "A Review on Processing and Tribological Properties of Metal Matrix Composites." *International Journal of Advancement in Engineering Technology, Management & Applied Sciences* 3: 302–10.

Dinaharan, I., N. Murugan, and Siva Parameswaran. 2011. "Influence of in Situ Formed ZrB2 Particles on Microstructure and Mechanical Properties of AA6061 Metal Matrix Composites." *Materials Science and Engineering: A* 528 (18): 5733–40. https://doi.org/10.1016/j.msea.2011.04.033

García-Rodríguez, S., B. Torres, A. Maroto, A. J. López, E. Otero, and J. Rams. 2017. "Dry Sliding Wear Behavior of Globular AZ91 Magnesium Alloy and AZ91/SiCp Composites." *Wear* 390–391 (October 2016): 1–10. https://doi.org/10.1016/j.wear.2017.06.010

Girish, B. M., K. R. Prakash, B. M. Satish, P. K. Jain, and Kameshwary Devi. 2011. "Need for Optimization of Graphite Particle Reinforcement in ZA-27 Alloy Composites for Tribological Applications." *Materials Science and Engineering: A* 530 (1): 382–88. https://doi.org/10.1016/j.msea.2011.09.100

Kala, Himanshu, K.K.S. Mer, and Sandeep Kumar. 2014. "A Review on Mechanical and Tribological Behaviors of Stir Cast Aluminum Matrix Composites." *Procedia Materials Science* 6: 1951–60. https://doi.org/10.1016/j.mspro.2014.07.229

Kumaran, S., M. Thirumalai, S. Uthayakumar, S. Aravindan, and S. Rajesh. 2016. "Dry Sliding Wear Behavior of SiC and B4C-Reinforced AA6351 Metal Matrix Composite Produced by Stir Casting Process." *Proceedings of the Institution of Mechanical Engineers, Part L: Journal of Materials: Design and Applications* 230 (2): 484–91. https://doi.org/10.1177/1464420715579302

Labib, F., H. M. Ghasemi, and R. Mahmudi. 2016. "Dry Tribological Behavior of Mg/SiCp Composites at Room and Elevated Temperatures." *Wear* 348–349: 69–79. https://doi.org/10.1016/j.wear.2015.11.021

Lim, C. Y.H., S. C. Lim, and M. Gupta. 2003. "Wear Behaviour of SiCp-Reinforced Magnesium Matrix Composites." *Wear* 255 (1–6): 629–37. https://doi.org/10.1016/S0043-1648(03)00121-2

Megahed, M., M. A. Attia, M. Abdelhameed, and A. G. El-Shafei. 2017. "Tribological Characterization of Hybrid Metal Matrix Composites Processed by Powder Metallurgy." *Acta Metallurgica Sinica (English Letters)* 30 (8): 781–90. https://doi.org/10.1007/s40195-017-0568-5

Patnaik, Amar, T. G. Mamatha, Sandhyarani Biswas, and Predeep Kumar. 2012. "Damage Assessment of Titania Filled Zinc-Aluminum Alloy Metal Matrix Composites in Erosive Environment: A Comparative Study." *Materials and Design* 36: 511–21. https://doi.org/10.1016/j.matdes.2011.11.054

Prasad, B. K., S. Das, A. K. Jha, O. P. Modi, R. Dasgupta, and A. H. Yegneswaran. 1997. "Factors Controlling the Abrasive Wear Response of a Zinc-Based Alloy Silicon Carbide Particle Composite." *Composites Part A: Applied Science and Manufacturing* 28 (4): 301–8. https://doi.org/10.1016/S1359-835X(96)00115-7

Radhika, N., and R. Raghu. 2015. "Dry Sliding Wear Behaviour of Aluminium Al-Si12Cu/TiB_2 Metal Matrix Composite Using Response Surface Methodology." *Tribology Letters* 59 (1). https://doi.org/10.1007/s11249-015-0516-3

Ramachandra, M., and K. Radhakrishna. 2007. "Effect of Reinforcement of Flyash on Sliding Wear, Slurry Erosive Wear and Corrosive Behavior of Aluminium Matrix Composite." *Wear* 262 (11–12): 1450–62. https://doi.org/10.1016/j.wear.2007.01.026

Ranganath, G., S. C. Sharma, and M. Krishna. 2001. "Dry Sliding Wear of Garnet Reinforced Zinc/Aluminium Metal Matrix Composites." *Wear* 250 (251): 1408–13. https://doi.org/10.1016/S0043-1648(01)00781-5

Rosso, M. 2006. "Ceramic and Metal Matrix Composites: Routes and Properties." *Journal of Materials Processing Technology* 175 (1–3): 364–75. https://doi.org/10.1016/j.jmatprotec.2005.04.038

Sabbaghian, M., M. Shamanian, H. R. Akramifard, and M. Esmailzadeh. 2014. "Effect of Friction Stir Processing on the Microstructure and Mechanical Properties of Cu-TiC Composite." *Ceramics International* 40 (8): 12969–76. https://doi.org/10.1016/j.ceramint.2014.04.158

Seah, K. H.W., S. C. Sharma, B. M. Girish, and S. C. Lim. 1996. "Wear Characteristics of As-Cast ZA-27/Graphite Particulate Composites." *Materials and Design* 17 (2): 63–67. https://doi.org/10.1016/S0261-3069(96)00033-7

Seetharam, R., S. Kanmani Subbu, M. J. Davidson, K. R. Ramkumar, and Pagidi Madhukar. 2022. "Influence of Reinforcement Particles on Dynamically Recrystallized Grain of Hot Upset Sintered Al-B4C Composites." *Journal of Materials Engineering and Performance*, no. Ref 24. https://doi.org/10.1007/s11665-022-06955-6

Shabani, Mohammadmehdi, Mohammad Hossein Paydar, Reza Zamiri, Maryam Goodarzi, and Mohammad Mohsen Moshksar. 2016. "Microstructural and Sliding Wear Behavior of SiC-Particle Reinforced Copper Matrix Composites Fabricated by Sintering and Sinter-Forging Processes." *Journal of Materials Research and Technology* 5 (1): 5–12. https://doi.org/10.1016/j.jmrt.2015.03.002

Sharma, Pardeep, Dinesh Khanduja, and Satpal Sharma. 2016. "Dry Sliding Wear Investigation of Al6082/Gr Metal Matrix Composites by Response Surface Methodology." *Journal of Materials Research and Technology* 5 (1): 29–36. https://doi.org/10.1016/j.jmrt.2015.05.001

Singh, Pawandeep, Vivudh Gupta, R. K. Mishra, and Balbir Singh. 2019. "Effect of Micro Filler Reinforcement on Mechanical and Wear Characteristics of Metal Matrix

Composites: A Review." *AIP Conference Proceedings*, 3 September 2019; 2148 (1): 030011. https://doi.org/10.1063/1.5123933

Sudarshan, M. K., and M. K. Surappa. 2008. "Dry Sliding Wear of Fly Ash Particle Reinforced A356 Al Composites." *Wear* 265 (3–4): 349–60. https://doi.org/10.1016/j.wear.2007.11.009

Suresha, S., and B. K. Sridhara. 2010. "Wear Characteristics of Hybrid Aluminium Matrix Composites Reinforced with Graphite and Silicon Carbide Particulates." *Composites Science and Technology* 70 (11): 1652–59. https://doi.org/10.1016/j.compscitech.2010.06.013

Tang, Feng, Xiaoling Wu, Shirong Ge, Jichun Ye, Hua Zhu, Masuo Hagiwara, and Julie M. Schoenung. 2008. "Dry Sliding Friction and Wear Properties of B4C Particulate-Reinforced Al-5083 Matrix Composites." *Wear* 264 (7–8): 555–61. https://doi.org/10.1016/j.wear.2007.04.006

Thirumalai Kumaran, S., and M. Uthayakumar. 2014. "Investigation on the Dry Sliding Friction and Wear Behavior of AA6351-SiC-B4C Hybrid Metal Matrix Composites." *Proceedings of the Institution of Mechanical Engineers, Part J: Journal of Engineering Tribology* 228 (3): 332–38. https://doi.org/10.1177/1350650113508103

Umanath, K., K. Palanikumar, and S. T. Selvamani. 2013. "Analysis of Dry Sliding Wear Behaviour of Al6061/SiC/Al2O3 Hybrid Metal Matrix Composites." *Composites Part B: Engineering* 53: 159–68. https://doi.org/10.1016/j.compositesb.2013.04.051

Ünlü, Bekir Sadik. 2008. "Investigation of Tribological and Mechanical Properties Al2O3-SiC Reinforced Al Composites Manufactured by Casting or P/M Method." *Materials and Design* 29 (10): 2002–8. https://doi.org/10.1016/j.matdes.2008.04.014

Veeresh Kumar, G. B., C. S. P. Rao, and N. Selvaraj. 2012. "Studies on Mechanical and Dry Sliding Wear of Al6061-SiC Composites." *Composites Part B: Engineering* 43 (3): 1185–91. https://doi.org/10.1016/j.compositesb.2011.08.046

Yang, J. B., C. B. Lin, T. C. Wang, and H. Y. Chu. 2004. "The Tribological Characteristics of A356.2Al Alloy/Gr(p) Composites." *Wear* 257 (9–10): 941–52. https://doi.org/10.1016/j.wear.2004.05.015

Yigezu, Belete Sirahbizu, M. M. Mahapatra, and P. K. Jha. 2013. "On Modeling the Abrasive Wear Characteristics of in Situ Al-12%Si/TiC Composites." *Materials and Design* 50: 277–84. https://doi.org/10.1016/j.matdes.2013.02.042

Yunus, Mohammed, and Rami Alfattani. 2023. "Assessment of Mechanical and Tribological Behavior of AA6061 Reinforced with B4C and Gr Hybrid Metal Matrix Composites." *Coatings* 13 (9), 1653. https://doi.org/10.3390/coatings13091653

Section III

Tribological Performance of Metal Matrix Composites

8 Influence of Nano-Sized Filler Reinforcement in Tribological Performance of Metal Matrix Composites

Vignesh Packkirisamy, Amith S. C.,
Ashokkumar Mohankumar, and
Arunkumar Thirugnanasambandam

8.1 INTRODUCTION

8.1.1 Overview of Metal Matrix Composites (MMCs) and Lightweight Composite Materials

Metal matrix composites (MMCs) are tailor-made materials comprised of a metallic matrix phase combined with one or more reinforcement phases, which are typically in the form of particulates, long or short fibers, or flakes. These advanced materials are chosen for their favorable strength-to-weight ratio, degradation resistance, and good mechanical and thermal properties. The secondary reinforcement phases are intentionally included in the metal matrix to enhance the mechanical and tribological performance of the monolithic metal, making MMCs highly preferrable for various applications requiring performance under multifaceted conditions.

Lightweight composite materials (LCMs), particularly utilizing aluminum and magnesium as metal matrices, have attracted considerable attention in aerospace, automobile, and medical applications. These specially tailored materials find applications where low density and high strength are required without compromising the structural integrity. The reinforcement phases to these matrices are typically ceramics or carbon-based nanostructures that can further improve the properties of MMCs, resulting in remarkable performance.

8.1.2 Importance of Tribological Performance in MMCs

Tribology, which includes the study of friction, wear, and lubrication, is essential for MMCs, especially in applications where parts are allowed to slide or roll over another under loading conditions. The wear resistance of MMCs plays a pivotal role in determining the longevity of the parts in service. Poor tribological behavior of

DOI: 10.1201/9781003530909-11

materials leads to material removal, increasing downtime, resulting in more service cost, particularly in aerospace and automobile industries, where components are continuously exposed to harsh ambient conditions and high mechanical stresses.

The tribological behavior of MMCs is governed by various factors, including type of metal matrix, the nature and distribution of filler materials, and working conditions. For instance, incorporation of hard ceramic filler particles into a soft metal matrix can enhance the wear resistance as well as modifying the frictional behavior of the composite. Understanding these properties is essential for the development of composites that can satisfy the stringent requirement of various engineering applications.

8.1.3 Role of Nano-Sized Filler Particles in Enhancing MMC Properties

Exceptionally advanced composites technology has been demonstrated by incorporating nano-sized filler materials into MMCs. Due to their small size and large surface area, nanofillers can change the characteristics of a metal matrix, leading to better mechanical as well as tribological performance. Such nanofillers consist of different types of materials, including CNTs (carbon nanotubes), graphene, nanoceramics like alumina (Al_2O_3), silica (SiO_2), tungsten carbide (WC), zirconia (ZrO_2), and others. Several mechanisms for improving MMC properties include grain refinement, improved load transfer from matrix to reinforcements that create stronger more homogeneous material and greater strength to the composite on its entirety upon the addition of these nanofillers in a metal matrix. In addition, it is possible for these nanofillers to lower matrix friction coefficients and enhance wear resistance via establishing a protective tribo-film at sliding contacts. The effectiveness of nano-sized fillers in MMCs depends on factors such as their dispersion within the matrix, the interfacial bonding between the filler and the matrix, and the overall volume fraction of the fillers. Achieving a uniform dispersion of nanofillers is particularly challenging due to their tendency to agglomerate, which can adversely affect the composite's performance. Therefore, advanced processing techniques and surface treatments are often required to fully realize the benefits of nano-sized fillers in MMCs.

8.1.4 Purpose of the Chapter

The aim of this section is a detailed overview on the role played by fillers that are nano-sized in improving tribological properties of metal matrix composites with an emphasis on both aluminum and magnesium matrices. This part will look at different types of nanofillers used for reinforcing MMCs, such as carbon nanotubes, graphene, and nano-ceramics. These will include aspects like how they affect the microstructure or performance characteristics among others.

8.2 ALUMINUM AND ITS ALLOY MATRIX COMPOSITES WITH NANOFILLERS

The major use of aluminum alloys (Al-Cu, Al-Si, Al-Mg, Al-Mg-Si, and Al-Zn series) are primarily in the field of aerospace, marine, and automobile applications. Among all these, Al-Mg-Si and Al-Zn alloys have structural applications for which wear properties can be further enhanced using nanoparticle reinforcements to form aluminum metal matrix nanocomposites. Nanofillers such as SiC, Al_2O_3, MWCNT, graphene, TiC, B_4C, etc. are used to refine the grain structures of aluminum alloy and increase the hardness of the developed nanocomposites; thus, the wear rate is reduced. Moreover, they form tough layers protecting against sliding wear. The factors influencing the wear properties of aluminum nanocomposites can be classified as follows: (1) concentration of nanoparticle reinforcements, (2) size of reinforcing particles, (3) composition of reinforcement, (4) fabrication methods, (5) tribology parameters (normal load, sliding speed, elevated temperature etc.), and (6) composition of the matrix. A few studies related to these influencing factors are given in Table 8.1.

8.2.1 Effect of Nanofillers on Microstructural Evolution and Tribological Properties of Aluminum MMCs

Pillari et al. (2024) discovered that adding graphene to an aluminum B319 alloy transformed the structure, allowing silicon to be avoided, resulting in more refined and fibrous Si structured phases. The mechanical and tribological characteristics of the alloy were enhanced by the addition of additional nucleation sites provided by graphene, resulting in a superior morphological structure. Adding 0.1 wt.% graphene increased the toughness of the B319 alloy by about 20%. Nonetheless, the hardness decreased beyond the 0.3 wt.% graphene limit, perhaps due to graphene aggregation and increased porosity. Pethuraj et al. (2022) observed that the pure Al with 5 wt.% Al_2SiO_5 and 5 wt.% B_4C hybrid nanocomposites had superior tensile, impact, and microhardness when compared to unreinforced and other single reinforced material. As reinforcement increased, wear rate decreased, with a 50.8% lower rate compared to Al_2SiO_5 composites. The scanning electron microscope (SEM-EDS) microstructures as shown in Figure 8.1(a–c) illustrate the presence of primary and secondary phases along with the uniformly dispersed nanofillers. The hybrid composites showed little abrasion wear, while the pure alloy had plastic and delamination wear behavior.

The Al_2O_3 nanoparticles were used in the study to develop 2024 aluminum matrix composite components using an ultrasonic-assisted semisolid stirring method (Jiang et al. 2018). Process variables like the volume percentage of Al_2O_3 nanoscale particles in semisolid slurries and the stirring temperature with duration are among the parameters that influence the metallurgical properties of rheoformed aluminum matrix composites. For a 5 vol.% Al_2O_3 nanoparticles distribution swirling at 620°C for 25 minutes, the rheoformed components were mostly stable at the bottom area. The results were the highest ultimate tensile strength (UTS) of 358 MPa and yield strength (YS) of 245 MPa. Comparing these values with the ones of an unreinforced matrix component, the gains obtained were 17.5% and 31.7%, respectively.

The process of homogeneous dispersion of Al_2O_3 nanoparticles across the aluminum matrix, which is an efficient barrier to the dislocation motion, is the main cause of improvement in mechanical properties, along with grain refinement, as shown in Figure 8.2(a–d). On top of that, there is the dislocation density that is big

TABLE 8.1
Summary of Mechanical and Tribological Values of Several Nanofiller-Reinforced Aluminum Matrix Composites

Source	Process	Material	Tensile UTS (MPa)	Microhardness (HV)	COF	Wear Rate (mm^3/Nm)
Pillari et al. (2024)	Powder metallurgy and stir casting. Dry sliding, linear reciprocating motion: L = 10 N, Sd = 300 m, F = 5 Hz, l = 10 mm	B319 Alloy	–	77	0.26	12.7×10^{-4}
		B319 – 0.1 wt.% graphene	–	93	0.25	10.7×10^{-4}
		B319 – 0.2 wt.% graphene	–	91	0.18	8.1×10^{-4}
		B319 – 0.3 wt.% graphene	–	91	0.17	6.1×10^{-4}
		B319 – 0.5 wt.% graphene	–	81	0.23	9.5×10^{-4}
		B319 – 0.7 wt.% graphene	–	80	0.25	9.7×10^{-4}
Pethuraj et al. (2022)	Stir casting. Dry sliding, Pin-on-disk: L = 10, 30, 50 N, Sd = 1000 m, S = 0.75, 1.5, 2.25 m/s	Pure Al	72	138	0.34–0.37	0.31–0.45
		Al – 5 wt.% Al_2SiO_5	90	220	0.3–0.35	0.29–0.43
		Al – 10 wt.% Al_2SiO_5	45	255	0.27–0.3	0.15–0.38
		Al – 5 wt.% $Al_2SiO_5 + B_4C$ each	48	240	0.2–0.22	0.1–0.23
Jiang et al. (2018)	Ultrasonic assisted semisolid stirring and rheoformed. Pin-on-disk: L = 30 N, S = 0.8 m/s, Sd = 1000 N	Al2024 – Al_2O_3 (1–10 vol.%)	~300 (Al2024) ~310–400 (Al2024 – Al_2O_3)	–	–	5.2 mg/Nm (Al2024) 3.4–1 mg/Nm (Al2024 – Al_2O_3)
Li et al. (2024)	Friction consolidation (FC) method. Pin-on-disk: L = 1 N, Sd = 120 m, F = 5 Hz	Al7075 T6 FC	526	181	0.45	–
		Al7075 – 12%TiB_2 T6 FC	566	218	0.55	–
		Al7075 – 24%TiB_2 T6 FC	672	270	0.6	–
		Al7075 T6	572	–	–	–

Note: L – applied load, Sd – sliding distance, F – reciprocating frequency, l – reciprocating length, S – sliding speed

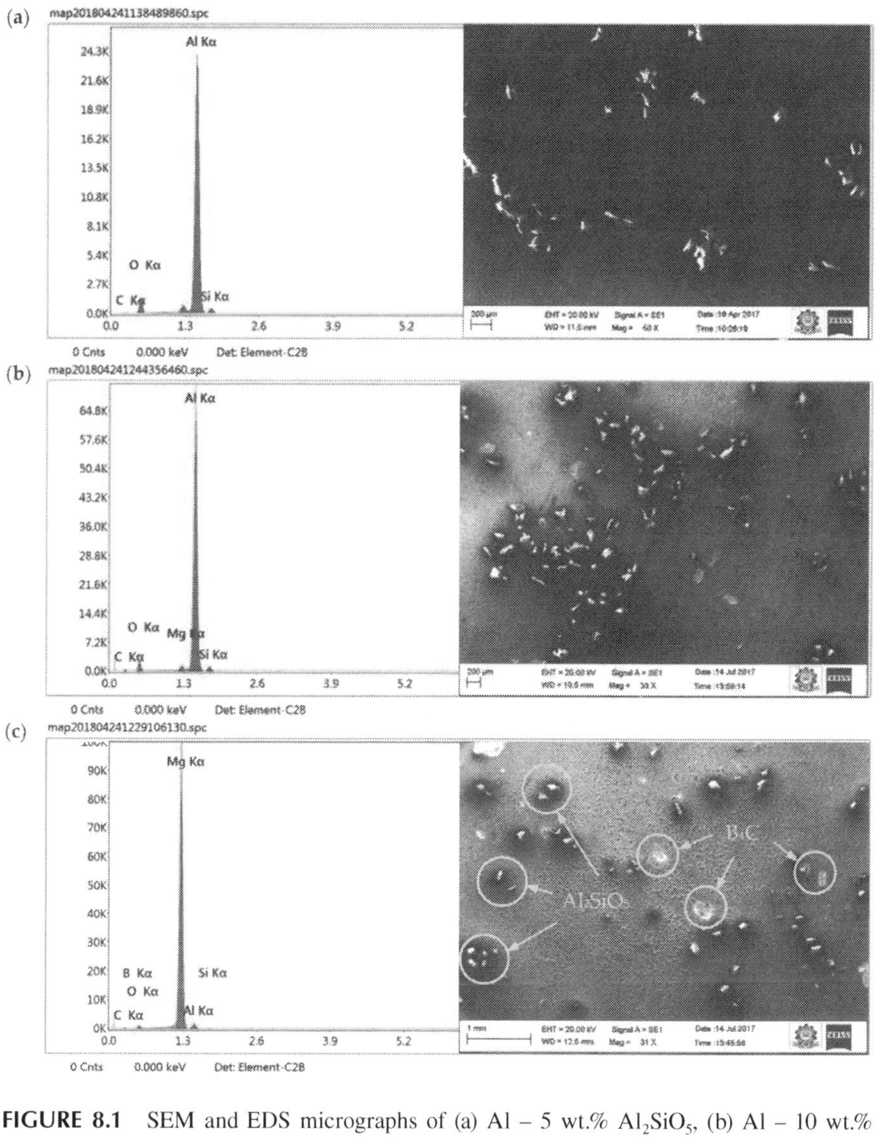

FIGURE 8.1 SEM and EDS micrographs of (a) Al – 5 wt.% Al_2SiO_5, (b) Al – 10 wt.% Al_2SiO_5, and (c) Al – 5 wt.% Al_2SiO_5 + 5 wt.% B_4C. (Reproduced with the permission of Pethuraj et al. 2022.)

and dislocation tangles are formed around the aluminum matrix, which is another factor for the development of mechanical properties. The increase in the volume fraction of Al_2O_3 nanoparticles in aluminum matrix composites leads to improved wear resistance. This is reflected in the shallower depth of ploughing grooves on the worn surface, which indicates better resistance to deformation and material removal during wear. The increase in hardness of the composite is primarily due to the presence of Al_2O_3, a hard ceramic material, within the softer aluminum matrix. Higher

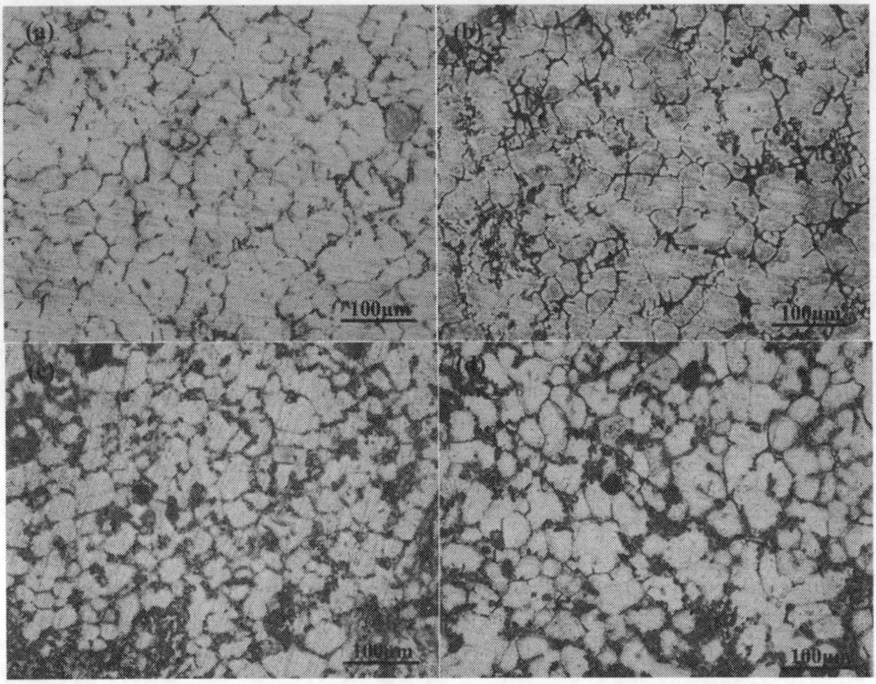

FIGURE 8.2 Optical microscope images of (a) AA2024, (b) AA2024/1% Al_2O_3, (c) AA2024/3% Al_2O_3, and (d) AA2024/5% Al_2O_3 aluminum composite. (Reproduced with the permission of Jiang et al. 2018.)

hardness reduces the material's susceptibility to abrasive wear, making it more resistant to penetration and deformation under contact stresses. This resistance is crucial in applications requiring continuous friction or sliding, as it extends the service life of components.

Microstructural analysis of the worn surface also reveals the presence of delamination, a combination of adhesion and delamination, as illustrated by Maddaiah et al. (2024) in Figure 8.3(a–e). Adhesive wear occurs when two contacting surfaces slide against each other, leading to material transfer or the generation of wear debris. The increase in hardness due to Al_2O_3 nanoparticles reduces the likelihood of material transfer, but under high stress conditions, the material may still experience localized failure in the form of delamination. This balance between these wear mechanisms highlights the complex nature of wear in metal matrix composites and emphasizes the importance of optimizing nanoparticle volume fraction for optimal performance.

The aluminum nanocomposites exhibit abrasion and oxidation at lower load conditions and sliding speed. The delamination of surface is observed due to the influence of higher load and sliding speed. The nanoparticles that are present in the matrix act as load bearing elements; thus, the sliding wear pin is resistant to remove the material from the surface. The rate of material removal is comparatively slower for aluminum nanocomposites when compared to the base alloy. The nanoparticles also enhance the wear resistance by influencing the material's surface as solid lubricant.

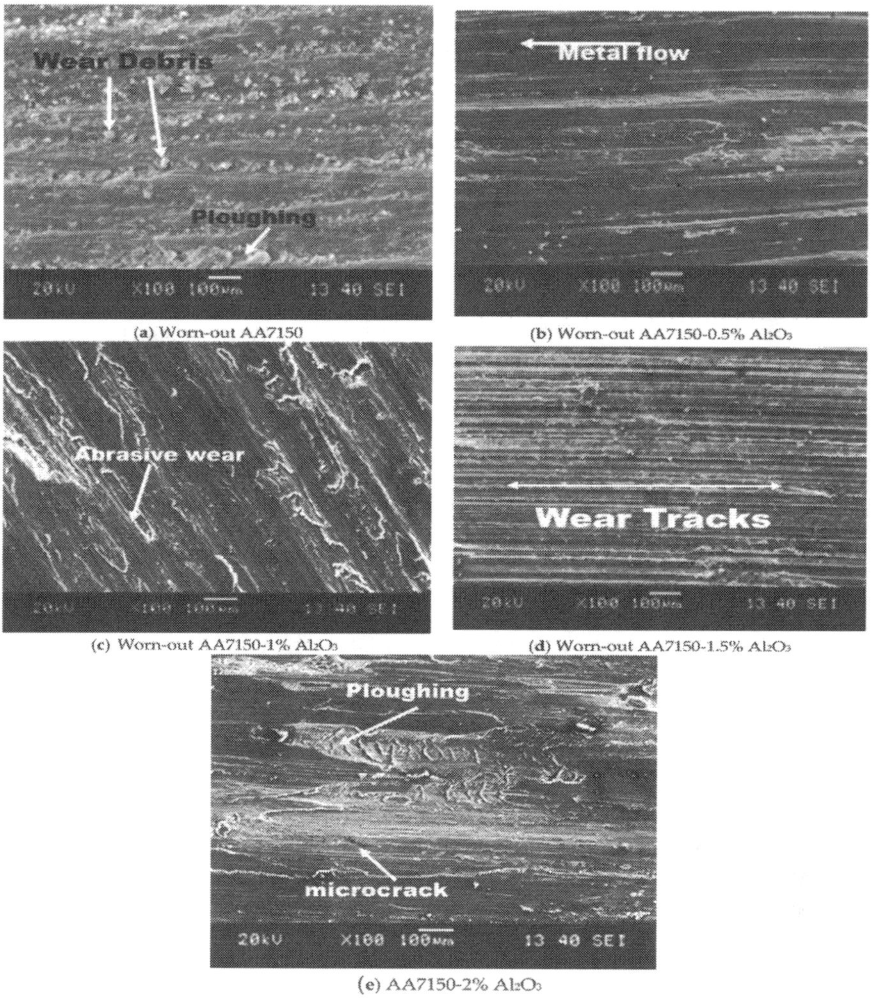

FIGURE 8.3 (a–e) SEM wear surface morphology of Al/Al_2O_3 nanocomposites. (Reproduced with the permission of Maddaiah et al. 2024.)

The heat generated at the point of wear does not affect the nanoparticles as they have high thermal coefficient. Alam et al. (2018) studied an A356 aluminum alloy reinforced with SiC particles ranging in size from 40 to 55 nm and a concentration of 1 to 5 wt.%. They discovered that the alloy's wear resistance improved as the SiC percentage rose. Specifically, increasing the SiC concentration improved the production of Fe-rich tribolayers, resulting in improved wear resistance. Furthermore, the friction coefficient of the base matrix alloy was reduced with increasing the load and SiC concentration, dropping from 0.62 to 0.48 at 5 wt.% and 30 N, indicating that an adhesive wear process was prevalent. Faisal and Kumar (2018) investigated an aluminum alloy called AA2199 that was reinforced with SiC particles with diameters ranging from 50 to 70 nm and a weight percentage between 0.5 and 2.5 wt.%.

Their results showed that as the SiC concentration rose, the reinforcement enhanced both hardness and tensile strength. Up to 1.5 wt.% SiC concentration, the wear and friction coefficient showed a decrease; beyond that, the attributes started to lose their superiority over the base alloy. The wear behavior of Al-2Cu alloy reinforced with B_4C particles, ranging in weight from 4 to 10% and with a particle size of 1 µm, was investigated by Alizadeh et al. (2017). According to their research, adding more B_4C improved the hardness and changed the wear process from delamination to abrasion. Furthermore, when the B_4C concentration rose, the alloy's wear rate also declined and the coefficient of friction (COF) reduced, highlighting the positive impacts of B_4C on the aluminum matrix's wear resistance. The hybrid reinforcement of the Al1030 alloy with B_4C and h-BN (hexagonal boron nitride) particles was studied by Harichandran and Selvakumar (2018). B_4C, whose particle size ranges from 2 to 6 µm, and h-BN, whose particle size ranges from 70 µm, were used in the study at weight percentages of 2% and 6%, respectively. The outcomes showed that the hybrid nanocomposite had wear resistance that was noticeably higher than that of the monolithic alloy and the nanocomposite that was only reinforced with B_4C. Additionally, the hybrid nanocomposite demonstrated a balanced increase in mechanical characteristics and wear resistance while maintaining improved ductility and impact energy.

8.3 MAGNESIUM AND ITS ALLOY METAL MATRIX COMPOSITES WITH NANOFILLERS

8.3.1 Evolution of Microstructure on Nanofiller-Reinforced Magnesium MMCs

The impact of nanofiller reinforcements on the material properties has been made apparent in the study of microstructure evolution in different magnesium nanocomposites. Subramani et al. (2022b) used SEM and EDS mapping analysis to study the effect of WS_2 nanotubes on AZ31 and their work indicated the growth of secondary β-$Mg_{17}Al_{12}$ formation as well as the uniform distribution of this phase (Figure 8.4). It was also mentioned that the nanotubes were responsible for cracking the Mg-Al alloys. However, detection by elemental mapping of these extra phases was not possible because WS_2 was at a very low weight fraction. Even if they were not detected in elemental maps, their presence in the microstructure was visible. In a different experiment, Banerjee et al. (2019) found that the addition of WC nanoparticles to AZ31 resulted in a more even distribution of the reinforcements and good junctions between the matrix and the reinforcement phase. WC nanoparticles' presence brought recrystallization and limited grain growth and the inclusion of finer grain structures in the nanocomposites. In spite of some agglomeration at a higher WC content, the microstructural integrity hitherto surpassed, as is confirmed by SEM and EDAX analysis. Jayaraman et al. (2016) performed the exploration of the influence of ball-milling on the AZ31 alloy with MWCNTs. They have shown that mechanical alloying can disperse the MWCNTs throughout that matrix efficiently while at a higher concentration, which leads to agglomeration. The microstructure of AZ31 and MWCNTs strengthened with a low amount of wt.%, the particles agglomerating with

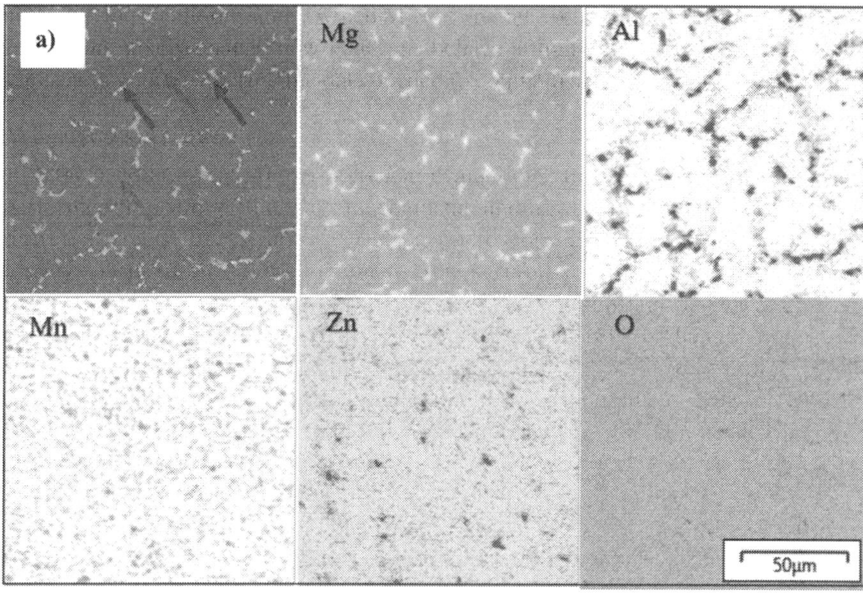

FIGURE 8.4 SEM microstructural image and EDS analysis of AZ31/WS_2 nanocomposite. (Reproduced with permission from Subramani et al. 2022b)

the smallest oxidation after the ball milling and sintering processes were detected. These studies collectively emphasize that nanofiller reinforcements like WS_2, WC, and MWCNTs contribute to the greatest improvement of the microstructural properties of AZ31-based composites, which leads to enhanced mechanical properties, wherein the challenge of agglomeration at high reinforcement contents is involved.

8.3.2 Effect of Nanofillers on the Tribological Behavior of Magnesium Nanocomposites

Nanofiller reinforcement in magnesium alloys has shown significant improvements in both microhardness and tribological properties, enhancing their potential for various industrial applications. The data compiled from multiple studies highlight how different nanofillers, such as Al_2O_3, MWCNT, GNPs, SiC, and WC, contribute to these enhancements. Table 8.2 demonstrates a summary of the mechanical and tribological test results of various magnesium matrix nanocomposites.

8.3.3 Microhardness

It is well-known fact that the addition of nanofillers brings the greatest improvement to the microhardness of magnesium composites. As per Azizieh et al. (2018), the microhardness of AZ31/Al_2O_3 nanocomposites was found to have risen up to the level of 70–90 HV while it was only 50 HV in the original AZ31 alloy. This rise is due to the particles of Al_2O_3 that add the matrix with the proper reinforcement of the

TABLE 8.2
Summary of Mechanical and Tribological Values of Several Nanofiller-Reinforced Magnesium Matrix Composites

Study	Material	Microhardness (HV)	COF	Wear Rate	Yield Strength (YS, MPa)	Ultimate Tensile Strength (UTS, MPa)	Wear Loss (mg)
Azizieh et al. (2018)	AZ31 base	50	0.29–0.31 (10 N) 0.25–0.29 (50 N & 90 N)	0.1 mm³/min (10 N) 0.16–0.2 mm³/min (50 N) 0.23–0.31 mm³/min (90 N)	–	–	–
	AZ31/Al_2O_3 nanocomposite	70–90	0.29–0.31 (10 N) 0.25–0.29 (50 N & 90 N)	0.1 mm³/min (10 N) 0.16–0.2 mm³/min (50 N) 0.23–0.31 mm³/min (90 N)	–	–	–
Turan et al. (2020)	AZ91	72.9±1.4	0.3–0.4 (250 mm/s)	0.006 mm³/m (250 mm/s)	–	–	–
	AZ91-MWCNT	82.7±1.6	–	–	–	–	–
	AZ91-MWCNT-GNPs	88.3±2.1	0.25 (250 mm/s)	0.004 mm³/m (250 mm/s)	–	–	–
Subramani et al. (2022a)	AZ31	40.44	–	–	83.87	175.3	1.56
	AZ31/0.2 SiC	50.1	–	–	101.63	132.6	1.28
	AZ31/0.5 SiC	50.75	–	–	108.61	143.1	1.25
	AZ31/1 SiC	51.37	–	–	109.07	157.2	1.15
Maddaiah et al. (2024)	AA7150	80–85	–	–	–	192	–
	AA7150-0.5% Al_2O_3	90–100	–	–	–	257	–
	AA7150-1% Al_2O_3	130–140	–	–	–	310	–
	AA7150-1.5% Al_2O_3	150–155	–	–	–	362	–
	AA7150-2% Al_2O_3	160–165	–	–	–	384	–

(Continued)

TABLE 8.2 *(Continued)*
Summary of Mechanical and Tribological Values of Several Nanofiller-Reinforced Magnesium Matrix Composites

Study	Material	Microhardness (HV)	COF	Wear Rate	Yield Strength (YS, MPa)	Ultimate Tensile Strength (UTS, MPa)	Wear Loss (mg)
Banerjee et al. (2019)	AZ31	60–70	0.6–0.7 (0.2 m/s, 40 N)	28–30 (10^{-12} Kg/N-m)	–	–	–
	AZ31/0.5 WC	80–82	–	–	–	–	–
	AZ31/1 WC	83–85	–	–	–	–	–
	AZ31/1.5 WC	95–97	–	–	–	–	–
	AZ31/2 WC	100–105	0.5–0.6 (0.2 m/s, 40 N)	5–6 (10^{-12} Kg/N-m)	–	–	–

element and push away the dislocation thereby achieving harder particles. Besides, Turan et al. (2020) also showed that with the impact of microhardness in AZ91 the MWCNT and GNPs were reinforced. The AZ91-MWCNT-GNP combinate showed a microhardness of 88.3±2.1 HV, greater and higher than the base AZ91 alloy's 73.6±0.2 HV. These results directly indicated the effect of MWCNT and GNPs on the ability of stress transfer. It can be concluded that the provision of both MWCNT and GNPs most probably results in the improvement of mechanical property in terms of improved load transfer capability. Subramani and others (2022a) reported that adding SiC nanoparticles in AZ31 metal made it much harder. The top hardness, 51.37 HV, was seen in the AZ31/1%SiC composite, while the base AZ31 metal had a hardness of 40.44 HV. The hard SiC aid with this betterment by improving the load bearing tendency of the nanocomposite. Banerjee and others (2019) also said that the AZ31/2% WC composite had a microhardness of 100–105 HV, which was a big jump from the base AZ31 metal's 60–70 HV. The WC nanoparticles made the matrix better by refined grain shape and made strong interfacial bonding with the matrix.

8.3.4 Tribological Behavior

Moreover, we can also say that when nanoparticles were added to nanocomposite materials their properties, including coefficient of friction (COF) and wear performance, improved significantly. Azizieh et al. (2018) reported that even under different load conditions, the COF for AZ31/Al_2O_3 remained constant, just like with pure AZ31 alloy. But the wear rate of the nanocomposite was always lower than that of its counterpart at higher loads, hence confirming to us that Al_2O_3 particles increase resistance against abrasive wear because they act as barriers according to Turan et al.

(2020), who sited reduction in COF for different compositions containing AZ91 alloy compared to plain AZ91 where the sliding speed at 250 mm/s yielded 0.25 while the base AZ91 ranged between 0.3–0.4, which suggest self-lubrication nature may be responsible due to formation of a protective film simply known as friction-reducing lubricant (FRL) during motion. Furthermore, the wear rate also was better, where AZ91-MWCNT-GNPs composite had 0.004 mm^3/m while base AZ91 contained 0.006 mm^3/m. A considerable reduction in wear rate for AZ31-WC nanocomposites was evidenced by Banerjee et al. (2019). For instance, an AZ31/2WC composite had a wear rate of 5–6 × 10^{-12} kg/N-m, which is much lower compared to a base alloy AZ31's 28–30 × 10^{-12} kg/N-m at a sliding speed of 0.2 m/s and under a load of 40 N. This enhancement can be explained by hardening WC particles that reduce the loss of material during sliding, thereby enhancing wear resistance. Again, there was drop in COF from about 0.6–0.7 for base AZ31 and down to 0.5–0.6 for AZ31/2 WC composite, indicating improved tribological performance due to the presence of WC.

8.3.5 Impact of Nanofiller Addition on Tribological Behavior of Magnesium Nanocomposites

The incorporation of nanofillers such as Al_2O_3, MWCNT, GNPs, SiC, and WC remarkably improves the wear behavior of magnesium-based nanocomposites. The included nanofillers enhance the wear resistance by refining the composite microstructure, triggering the formation of a protective layer, and giving solid lubrication. The enhanced wear resistance was mainly due to the hard nature of the nanofillers, which reduce material susceptibility to wear. Additionally, the reduction in COF with the inclusion of lubricating nanofillers like GNPs contributes to smoother sliding and reduced friction. The combination of these effects results in magnesium nanocomposites that exhibit not only higher hardness but also superior tribological performance, making them suitable for applications requiring high wear resistance and low friction.

8.4 CONCLUSIONS AND FUTURE SCOPE

This book chapter has highlighted the substantial impact of nanofiller reinforcements on the microhardness, tribological conduct, and mechanical performance of lightweight aluminum and magnesium-based nanocomposites. By integrating various nanofillers along with Al_2O_3, MWCNT, GNPs, SiC, WC, and B_4C into those matrices, it remarkably improves the microhardness and wear resistance, as observed. These results are attributed to the potential of nanofillers to refine grain structure, promote the formation of protective phases during sliding, and offer stable lubrication, all of which make contributions to the stepped-up durability and performance of the composites underneath various working situations.

Similarly, the evaluation of tribological studies confirmed that nanofillers not only reduce material loss during sliding but also lower friction coefficients, thereby extending the service life of components made from these materials. The findings presented in this chapter underscore the ability of lightweight aluminum- and magnesium-based nanocomposites in a wide variety of industries, inclusive of aerospace, automobile, and biomedical engineering, in which the combination of lightweight and enhanced overall performance is of major interest.

8.4.1 Future Scope

The future scope of research in lightweight MMCs includes optimizing nanofiller content, distribution, and particle size to achieve tailored mechanical properties. Exploring the synergistic effects of multiple nanofillers, such as MWCNTs and GNPs, offers potential for further enhancing material characteristics. Additionally, investigating the long-term stability and performance of these nanocomposites under various environmental conditions, including temperature, humidity, and corrosion, is essential for their application in industries like aerospace and biomedical engineering. Improving corrosion resistance, particularly for biomedical implants, remains a critical research focus, ensuring biocompatibility and controlled degradation rates for safe and effective use.

REFERENCES

Alam, M. T., Arif, S., & Ansari, A. H. (2018). Wear behaviour and morphology of stir cast aluminium/SiC nanocomposites. *Materials Research Express*, 5(4), 045008.

Alizadeh, A., Maleki, M., & Abdollahi, A. (2017). Preparation of super-high strength nanostructured B4C reinforced Al-2Cu aluminum alloy matrix composites by mechanical milling and hot press method: Microstructural, mechanical and tribological characterization. *Advanced Powder Technology*, 28(12), 3274–87.

Azizieh, M., Larki, A. N., Tahmasebi, M., Bavi, M., Alizadeh, E., & Kim, H. S. (2018). “Wear behavior of AZ31/Al2O3 magnesium matrix surface nanocomposite fabricated via friction stir processing.” *Journal of Materials Engineering and Performance*, 27(4), 2010–17. https://doi.org/10.1007/s11665-018-3277-y

Banerjee, S., Poria, S., Sutradhar, G., & Sahoo, P. (2019). Dry sliding tribological behavior of AZ31-WC nano-composites. *Journal of Magnesium and Alloys*, 7(2), 315–27. https://doi.org/10.1016/j.jma.2018.11.005

Faisal, N., & Kumar, K. (2018). Mechanical and tribological behaviour of nano scaled silicon carbide reinforced aluminium composites. *Journal of Experimental Nanoscience*, 13(sup1), S1–S13.

Harichandran, R., & Selvakumar, N. (2018). Microstructure and mechanical characterization of (B4C+h-BN)/Al hybrid nanocomposites processed by ultrasound assisted casting. *International Journal of Mechanical Sciences*, 144, 814–26.

Jayaraman, J., Kuppusamy, R., & Rao, H. (2016). Investigation on wear properties of AZ31-MWCNT nanocomposites fabricated through mechanical alloying and powder metallurgy. *Science and Engineering of Composite Materials*, 23(1), 61–66. https://doi.org/10.1515/secm-2014-0056

Jiang, J., Xiao, G., Che, C., & Wang, Y. (2018). Microstructure, mechanical properties and wear behavior of the rheoformed 2024 aluminum matrix composite component reinforced by Al_2O_3 nanoparticles. *Metals*, 8(6), 460.

Li, X., Das, H., Pole, M., Li, L., Soulami, A., Grant, G. J., & Efe, M. (2024). Exceptional strength and wear resistance in an AA7075/TiB2 composite fabricated via friction consolidation. *Materials & Design*, 242, 113006.

Maddaiah, K. C., Kumar, G. B. V., & Pramod, R. (2024). Studies on the mechanical, strengthening mechanisms and tribological characteristics of AA7150-Al_2O_3 nano-metal matrix composites. *Journal of Composites Science*, 8(3), 97. https://doi.org/10.3390/jcs8030097

Pethuraj, M., Uthayakumar, M., Rajesh, S., Abdul Majid, M. S., Rajakarunakaran, S., & Niemczewska-Wójcik, M. (2022). Dry sliding wear studies on sillimanite and B_4C reinforced aluminium hybrid composites fabricated by vacuum assisted stir casting process. *Materials*, 16(1), 259.

Pillari, L. K., Lessoway, K., & Bichler, L. (2024). Reciprocating dry sliding friction and wear behavior of B319 aluminum alloy-graphene composites. *Tribology International*, 192, 109334.

Subramani, M., Huang, S.-J., & Borodianskiy, K. (2022a). Effect of SiC nanoparticles on AZ31 magnesium alloy. *Materials*, 15(3), 1004. https://doi.org/10.3390/ma15031004

Subramani, M., Huang, S.-J., & Borodianskiy, K. (2022b). Effect of WS2 nanotubes on the mechanical and wear behaviors of AZ31 stir casted magnesium metal matrix composites. *Journal of Composites Science*, 6(7), 182. https://doi.org/10.3390/jcs6070182

Turan, M. E., Zengin, H., & Sun, Y. (2020). Dry sliding wear behavior of (MWCNT + GNPs) reinforced AZ91 magnesium matrix hybrid composites. *Metals and Materials International*, 26(4), 541–50. https://doi.org/10.1007/s12540-019-00338-8

9 Influence of Fillers on High-Temperature Tribology of Metal Matrix Composites

Shrishail B. Sollapur, Rajesh Kamalkishor Mantri, Devakant Baviskar, Yugesh A. Kharche, and Santosh R. Shekokar

9.1 INTRODUCTION

Within the field of engineering materials, the search for solutions capable of withstanding extreme conditions is constant. Applications that operate at high temperatures, such as those in aerospace propulsion, automotive engines, and power production, require materials that can retain their structural integrity, resistance to wear, and low friction characteristics even under high heat. Metal matrix composites (MMCs) have emerged as a promising option, offering a unique combination of metallic attributes like malleability and strength alongside the high stiffness of reinforcing materials. However, the ability of MMCs to withstand high temperatures without significant tribological issues is a major concern, as it directly affects their efficiency, longevity, and safety in various high-temperature environments (Chawla & Chawla, 2013). MMCs typically consist of a metal or alloy matrix reinforced with elements such as particles, whiskers, or fibers. Common matrix materials include aluminum, magnesium, and titanium, while the reinforcing components are often ceramics like silicon carbide (SiC) and alumina (Al_2O_3). The selection of materials depends on the specific needs of the application, such as the need for high strength-to-weight ratio or thermal conductivity. In high-temperature conditions, MMCs encounter several unique challenges, including softening of the matrix, deterioration of the interface between the matrix and reinforcement, and oxidation, all of which can significantly alter their tribological behavior.

To improve the tribological performance of MMCs at high temperatures, researchers are investigating the use of various additives known as fillers. These additives, also referred to as solid lubricants or self-lubricating materials, are integrated into the MMC structure to enhance its friction and wear resistance, especially at elevated temperatures. According to Ibrahim et al. (1991), fillers can be introduced into composites through various approaches: as an additional component in the composite, as a coating on the reinforcement particles, or as a phase formed during the manufacturing process. Several types of fillers have shown potential in enhancing

DOI: 10.1201/9781003530909-12

the tribological properties of MMCs. Graphite, with its layered structure and weak bonds between layers, offers superior solid lubrication, though its effectiveness decreases above 400°C due to oxidation. Molybdenum disulfide (MoS_2) provides low friction up to approximately 400°C in air, but its susceptibility to oxidation, which results in the formation of abrasive MoO_3, is a limitation. Hexagonal boron nitride (h-BN), with its structural similarity to graphite, can provide lubrication up to 900°C. Its chemical inertness and resistance to oxidation make it an ideal choice for applications requiring high-temperature performance.

Additional noteworthy components are calcium fluoride (CaF_2) and barium fluoride (BaF_2), which create films with low shear strength at elevated temperatures, offering lubrication capabilities up to 900°C, especially in materials with nickel-based MMCs. Prasad and Asthana (2004) explain that soft, noble metals like silver (Ag) and gold (Au) do not form stable oxides, which help maintain their lubricating properties even at elevated temperatures. They are frequently utilized in bearings designed for high-temperature applications. Tungsten disulfide (WS_2), akin to MoS_2, exhibits low friction and superior resistance to oxidation, retaining its properties up to 600°C. The positive impact of these additives on the high-temperature performance of MMCs can be explained through various mechanisms. Many additives form a thin, continuous layer at elevated temperatures. This layer, possessing low shear strength, facilitates smooth sliding between surfaces, thereby reducing friction. Soft metal additives like silver can fill in irregularities and micro-cracks during motion, acting as a "repair" mechanism that smoothens the surface and decreases wear. Hard ceramic additives, when present in small amounts, can smooth the contact surface by removing irregularities, leading to reduced friction.

Hard additives also provide structural support, bearing a portion of the applied load. This is especially advantageous at high temperatures when the matrix softens. Certain additives, such as boron nitride or glassy carbon, create a protective layer that blocks oxygen from reaching the matrix or the interface between the matrix and reinforcement, thus preventing oxidation. Fillers with high thermal conductivity, like graphite or SiC, enhance the dissipation of frictional heat. At elevated temperatures, MMCs can experience various forms of wear, including abrasion, adhesion, delamination, and oxidation wear. Additives significantly affect these wear processes. While hard ceramic reinforcements in MMCs resist abrasion, they may become abrasive if they fracture. Soft additives like graphite can surround these hard particles, diminishing their abrasive effect. At high temperatures, the matrix can soften and become more susceptible to adhesion. Additives like h-BN create a protective layer that prevents direct metal-to-metal contact, thereby reducing adhesion. According to Rohatgi and Tabandeh-Khorshid (2014), delamination wear arises when subsurface cracks propagate parallel to the surface. Soft additives mitigate this by absorbing shear stress, thereby reducing subsurface deformation, while hard additives counteract it by deflecting crack propagation. High temperatures accelerate oxidation, leading to the formation of brittle oxides that are prone to spallation. Some additives, such as silver or glassy carbon, offer a physical barrier against oxygen, while others, like h-BN, are chemically inert.

However, adding fillers to MMCs poses difficulties, especially in ensuring even spread and keeping the interface strong. The usual ways of making these materials involve powder metallurgy, which sees powders mixed, pressed, and heated; stir casting, where fillers are mixed into the liquid metal; and in-situ methods, where fillers develop within the metal through chemical reactions. Each approach has its advantages and disadvantages, influencing how the fillers are distributed and how well they bond. Obstacles in the production process include poor interaction between fillers and the metal, resulting in weak connections; clumping of small filler particles; mismatched temperatures leading to stress; and possible chemical reactions at high temperatures. Despite these hurdles, the improved high-temperature lubrication properties of MMCs with added fillers have opened up new opportunities. Erdemir and Donnet (2006) state that in the aerospace industry, MMCs reinforced with fillers such as hexagonal boron nitride (h-BN) or calcium fluoride (CaF_2) are employed in turbine components to reduce friction at temperatures exceeding 600°C. Similarly, in the automotive sector, brake rotors made from MMCs containing graphite exhibit superior performance under high-temperature braking conditions. In the field of power generation, bearings created from silver-filled MMCs keep friction low at temperatures up to 800°C.

Looking forward, there are several research areas being explored. These include incorporating nanoscale fillers such as graphene for improved dispersion, mixing various fillers to achieve lubrication over a broader temperature range, and drawing inspiration from natural surfaces that lubricate well to design new filler structures. The use of computational modeling is also being considered to forecast and enhance the behavior of fillers at high temperatures.

9.2 FILLERS IN MMC

9.2.1 Role of Fillers in Metal Matrix Composites

Fillers are essential in MMCs as they boost different characteristics of the materials to fulfill particular engineering needs. They greatly enhance the material's mechanical attributes like tensile strength, hardness, and resistance to wear. Additionally, fillers can improve the material's ability to conduct heat and its stability, making it ideal for use in high-temperature environments. Moreover, fillers contribute to reducing the composite's weight while ensuring or even enhancing its structural strength. By choosing the right fillers, MMCs can be customized for specific uses in the aerospace, automotive, and electronics sectors.

9.2.2 Fillers Types and Properties

Additives in MMCs boost the characteristics of the material and can be categorized into various groups according to their makeup and structure. Ceramic additives, such as silicon carbide and aluminum oxide, enhance durability and resistance to heat. Metallic additives, such as tungsten and titanium, improve toughness and electrical conductivity. Moreover, additives based on carbon, like carbon nanotubes and graphite, provide better rigidity and heat transfer. Every type of additive adds unique features to the MMC, customized for particular uses and performance as shown in Figure 9.1.

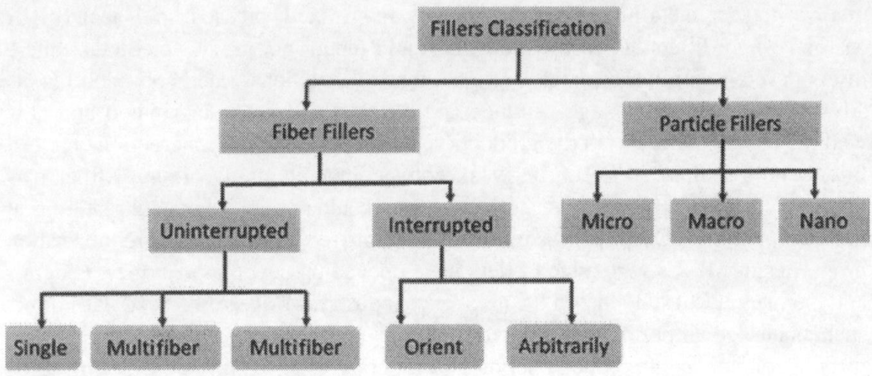

FIGURE 9.1 Classification of fillers.

9.2.2.1 Graphite

Adding graphite fillers to MMCs brings several benefits due to its outstanding characteristics. Graphite is recognized for its high ability to conduct heat and its excellent lubricating qualities, which reduce friction and wear in composite materials. By adding graphite fillers to MMCs, their ability to manage heat is improved, making them suitable for uses that require efficient heat dissipation, such as in electronic devices and high-performance engines. Moreover, graphite's lightness helps decrease the overall weight of the composite while keeping its strength. The use of graphite fillers also makes MMCs easier to work with, making the processing and shaping of these materials simpler for various industrial purposes.

9.2.2.2 Molybdenum Disulfide (MoS_2)

Scharf et al. (2013) describe molybdenum disulfide (MoS_2) as a solid lubricant that provides low friction up to approximately 400°C in air and even higher in non-reactive environments. However, its drawback lies in its susceptibility to oxidation at elevated temperatures, leading to the formation of abrasive molybdenum trioxide (MoO_3).

9.2.2.3 Hexagonal Boron Nitride (h-BN)

Wu et al. (2018) highlight that hexagonal boron nitride (h-BN), with a structure akin to graphite, offers lubrication up to 900°C. Its chemical stability and resistance to oxidation make it an ideal candidate for applications demanding high-temperature performance.

9.2.2.4 Calcium Fluoride (CaF_2) and Barium Fluoride (BaF_2)

Muratore et al. (2007) note that fluoride compounds form films with low shear strength at elevated temperatures, providing effective lubrication up to 900°C. They are particularly effective when incorporated into MMCs containing nickel.

9.2.2.5 Silver (Ag) and Gold (Au)

Soft, noble metals that do not form stable oxides, silver and gold retain their lubricating abilities at high temperatures. Wang et al. (2014) states that these materials are commonly utilized in the manufacturing of bushings and bearings for high-temperature applications.

9.2.2.6 Tungsten Disulfide (WS_2)

Lansdown (1999) explains that, similar to MoS_2, tungsten disulfide (WS_2) provides low friction but has superior resistance to oxidation, retaining its lubricating properties up to 600°C.

9.2.2.7 Glassy Carbon

Wu et al. (2010) describe an amorphous form of carbon that functions as a lubricant and as a barrier to solid-state diffusion, helping to preserve the interface between the matrix and reinforcement at high temperatures.

9.3 PROCESSING METHODS

Figure 9.2 illustrates the methods by which fillers affect the performance of MMCs at high temperatures through the study of tribology. Panin et al. (2022) explain that fillers modify the tribological properties by redistributing the load across various types of contacts, such as point, line, and area contacts. They contribute to the formation of a protective transfer film layer on the contact surfaces, which reduces wear and friction. The formation of this film is affected by factors such as the filler type, temperature, speed, and the conditions around the contact. Additionally, the hardness, roughness, and chemical makeup of the fillers play a significant role in the effectiveness of the transfer film, thereby improving the high-temperature performance of the MMCs.

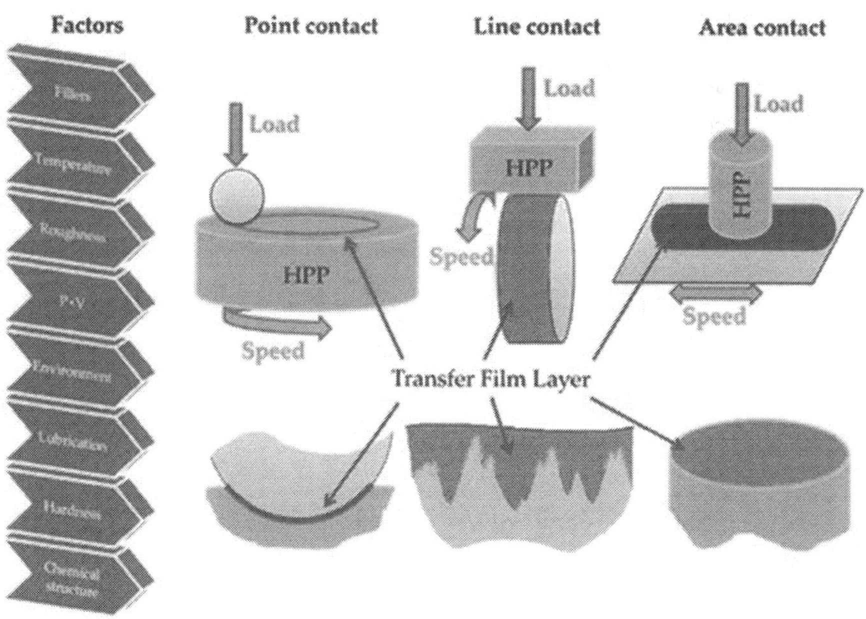

FIGURE 9.2 Fillers influence high-temperature tribology in metal matrix composites (MMCs) (Kurdi et al., 2020).

The positive impacts of fillers on the high-temperature performance of MMCs can be explained through various processes.

9.3.1 Creation of a Lubricating Layer

At elevated temperatures, numerous fillers (such as graphite, MoS_2, and h-BN) generate a thin, uniform layer at the interface. This layer possesses a low resistance to shear, facilitating smooth movement between the surfaces and decreasing friction.

9.3.2 Repairing Action

Prasad (2007) states that certain fillers, particularly malleable metals like silver and gold, can fill irregularities and small cracks during motion, enhancing the material's performance. This "repairing" action smoothes the interface, lessening stress points and reducing wear.

9.3.3 Surface Smoothing Effect

Rigid ceramic fillers like Al_2O_3 or SiC, when present in small amounts, can smooth the interface by eliminating rough spots. Prasad et al. (1993) explain that this process leads to a more uniform surface and a reduction in friction.

9.3.4 Bearing Capacity

Rigid fillers can also support a part of the applied force, lessening the strain on the surrounding material. Miracle (2005) notes that this effect is particularly beneficial at high temperatures, where the surrounding material tends to become softer.

9.3.5 Protection from Oxidation

Kaushik et al. (2022) use fillers such as glassy carbon or boron nitride (BN) to create a protective layer that prevents oxygen from reaching the surrounding material or the interface between the material and reinforcement, effectively inhibiting oxidation.

9.3.6 Improved Heat Dissipation

Manivannan et al. (2020) highlight those fillers with high thermal conductivity, such as graphite or SiC, play a key role in the more efficient dissipation of frictional heat, thereby reducing the formation of localized hotspots.

9.4 INFLUENCE ON HIGH-TEMPERATURE WEAR MECHANISMS ON MMCs

In elevated temperatures, MMCs can experience a variety of mechanisms of wear, such as friction, sticking, separation, and rusting. The presence of fillers greatly impacts these wear processes:

9.4.1 Friction Wear

The tough ceramic particles added to MMCs can withstand friction, yet they might also act as abrasive particles if they detach. Vencl et al. (2010) suggest that soft fillers, such as graphite or MoS_2, can surround tough particles, thereby reducing their abrasive effect.

9.4.2 Sticking Wear

When temperatures are high, the material of the MMC becomes more susceptible to sticking. Aranganathan and Bijwe (2016) explain that fillers like h-BN or CaF_2 form a protective layer that prevents direct contact between metals, effectively reducing sticking.

9.4.3 Separation Wear

This happens when cracks spread out from the inside, leading to wear debris that looks like sheets. Suh (1977) explains that fillers can affect crack propagation in two ways: soft fillers absorb the forces that drive crack growth, reducing their depth, while hard fillers deflect the cracks, preventing their further expansion.

9.4.4 Rusting Wear

Elevated temperatures speed up the rusting process, creating hard, brittle oxides that can easily break off. Stott (1998) notes that certain fillers, such as silver or glassy carbon, block oxygen from reaching the material, while others, like h-BN, are non-reactive and do not accelerate the rusting process.

9.5 EFFECT OF FILLERS ON HIGH-TEMPERATURE TRIBOLOGY OF METAL MATRIX COMPOSITES

9.5.1 Testing Methods and Tribology

The research of tribology, especially when focusing on high-temperature uses for MMCs, demands thorough testing techniques to examine the effects of fillers on wear, friction, and the composite's overall functionality. Tjong and Ma (1997) emphasize that as MMCs are increasingly used in high-temperature environments, it is crucial to understand how different fillers impact the tribological properties of these composites. Below is a summary of the main testing techniques used in tribological research of MMCs, along with how these techniques aid in evaluating the performance of different fillers.

9.5.1.1 Pin-on-Disk Testing

The pin-on-disk test is a widely employed technique for assessing the lubrication characteristics of MMCs. In this procedure, a pin crafted from the composite material is pressed onto a disk that is in motion, while a predetermined force is applied,

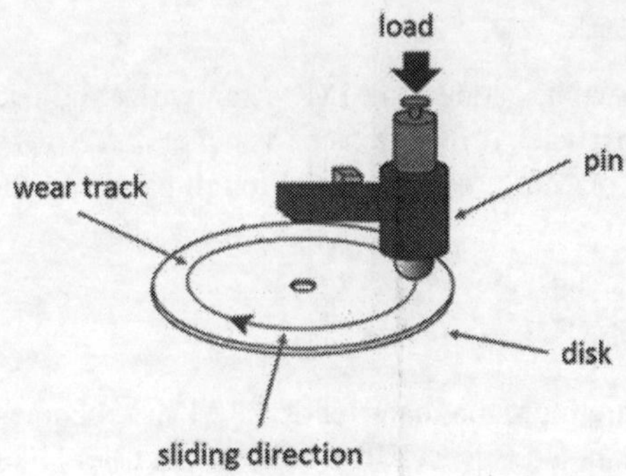

FIGURE 9.3 Pin-on-disk tester layout (Kaushik et al., 2022).

as shown in Figure 9.3. The test determines the resistance to sliding friction and the rate at which the material wears as the pin moves across the disk. When performed at higher temperatures, this test aids in evaluating the material's resistance to wear at elevated temperatures and the efficiency of the additives in minimizing friction and wear. Various additives can be subjected to the same conditions to compare their effects on the material's sliding friction.

9.5.1.2 Ball-on-Disk Testing

The ball-on-disk experiment employs a round indenter (usually a tough steel or ceramic ball) rather than a pin (Figure 9.4). This method is especially effective for examining the interaction mechanics and friction characteristics of MMCs across a range of temperatures. The arrangement of the ball-on-disk experiment facilitates the exploration of various tribological features, including the creation of transfer films, which play a crucial role in the material's resistance to wear. Through studying how additives affect the formation and durability of these films at elevated temperatures, scientists can identify the best additives to improve the functionality of MMCs.

9.5.1.3 High-Temperature Tribometers

Tribological performance evaluation tools for high temperatures are specifically crafted to assess how materials perform in terms of friction, wear, and material transfer at elevated temperatures. (Hashim et al., 1999) explain that these tools are capable of simulating the extreme conditions that materials encounter in real-world applications, such as those in aerospace or automotive engines. High-temperature tribological evaluation tools are able to measure friction, wear, and material transfer in controlled

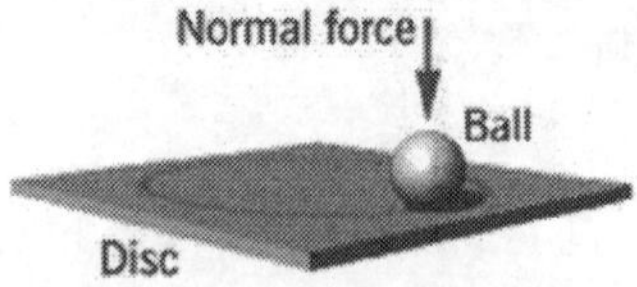

FIGURE 9.4 Ball-on-disk tester (Kaushik et al., 2022).

settings that mimic the conditions under which MMCs operate. When fillers are added to MMCs, they can greatly change how materials behave at high temperatures, and these tools are essential for gathering the data needed to comprehend these alterations.

9.5.1.4 Wear Track Analysis

Rajan et al. (2007) explain that after tribological evaluations, the analysis of wear tracks is performed to examine the wear mechanisms and assess the effectiveness of the additives. Methods like scanning electron microscopy (SEM) and energy-dispersive X-ray spectroscopy (EDS) are employed to scrutinize the surfaces of the composites that have experienced wear. Tjong and Ma (2000) highlight that SEM provides detailed images of wear paths, allowing scientists to observe changes in the microstructure, crack formation, and material displacement. EDS can determine the chemical makeup of the debris from wear and the films formed during transfer, shedding light on how additives help decrease wear and improve the composite's performance under high-temperature conditions.

9.5.1.5 Hardness Testing

Voevodin et al. (1999) emphasize that the toughness of MMCs is crucial in predicting their wear resistance, particularly when subjected to high temperatures. To evaluate the effect of fillers on the material's toughness, hardness tests like the Vickers or Rockwell hardness tests are performed both before and after conducting tribological tests as shown in Figure 9.5. Fillers, especially those made of ceramics or nanoparticles, can greatly enhance the toughness of the composite, thus boosting its resistance to wear. These hardness tests enhance the tribological tests by offering a deeper insight into the impact of fillers on the composite's overall toughness.

FIGURE 9.5 Hardness tester.

9.5.1.6 Thermogravimetric Analysis (TGA)

Thermogravimetric analysis (TGA) is a technique employed to examine the heat stability and deterioration characteristics of MMCs. Mortensen and Llorca (2010) explain that thermogravimetric analysis (TGA) measures the weight loss of a substance when exposed to heat, providing valuable insights into the heat resistance of the composite and its additives. This test is especially crucial for applications operating at high temperatures, where the deterioration of the matrix or additives can greatly impact their lubrication capabilities. Through the analysis of TGA data, scientists can choose additives that boost the heat resistance of MMCs, thus enhancing their functionality in high-temperature environments.

The mix of these evaluation techniques enables a thorough assessment of the impact of fillers on the high-temperature performance of metal matrix composites. Jiang and Wang (2015) explain that through pin-on-disk and ball-on-disk experiments, scientists can assess wear and friction levels under conditions that closely replicate real-world use cases. High-temperature tribometers yield information on the performance of these materials under severe conditions, while the examination of wear tracks and hardness provide understanding into the changes at the microscopic level and the mechanical characteristics that enhance the material's toughness. You et al. (2004) and Lloyd (1994) note that TGA also aids in understanding the thermal stability of composites and the behavior of their fillers under various conditions, in precise choice of evaluation techniques and the in-depth examination of the outcomes are crucial for refining the application of fillers in metal matrix composites. These steps assist in pinpointing the most efficient fillers that improve the tribological efficiency of composites at elevated temperatures, guaranteeing their dependability and endurance in challenging industrial settings.

9.5.2 Tribological Properties of Filler Composites

DellaCorte and Edmonds (1996) emphasize that fillers are crucial for enhancing the lubrication and wear resistance of MMCs when exposed to high temperatures. Adding fillers can greatly affect the wear resistance, how materials slide against each other, and their overall strength. Meng et al. (2018) explain that at high temperatures, MMCs often face challenges with lubrication and wear, as the metal may become softer and certain reinforcing materials may weaken. However, the right choice of fillers can help overcome these issues, ensuring the material stays strong and functional even under extreme conditions. Common fillers include ceramics, carbides, oxides, and natural substances, all of which are used to boost the ability of MMCs to withstand high temperatures. Agarwal et al. (2014) highlight those ceramic fillers, such as silicon carbide (SiC) and aluminum oxide (Al_2O_3), significantly improve wear resistance due to their hardness and ability to maintain stability at high temperatures. These materials can handle a lot of heat without breaking down, which lowers the rate of wear and keeps friction to a minimum even when the composite is put under tough conditions.

Furthermore, using solid lubricants such as graphite or molybdenum disulfide (MoS_2) as fillers in MMCs can greatly reduce friction, leading to better overall performance at high temperatures. These lubricants create a protective layer on the composite's surface during use, which reduces the contact between metal surfaces and decreases wear. The size and distribution of the fillers also play a big role in the tribological properties. Small, evenly spread particles can create a more uniform composite structure, which helps distribute the load more evenly across the material and prevents certain areas from wearing out faster. Tomala et al. (2017) emphasize that the filler's ability to bond effectively with the metal matrix is crucial. Strong bonding ensures the filler particles are well integrated with the matrix, thereby enhancing the composite's wear resistance and thermal stability. The ability of filler composites to resist wear and friction at high temperatures is crucial for their use in demanding situations. Important tribological characteristics include wear resistance, the coefficient of friction, and the capacity to form a protective tribo-layer that reduces the direct contact between surfaces.

9.5.2.1 Wear Resistance

The durability against wear in a composite material is closely linked to the toughness and ability to withstand heat of the additives. Ceramic additives such as SiC and Al_2O_3 are recognized for their superior durability in resisting wear due to their ability to withstand harsh forces and keep their shape at elevated temperatures. Moghadam et al. (2014) explain that by incorporating these additives, the material's resistance to wear-induced material loss is enhanced, which ultimately extends the composite's lifespan under high levels of stress.

9.5.2.2 Friction Coefficient

The coefficient of friction in filler composites can be adjusted by choosing fillers that have lubricating characteristics, like graphite or MoS_2. Rashad et al. (2015) explain that these substances help maintain a low friction coefficient at high temperatures by forming a solid lubricating film on the composite's surface. This film decreases the forces acting between the surfaces, resulting in reduced friction and, as a result, less heat produced when there is sliding contact.

9.5.2.3 Tribo-Layer Formation

The tribological efficiency of filler composites is significantly influenced by the creation of a tribo-layer during use. This layer serves as a shield that blocks direct contact between metals, thus lowering wear and friction. Liu et al. (2014) and Fan and Kalia (1995) note that the effectiveness of this layer depends on the type of filler used and its interaction with the matrix material at elevated temperatures. Filler types that aid in forming a stable and cohesive tribo-layer are highly sought after in situations involving high temperatures and mechanical forces. The choice of fillers greatly affects the tribological characteristics of metal matrix composites, especially at high temperatures. White et al. (2001) explain that by selecting the right combination of fillers, it is possible to enhance wear resistance, reduce friction, and promote the formation of protective layers, thereby improving the composite's durability and performance under harsh conditions.

9.6 ENVIRONMENTAL BENEFITS

Adopting natural components in composite materials brings significant environmental benefits, particularly when looking at how these materials perform at high temperatures in MMCs. Substances like coconut coir, rice husk, and wheat husk reduce our reliance on synthetic fibers, which are often derived from non-renewable sources, and also contribute to reducing industrial waste. By incorporating agricultural by-products into composite materials, manufacturers can minimize environmental damage, address ecological concerns, and meet stringent environmental regulations. For example, coconut coir has shown remarkable mechanical properties in glass/epoxy composites, demonstrating that natural materials can compete with traditional synthetic additives. Similarly, mixing ground walnut shells, another type of organic waste, into epoxy composites has led to improved hardness and stiffness, albeit with a slight reduction in impact and tensile strength. These examples illustrate how natural components can enhance certain mechanical properties while also promoting environmental sustainability. Furthermore, adding bamboo, E-glass, and coconut shells to polyester composites has not only increased the materials' fatigue resistance and tensile strength but has also promoted the use of renewable resources in engineering projects. By choosing natural components over synthetic ones, the engineering sector can reduce its environmental footprint, enhance sustainability, and foster the development of eco-friendly materials that meet the demands of modern industries.

9.7 FUTURE TRENDS AND RECENT BREAKTHROUGHS IN THE HIGH-TEMPERATURES TRIBOLOGY OF METAL MATRIX COMPOSITES

The development of friction studies at high temperatures in MMCs is being shaped by advancements in material science and engineering, aiming to improve performance, durability, and sustainability. Sahoo et al. (2011) highlight that as industries like automotive, aerospace, and energy require materials capable of withstanding extreme conditions, including high temperatures and significant mechanical stresses, researchers and manufacturers are increasingly focusing on innovative methods to improve the friction properties of MMCs. This includes the development of nanocomposites, composite materials, and environmentally friendly alternatives, as well as the implementation of new processing techniques and the exploration of emerging technologies.

9.7.1 Nanocomposites and Hybrid Materials: A Promising Future

Nanotechnology is playing a crucial role in enhancing the tribological properties of MMCs, especially when exposed to high temperatures. Nanocomposites, which incorporate nanoparticles into the metal framework, show promising results in increasing resistance to wear, reducing friction, and improving mechanical performance. By adding nanoparticles like carbon nanotubes (CNTs), graphene, and other nano-sized ceramics, these particles are uniformly spread throughout the matrix, reinforcing the material and facilitating the formation of a tribo-protective layer that minimizes the direct interaction between metals during operation. Ye et al. (2024) note that hybrid nanocomposites, which combine different types of nanoparticles or a blend of micro- and nano-sized reinforcements, are gaining significant attention for their potential to enhance material properties. These materials gain from additive effects, where the characteristics of different reinforcements complement each other, leading to enhanced thermal stability, increased load-bearing capacity, and improved wear resistance at elevated temperatures. For instance, a mixture of ceramic nanoparticles and CNTs can provide both hardness and flexibility, rendering the composite more resistant to cracks and wear.

9.7.2 Eco-Friendly Fillers and Sustainable Practices

As concerns about the environment grow and rules become stricter, the push for eco-friendly materials in making MMCs is getting stronger. The use of natural fillers, which are sourced from farm or biological waste, is increasing as a more sustainable choice than traditional synthetic fillers. Substances such as coconut coir, rice husk, and bamboo are not only sustainable but also break down easily, making them attractive for reducing the environmental impact of MMCs. Fragassa et al. (2024) highlight that natural fillers have demonstrated potential in improving specific tribological properties, such as wear resistance and friction, especially when combined with other additives in mixed-matrix composites. The emphasis on being eco-friendly extends beyond just choosing fillers. The entire life cycle of the composite, from its manufacturing to its disposal at the end of its life, is being thoroughly reviewed to minimize its environmental footprint. The recycling of MMCs, though challenging, is being pursued with the aim of finding ways to efficiently recover valuable materials and reduce waste. Additionally, advancements in green chemistry are paving the way for more sustainable manufacturing processes that lessen the use of toxic chemicals and lower the energy required for production.

9.7.3 Additive Manufacturing and Advanced Processing Techniques

Additive manufacturing, also referred to as 3D printing, is revolutionizing the production of materials, particularly those capable of withstanding elevated temperatures. This technology enables the fabrication of complex shapes with precise control over the distribution of reinforcements throughout the material. This level of accuracy allows for the development of materials that can alter their characteristics from one section to another, tailored to meet specific requirements for reducing friction and enhancing durability. For example, a component might feature a tougher exterior to resist wear from friction and a more robust interior to protect against impacts. Recent progress in 3D printing techniques, such as selective laser melting (SLM) and direct energy deposition (DED), has demonstrated the feasibility of producing materials that are more adept at withstanding high temperatures for their designated applications. Ngo (2018) explain that these approaches enable the rapid production of materials with customized structures that deliver enhanced performance under demanding conditions. In addition to 3D printing, other advanced techniques are being explored. Spark plasma sintering (SPS), for instance, is gaining traction for its capability to produce dense materials with small structures efficiently. SPS operates by applying a short electric current and pressure to bond the material, thereby enhancing the connection between the material and the reinforcements. This method is particularly effective in increasing resistance to wear and improving the material's ability to withstand higher temperatures.

9.7.4 Tribological Testing and Modeling: Pushing the Boundaries

As the demand for advanced MMCs grows, the need for more intricate tribological testing and modeling methods also escalates. Traditional testing methods, though reliable, may not fully capture the complex interactions occurring at high temperatures and under various stress conditions. Kurdi et al. (2020) state that to address this challenge, researchers are developing advanced tribometers that accurately simulate extreme conditions, thereby providing a deeper understanding of wear mechanisms and the frictional properties of MMCs. The application of computational modeling is increasingly vital in the development and improvement of MMCs. Methods such as finite element analysis (FEA) and molecular dynamics (MD) simulations are used to predict the performance of MMCs under different conditions, aiding in the modification of their composition and processing techniques. These models help identify potential failure areas and guide the creation of stronger and more efficient MMCs for use in high-temperature settings.

9.8 CONCLUSION

The incorporation of fillers into metal matrix composites represents a significant advancement in the field of high-temperature tribology. By carefully selecting and integrating these fillers, researchers have been able to engineer MMCs that maintain low friction and high wear resistance even in extreme temperature environments. The fillers achieve this through a variety of mechanisms: forming lubricating films, mending and polishing surfaces, bearing loads, and protecting against oxidation. As technology pushes the boundaries of temperature and stress in applications like hypersonic flight, deep geothermal energy, and nuclear fusion, the demand for even more robust materials grows. The future of high-temperature tribology in MMCs lies in nanoscale engineering, multifunctional fillers, bio-inspired design, and computational optimization. With continued research and innovation, filler-enhanced MMCs promise to unlock new frontiers in high-temperature applications, driving progress across industries and shaping the technologies of tomorrow.

REFERENCES

Agarwal, K., et al. (2014). Development of self-lubricating Ni-based composites for high-temperature applications: A review. Critical Reviews in Solid State and Materials Sciences, 39(3), 179–196.

Aranganathan, N., & Bijwe, J. (2016). Comparative performance evaluation of NAO friction materials containing different classes of graphite. Wear, 358, 17–22.

Chawla, N., & Chawla, K. K. (2013). Metal Matrix Composites. Springer Science & Business Media.

DellaCorte, C., & Edmonds, B. J. (1996). NASA PS400: A new high temperature solid lubricant coating for high temperature wear applications. NASA Technical Memorandum, 107056.

Erdemir, A., & Donnet, C. (2006). Tribology of diamond-like carbon films: Recent progress and future prospects. Journal of Physics D: Applied Physics, 39(18), R311.

Fan, J., & Kalia, R. K. (1995). Atomic-scale simulation of metal-ceramic interfaces. MRS Bulletin, 20(2), 18–21.

Fragassa, C., Vannucchi de Camargo, F., & Santulli, C. (2024). Sustainable biocomposites: Harnessing the potential of waste seed-based fillers in eco-friendly materials. Sustainability, 16(4), 1526.

Hashim, J., et al. (1999). Metal matrix composites: Production by the stir casting method. Journal of Materials Processing Technology, 92, 1–7.

Ibrahim, I. A., Mohamed, F. A., & Lavernia, E. J. (1991). Particulate reinforced metal matrix composites—a review. Journal of Materials Science, 26(5), 1137–1156.

Jiang, J., & Wang, R. H. (2015). Particle size-dependent interface adhesion energy and its implications for the mechanical behavior of metal matrix nanocomposites. Scripta Materialia, 106, 51–54.

Kaushik, J., et al. (2022). A review on application and optimization processes used for wear testing machine (pin on disc apparatus). Materials Today: Proceedings, 64 (Part 3). https://doi.org/10.1016/j.matpr.2022.04.735

Kurdi, A., Alhazmi, N., Alhazmi, H., & Tabbakh, T. (2020). Practice of simulation and life cycle assessment in tribology—A review. Materials, 13(16), 3489.

Lansdown, A. R. (1999). Molybdenum Disulphide Lubrication. Elsevier.

Liu, M., et al. (2014). Nacre-inspired design of mechanical stable coating with underwater superoleophobicity. ACS Nano, 8(2), 1323–1330.

Lloyd, D. J. (1994). Particle reinforced aluminium and magnesium matrix composites. International Materials Reviews, 39(1), 1–23.

Manivannan, J., et al. (2020). The influence of heat treatment on the thermal and electrical conductivity of Al-SiC composites. Journal of Alloys and Compounds, 822, 153425.

Meng, F., et al. (2018). Graphite self-lubricating composites: A review. Reviews on Advanced Materials Science, 53(1), 76–90.

Miracle, D. B. (2005). Metal matrix composites – From science to technological significance. Composites Science and Technology, 65(15–16), 2526–2540.

Moghadam, A. D., et al. (2014). Functional metal matrix composites: Self-lubricating, self-healing, and nanocomposites-an outlook. JOM, 66(6), 872–881.

Mortensen, A., & Llorca, J. (2010). Metal matrix composites. Annual Review of Materials Research, 40, 243–270.

Muratore, C., et al. (2007). CaF_2-coated carbons as lubricious adaptive nanocomposite materials for MEMS devices. Tribology Letters, 25(2), 113–117.

Panin, S. V., Alexenko, V. O., & Buslovich, D. G. High performance polymer composites: A role of transfer films in ensuring tribological properties—A review. Polymers 2022, 14, 975. https://doi.org/10.3390/polym14050975.

Prasad, B. K. (2007). Investigation into sliding wear performance of zinc-based alloy reinforced with SiC particles in dry and lubricated conditions. Wear, 262(3–4), 262–273.

Prasad, S. V., et al. (1993). Tribology of aluminum metal-matrix composites: Lubrication by graphite. Wear, 169(1), 59–69.

Prasad, S. V., & Asthana, R. (2004). Aluminum metal-matrix composites for automotive applications: Tribological considerations. Tribology Letters, 17(3), 445–453.

Rajan, T. P. D., et al. (2007). Developments in solidification processing of functionally graded aluminium alloys and composites by centrifugal atomisation and casting. Transactions of the Indian Institute of Metals, 60(2–3), 159–174.

Rashad, M., et al. (2015). Effect of graphene nanoplatelets addition on mechanical properties of pure aluminum using a semi-powder method. Progress in Natural Science: Materials International, 25(5), 460–470.

Rohatgi, P. K., & Tabandeh-Khorshid, M. (2014). Self-lubricating metal matrix composites: A review. Tribology Online, 9(5), 189–200.

Sahoo, P. (2011). Optimization of turning parameters for surface roughness using RSM and GA. Advances in Production Engineering & Management, 6(3).

Scharf, T.W., Prasad, S.V., 2013. "Solid Lubricants: A Review," Journal of Materials Science, 48(2), 511–531.

Stott, F. H. (1998). The Role of Oxidation in the Wear of Alloys. Tribology International, 31(1–3), 61–71.

Suh, N. P. (1977). An overview of the delamination theory of wear. Wear, 44(1), 1–16.

Tjong, S. C., & Ma, Z. Y. (1997). Microstructural and mechanical characteristics of in situ metal matrix composites. Materials Science and Engineering: R: Reports, 29(3), 49–113.

Tjong, S. C., & Ma, Z. Y. (2000). High-temperature creep behaviour of powder-metallurgy aluminium composites reinforced with SiC particles of various sizes. Composites Science and Technology, 60(7), 1141–1150.

Tomala, A., et al. (2017). WS_2 nanotubes as efficient lubricant additives for hot forming operations. Lubricants, 5(2), 9.

Vencl, A., Rac, A., & Mrdak, M. (2010). Characterization of the plasma spray coatings by abrasive wear tests. Tribology in Industry, 32(1), 23–28.

Voevodin, A. A., et al. (1999). Nanocrystalline WC/amorphous diamond-like carbon composites with enhanced mechanical properties for engineering applications. Diamond and Related Materials, 8(9), 1524–1527.

Wang, H., Liu, Y., & Sun, J. (2014). Development and application of self-lubricating metal matrix composites: A review. Materials & Design, 55, 496–505.

White, S. R., et al. (2001). Autonomic healing of polymer composites. Nature, 409(6822), 794–797.

Wu, J. H., et al. (2010). The effect of glassy carbon on the friction and wear characteristics of Ni3Al-based alloys. Wear, 269(1–2), 48–56.

Wu, Y., et al. (2018). Hexagonal boron nitride as a multifunctional additive in matrix materials: A review. Materials Today Communications, 17, 82–95.

Ye, W., Shi, Y., Zhou, Q., Xie, M., Wang, H., Bou-Saïd, B., & Liu, W. (2024). Recent advances in self-lubricating metal matrix nanocomposites reinforced by carbonous materials: A review. Nano Materials Science, 6, 701–713.

You, C. P., et al. (2004). The effect of interfacial heat transfer coefficient on the high-temperature sliding wear behavior of aluminum matrix composite reinforced with granulated carbon fibers. Carbon, 42(2), 359–369.

10 Influence of Fillers in the Tribological Performance of Self-Lubricating Metal Matrix Coating

G. J. Naveen, P. Sampathkumaran, Nagaraja C. Reddy, E. R. Babu, N. R. Thyagaraj, and S. Seetharamu

10.1 INTRODUCTION

The basis of modern engineering research and fillers play a very important role in the tribological behavior of self-lubricating metal matrix coatings for machines with tightly packed components intended for high efficiency and extended life. Normally they are fixed onto a metal matrix from which solid lubricant fillers contained in it can intermittently get freed to reduce dirt and rubbing in a few modern fields, for example, substantial gear, automobile, and aviation (Raj et al. 2024; Song et al. 2024; Yang et al. 2024). Metal matrix components show exceptional thermal endurance while keeping up their mechanical power through various base metals such as cobalt, nickel, and aluminum and their reliable structure stops breakdown during actual usage. These coatings produce self-lubricating properties thanks to the integrated fillers including graphite, molybdenum disulfide (MoS_2), and hexagonal boron nitride (h-BN). The tribological behavior of the coating changes significantly with these fillers because they improve the coating's performance during high-speed operations by decreasing friction and wear (Malaki et al. 2019; Vo et al. 2023; Ye et al. 2024).

The preparation of solid lubricant fillers in the metal matrix results in higher life of the basic components as a whole; sacrifice of those forms a high wear resistant and low coefficient of friction in comparison to clean bearings. This creates a transfer film to provide lubrication and eliminate metal-to-metal contact, which causes wear. For this reason, filler selection is often more influenced by application domains such as the operating temperature range, climatic influences if any, and mechanical stresses. Graphite keeps its excellent lubrication trait up to modest temperatures but its dry lubrication properties end when temperatures rise. Top aerospace companies select h-BN systems because they stay chemically stable at high temperatures yet MoS_2 works best when combined with vacuum-based equipment (Ay 2025; Beyanagari et al. 2024; Mukhtar et al. 2024; Xing et al. 2024).

DOI: 10.1201/9781003530909-13

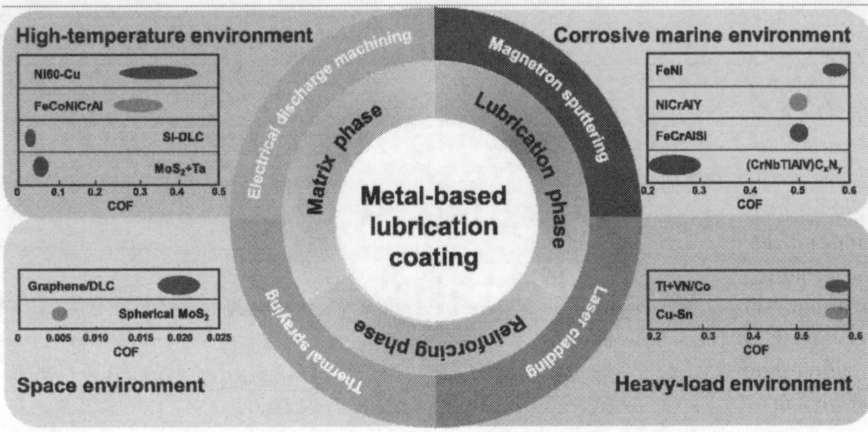

FIGURE 10.1 The composition, methods of manufacture, and essential functions of metal-based lubricants (Wu et al. 2024).

Many experts from materials science mechanical engineering and surface chemistry team up to study coating features that improve friction performance. Several advanced testing methods such as X-ray diffraction, transmission electron microscopy (TEM), and scanning electron microscopy (SEM) study how these coatings are made up internally. Actual components and test surfaces (pin-on-disk, ball-on-flat) display resulting coating effects while generating data under realistic working conditions. These tests display how fillers interact with their matrix system while showing their effect on both testing surfaces and the oil film formation process (Gao et al. 2024; Jakovljevic and Landek 2023; Jansons et al. 2022; Singh et al. 2024a; Wood and Lu 2024; Xu and Zhang 2024).

The coatings have also been developed in part due to a growing demand for environmentally friendly replacements for liquid lubricants, which can be hazardous to the ecosystem and human health by way of leaks, evaporation, and disposal problems. Such self-lubricating metal matrix coatings will not only be effective by itself, but also minimize its harmful effects on nature by reducing its maintenance, which highlights its sustainability perspective. Formulations of established coatings

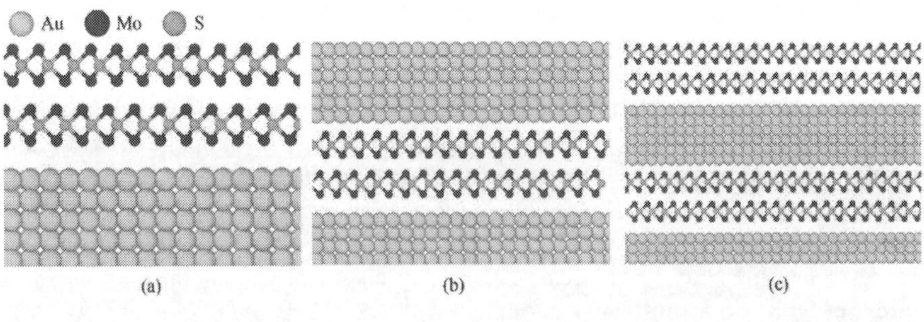

FIGURE 10.2 Several films that are bilayer or multilayer and made of two components, Au and MoS_2: (a) MoS_2/Au, (b) Au/MoS_2/Au, and (c) MoS_2/Au/MoS_2/Au (Zhou et al. 2023).

are non-trivial to develop and characterize, necessitating a thorough understanding of both the materials and the operational environments they are intended to protect against. These coatings will perform better under the conditions of some factors like compatibility of the filler with the matrix, the size and distribution of the filler particles, and the integrity of the coating on the substrate (Bernasconi et al. 2016; Li et al. 2024a; Ling et al. 2024; Ren et al. 2021; Ye et al. 2024).

Furthermore, the use of high-throughput experimental infrastructure along with the cultivation of computational materials science will augment the development of self-lubricating metal matrix coatings tuned to specific tribological properties. They are widely used for modeling and simulating the different loading conditions and matrix-filler combinations to predict performance before the actual manufacture. High-throughput tests allow for rapid screening of large numbers of combinations of materials and processing parameters while giving an indication of the most promising candidates for detailed characterization. This streamlined method considerably decreases the time and cost of developing fresh coatings and enables confining solutions to particular industrial needs (Baiocco et al. 2024; Chen and Zhao 2024; Ning et al. 2024).

Self-lubricating metal matrix coatings have received significant interest due to their ability to combine desirable properties of both metals and polymers, making them suitable for a range of applications where lubrication is critical. Such coatings help reduce wear and friction, increasing the service life and operational efficiency of machines and hence contribute to economy as well as support sustainability targets. Advancements in the discipline will result in increasingly better solutions as material science and manufacturing technology advance. These high-performance, sustainable materials are the proof for a part of the mystery sought by industries. It needs more research and development to resolve such challenges. This research will deepen insight in the way materials behave alongside constructing specific technologies for radically improving business practices for machine care and agile performance (Baig and Samad 2021; Daneshifar et al. 2019; Hao et al. 2019; Siddiqui and Mittal 2023; Tian et al. 2023).

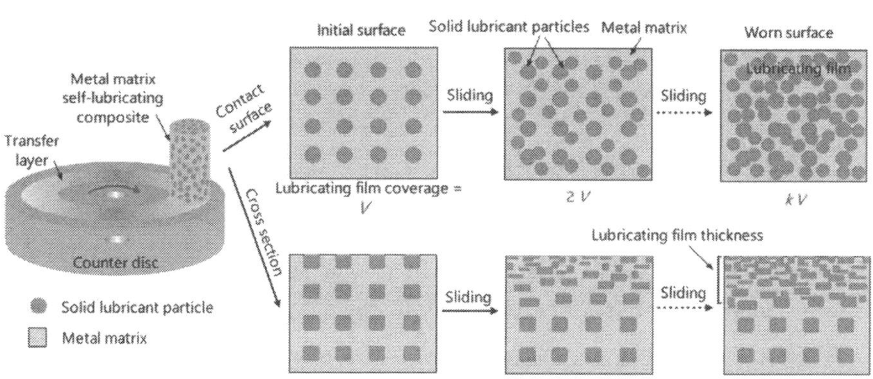

FIGURE 10.3 Diagram illustrating the lubricating film's development on the MMSC's worn surface (Xiao et al. 2020).

10.2 MATERIALS AND METHODS

A broader range of advanced materials and methods are used to study the fillers affecting the tribological behavior of self-lubricating metal matrix coatings as they were able to provide insightful data in the way different fillers helped improve the wear resistance and/or lower the coefficient of friction in a number of industrial applications. Metal matrix, normally employed with strong and thermal-resistant alloys such as cobalt, nickel, and aluminum. Graphite MoS_2 and h-BN wear-resistant materials enter the matrix when used as natural lubricants. The material breakdown shows graphite works best for low speeds while MoS_2 reliably stands shear load and h-BN resist heat and oxidation. Each filler has distinctive advantages (Li et al. 2024b; Singh Gautam et al. 2023, Kumar et al. 2023; Zhao et al. 2024).

Choosing the matrix and filler materials is the first step of an additive process relative to the intended application. The matrix and filler materials are mixed, ground, and sintered to form the composite materials. This is a process of powder metallurgy. The proper use of ball mills, for example, will resolve the problem of dispersing filler homogeneously in the matrix that will improve the mechanical bonding and integrity of the coating. To form a strong and low porosity coating that guarantees good adhesion to the substrate, the powder mixture is melted and then sprayed on the substrate using a plasma torch. Substrates are then covered with thermal spraying techniques, such as plasma spraying (Ginting 2022; Mihoob et al. 2023; Neves et al. 2007).

Controlled wear and friction property investigations including pin-on-disk, ball-on-flat, and reciprocating wear tests can be done in accordance to standardized tribological tests of the coatings. These experiments provide a systematic comparison of tribological performance between different coating formulations under representative mechanical and thermal loading conditions. Techniques like TEM and SEM are also utilized to investigate wear mechanisms, providing insight to wear processes from a surface perspective and microstructural transformations as well as the morphology of wear tracks. The phase composition of the coatings is also determined by X-ray diffraction (XRD), which helps to discover changes or reactions that can occur during the wear tests. Research describes the activity of fillers during wear testing and explains the development of a smooth lubricious transfer film on contact areas. We use both new and standard surface analysis tools to study our samples (Raman spectroscopy) in order to verify the obtained results; the resulting coating is evaluated for their ability to reduce friction and protect substrate from wear. Understanding how filler and polymer alloys interact with each other plus the opposing surface defines the best tribological performance of the coatings.

Modern engineering tools help detect how different filler materials respond inside metal during operation. Simulation techniques like Molecular Dynamics and Finite Element Analysis show the real failure mechanisms on tiny filler surfaces plus reveal how temperature spreads during stress tests. The simulations help researchers cut development expenses and identify optimized processing conditions before physical production. Exploring how fillers change coated metal properties from advanced materials with these techniques advances our tribological knowledge while driving environmental-friendly next-generation coating design. This research helps organizations better understand component life spans while reducing maintenance expenses. The research creates essential material science knowledge for tribological engineering which needs modern technology advancements (Abdelbary and Chang 2023; Domínguez García et al. 2023; Gee et al. 2022; Nazir et al. 2023; Ray et al. 2023; Zhang et al. 2024).

10.3 RESULTS

Understanding how filler and polymer alloys interact with each other plus the opposing surface defines the best tribological performance of the coatings. Modern engineering tools help detect how different filler materials respond inside metal during operation. Simulation techniques like MD and FEA show the real failure mechanisms on tiny filler surfaces plus reveal how temperature spreads during stress tests. The simulations help researchers cut development expenses and identify optimized processing conditions before physical production. Exploring how fillers change coated metal properties from advanced materials with these techniques advances our tribological knowledge while driving environmental-friendly next-generation coating design. This research helps organizations better understand component life spans while reducing maintenance expenses. The research creates essential material science knowledge for tribological engineering which needs modern technology advancements. A case in point is graphite, which finds incredibly vast usage as a high-temperature lubricant owing to its excellent thermal conductivity as well as low friction characteristics. On the basis of its multilayer structure, MoS_2 exhibits superior lubrication properties primarily in low oxygen or in vacuum atmosphere where lubrication performance of graphite is declined. Because PTFE has an extremely low coefficient of friction and remarkable chemical resistance, it is used in corrosive environments where other fillers would fail. Along with natural characteristics, the incorporation of these fillers into the matrix also determines their performance. The tribological behavior of the whole coating has also reported to depend on the dispersing technique used to embed fillers (e.g., thermal spraying or powder metallurgy) and therefore the uniform distribution of fillers within the matrix. The homogeneous distribution of lubricating attributes across coating surfaces helps achieve standardized performance while minimizing specific wear areas. An improved dispersion and increased mechanical interlocking with the matrix occur when using fine spherical filler particles which enhance both wear resistance and mechanical integrity of coatings. Another consideration with respect to filler particles is their size and shape. Recently fillers of nano size have been explored as they can provide higher surface coverage and integration and thus leading to better tribological performance due to their high surface area-to-volume ratio. Too low filler content will be unable to deliver sufficient lubrication and too high filler content will result in a decrease in mechanical strength of the matrix. There are many materials out there, but explaining their specific additive interactions and how they interpret performance is the other elephant in the room. A promising pathway to tune the mechanical wear vs. lubrication balance and hence wear behavior of the coating could be through hybridization, that is mixing of hard and soft particle component (for example, diamond and graphite particles). Various tribological behaviors of coatings are also influenced by other operational parameters such as temperature, speed, and load. Overloading can lead to delamination or deformation of the coatings, exposing the base material to wear. Hence, to prevent/treat failure modes in an application, its mechanics need to be understood as the first step, and hence the mechanical thresholds of the fillers as well as that of the matrix need to be explored. Additionally, environmental conditions like humidity and corrosive chemicals can hinder the metal matrix and the fillers, which means that materials should be selected carefully based on expected working conditions. Current activities in this area of investigation encompass advanced simulation approaches to assess wear mechanisms and simulate the performance of the filler-matrix metal matrix coatings. These latter models allow for enhanced designs, to be tuned for usage, by optimizing both the fill, and the distribution of that fill in the matrix. Thus, fillers with a significant increase in tribological properties of metal matrix coating and proper selection of fillers is vital to maximize the tribological performance. Fillers play a vital role in extending the longevity and effectiveness of mechanical components in various industries. The reason as to why is because they improve the lubricating properties of the coatings and structural strength of the aforementioned coatings, because at the start of the day equipment need to be functioning reliable as well as affordably (Aly et al. 2012; Gangwani et al. 2024; Joshi et al. 2024; Keshavanarayana et al. 2014; Kumar and Bijwe 2013; Lin et al. 2023; Liu et al. 2024; Skarvelis and Papadimitriou 2009).

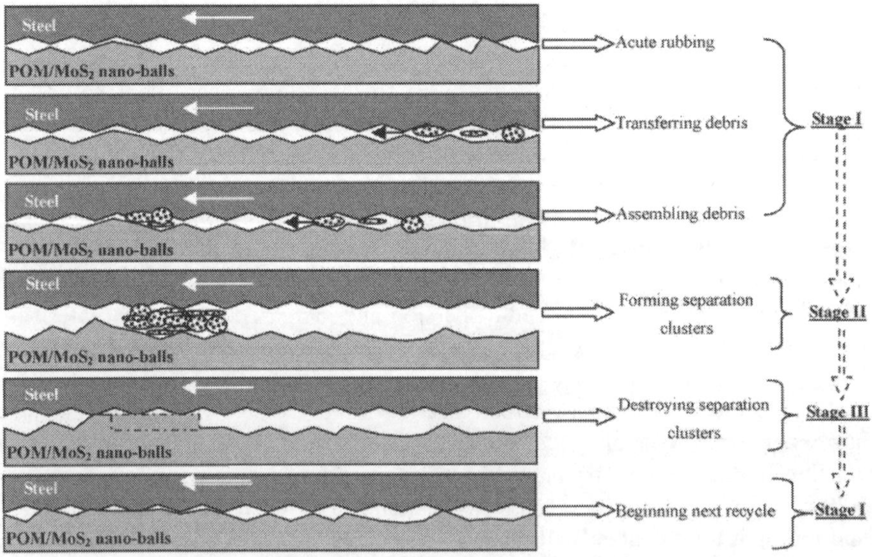

FIGURE 10.4 Schematic illustration of the mechanism of debris' assembling (Hu et al. 2009).

10.4 DISCUSSION

Experimental variables have been used to study, to a great extent, the behavior of self-lubricating metal matrix coatings and also self-lubricating metal matrix composites, with fillers being added through additional processing routes as alternative controlled results, in this very important field of research focused in a better understanding of self-lubricating metal matrix materials, to yield a better way of making metal tribological materials as they were to be used in industries. MoS_2 and h-BN and graphite enter metallic alloy systems to produce hard yet thermally stable structures including nickel aluminum cobalt as the base components and filler breakthroughs in material structure engineering. These fillers are primarily used to lessen wear and friction between various mechanical components each time they enter into motion, increasing the life span of the start-up and its efficiency. Processing graphite using methods that involve or tolerate higher temperatures enables applications, but oxidation limits its use at high temperatures. h-BN is a well-regarded material for high-temperature stability and chemical inertness, but is only appropriate for dry tribology and thermal barrier coatings in extremely high-temperature applications as it is not a suitable lubricant due to its weak functionalization and friction reduction properties, while MoS_2 will allow for good lubrication under vacuum or inert conditions, making it appropriate for engineering steward spectrum for aircraft applications (Singh et al. 2024b; Shi et al. 2024; Smoleń et al. 2024).

Controlled testing using direct wear/friction measurement is the primary tool to characterize the tribological performance of these coatings. These fewer common tests generally replicate the types of forces coatings might be exposed to under real-world industrial conditions. A key aspect of the effectiveness of a filler is its ability to create a solid, lubricative layer between two frictional surfaces to prevent wear, by preventing mechanical engagement. Therefore, although transfer film is very important to protect the inner metal surface from being worn and machine-destroyed on the surface. Filler selection is based on factors that include the operating temperature, mechanical stress, and climatic conditions as they determine how efficiently the filler functions in the expected service life of the coating (Cotell and Hirvonen 1996; Sahoo et al. 2022; Singh and Sinha 2021).

Optimization of tribological performance has thus been the focus of recent studies in this area. Recent advances in this domain includes enhancement of adhesion and elimination of interfacial separation, progress in controlling the size, distribution, and volume density of fillers in matrix. Recently, the two-dimensional (2D) fillers with nanoscale sizes were found to be more effective than the common microscale fillers, and thus revolutionized the whole field of nanotechnology. For instance, MoS_2 and h-BN nanoparticles have been demonstrated to offer a more homogeneous and effective lubrication film than bigger particles, leading to lower friction coefficients and wear rates (Bukvić et al. 2024; Moustafa and Taha 2023; Ziyamukhamedova et al. 2024).

Modern hybrid coating solutions that incorporate various fillers into single protective layers show evidence of improved performance properties based on user application needs. A coating developed from graphite and MoS_2 fillers demonstrates better operational characteristics over diverse conditions because the frictional properties of graphite match the load-bearing capability of MoS_2. Striking the right balance of different fillers within hybrid coatings remains a challenging task depending on the required level of understanding about material behavior during component usage (Lu et al. 2023; Voevodin et al. 1999).

Metal matrix coatings which are self-lubricating would enhance the tribological performance, but would forgo the application of liquid lubricants which is another advancement towards an bio-friendly lubrication concept, eliminating the risks regarding leakage of liquid lubricants and/or unconscientious disposal. Such coatings make them more attractive for companies that are involved in the expansion of the environmental footprint and improving efficiency by removing the need for regular re-lubrication and thus help to reduce downtimes and costs. The future of research into this area, boosted by continued advances in material synthesis, computer modeling, and experimental methodology that will allow ever more clever repairs, looks bright. In this modern design-forward view, computational modeling approaches in particular hold the promise of accelerating coating designers because they have been shown to be very effective for predicting properties related to how new arrangements of fillers interact with metal matrices in different ways. Similarly, increased development multilayered characterization techniques including in-situ tribological studies and online surface analysis are expanding our understanding of the performance of these coatings under operational stressors, and guiding design for materials which would survive harsh environments more effectively (Somya et al. 2024; Sukumaran et al. 2024; Vafaeenezhad and Eslami-Farsani 2024).

The importance of fillers in the improvement of tribological properties for self-lubricating metal matrix coatings is difficult to overstate. By significantly decreasing wear and friction, they extend the longevity and efficiency of critical components in a variety industry contributing to effectiveness sustainability and progression with technology systems. In conclusion, it is necessary to further investigate these advanced materials to satisfy the increasing requirements of modern industry. It is inevitable that new or improved coating systems, characterized by high efficiency and durability will be developed as a consequence of the modernization of production processes and innovation in material structure, eventually pointing out to one way for setting industry benchmarks into those coatings with large demand/high concentration level (Jakab et al. 2021; Pan et al. 2022).

10.5 CONCLUSION

Nonetheless, the role of fillers with respect to tribological performance of self-lubricating metal matrix coatings is a key factor to consider in material engineering that could contribute to enhancing the dry sliding wear and frictional response of mechanical systems. Fillers in metal matrices such as graphite, molybdenum disulfide (MoS_2), and hexagonal boron nitride (h-BN) can define coating functionalization capabilities by lowering the coefficient of friction and improving wear resistance. There is a broad array of functions that need improving efficiency and durability hugely for traditional, automotive, and aerospace machinery applications. This action of fillers is mediated primarily by the generation of a lubricious transfer film on contacting interfaces. This film serves as a buffer that prevents metal-to-metal contact allowing for wear reduction. The success of these fillers depends on the physical and chemical properties of these fillers, the intended environment, and compatibility with a metallic matrix. For applications, they are best as fillers, depending on needs. For example, MoS_2 has better high-temperature stability, which is suitable for high-performance application when the atmospheric environ conditions obtain reparation under vacuum or inert circumstances, while due to its excellent lubrication properties, graphite is generally selected in the medium-temperatures platforms. The cultural importance of h-BN comes from the special chemical inertness and thermal stability that h-BN contributes. In addition to the selection of appropriate fillers, formulation of such coatings also includes manipulation of size, geometry, dispersion, and loading levels of the fillers in the matrix. This is crucial for its uniform performance under different load conditions, which is possible only when fillers are properly encapsulated and evenly dispersed inside the matrix.

Furthermore, the preparation of nano-sized fillers has been accomplished in recent years by putting the principles of nanotechnology into practice and these fillers have shown higher performance than micro-based fillers in terms of wear and friction reduction owing to the larger surface area and interactions with the matrix and counter surfaces. Exponentially growing research advances targeted to damage-reduction of friction and wear losses have the potential in use of nanomaterials or fillers. Hybrid fillers, which are a mixture of two or more different types of fillers, have also been utilized as a versatile method to achieve improved tribological properties through the synergistic effects of the benefits of each type of filler. This results in a hybrid approach that significantly lowers wear and friction thus extending both the longevity of the coating and its component protection. Computer modeling and simulation also emerged as vital tools in helping estimate how the coatings performed when subjected to service stresses and how fillers behaved within the metal matrix. This provides an initial guide to the optimal filler and matrix combinations for an application and saves time and money wasted on trial-and-error approaches. Also, further developments in manufacturing technologies like plasma spraying and laser cladding have allowed the coatings to be applied with more accuracy and precision thereby leading to better adhesion and uniformity of covering along with lesser defects. Additionally, these coatings are green; because the self-lubricating metal matrix coatings eliminate the use of liquid lubricants, this can be dangerous to the environment sometimes. This minimizes the requirement for ongoing upkeep and component replacement, which is wasteful and has an effect on production and operating processes. This relevance of these coatings to sustainability fits in with global efforts to become more energy-efficient and reduce the deleterious emissions produced from industrial processes.

In summary, the strategic inclusion of fillers into metal matrix coatings is a breakthrough in the field of wear science and tribology. The addition of fillers improves the wear resistance and/or lubricating performance of the coatings; therefore, increasing the lifetime and performance of mechanical components. The first paper describes the importance of research and development in the field of coating technology in order to ensure that the coating technologies develop and meet the future needs of modern engineering applications. As the industry witnesses advancements, the performance and dependability of self-lubricating metal matrix coatings will be transformed through the amalgamation of enhanced materials, patented manufacturing processes, and creative design principles, underscoring systems the pivotal part played by self-lubricating coatings in tactile development of modern mechanical devices. This in-depth study of the filler effect not only contributes immensely to the field of tribology but also lays the groundwork for future classes of innovations that will push the boundaries of both materials' science and engineering.

REFERENCES

Abdelbary, A., & Chang, L. (2023). 9 - Tribology testing, measurements, and evaluation. In A. Abdelbary, & L. Chang (Eds.), *Principles of engineering tribology* (pp. 391–453). Academic Press. https://doi.org/10.1016/B978-0-323-99115-5.00010-4

Aly, A. A., Zeidan, E.-S. B., Alshennawy, A. A., El-Masry, A. A., & Wasel, W. A. (2012). Friction and wear of polymer composites filled by nano-particles: A review. *World Journal of Nano Science and Engineering*, *2*(1), 1–10.

Ay, G. M. (2025). Solid lubricants against wear formation in aerospace components. In S. Gürgen (Ed.), *Failure in aircraft materials: Sustainable aviation* (pp. 1–X). Springer. https://doi.org/10.1007/978-3-031-65850-1_1

Baig, M. M. A., & Samad, M. A. (2021). Epoxy/epoxy composite/epoxy hybrid composite coatings for tribological applications-A review. *Polymers*, *13*(2), 179. https://doi.org/10.3390/polym13020179

Baiocco, G., Menna, E., Mingione, E., Rubino, G., & Ucciardello, N. (2024). Effect of process parameters and film stratification on morphology and performance of auto-lubricating Ni-GnP electroplated coated steel. *Engineering Failure Analysis*, *161*, 108223. https://doi.org/10.1016/j.engfailanal.2024.108223

Bernasconi, R., Tucci, A., Magagnin, L., & Nobili, L. (2016). Metal matrix composites as environment-friendly protective coatings. *Materials Science Forum*, *879*, 1387–1392. https://doi.org/10.4028/www.scientific.net/msf.879.1387

Beyanagari, S. R., Sivalingam, A., Kandasamy, J., & Katiyar, J. K. (2024). Influence of 2D solid lubricants on mechanical and tribological behaviour of Al 7XXX series metal matrix composites: A comprehensive review. *Tribology - Materials, Surfaces & Interfaces*, *18*(3), 145–170. https://doi.org/10.1177/17515831241265341

Bukvić, M., Gajević, S., Skulić, A., Savić, S., Ašonja, A., & Stojanović, B. (2024). Tribological application of nanocomposite additives in industrial oils. *Lubricants*, *12*(1), 6. https://doi.org/10.3390/lubricants12010006

Chen, B., Zhao, Y., Yang, H., & Zhao, J. (2024). Process parameters optimization and numerical simulation of AlCoCrFeNi high-entropy alloy coating via laser cladding. *Materials*, *17*(17), 4243. https://doi.org/10.3390/ma17174243

Cotell, C. M., & Hirvonen, J. K. (1996). Effect of ion energy on the mechanical properties of ion beam assisted deposition (IBAD) wear resistant coatings. *Surface and Coatings Technology*, *81*(1), 118–125. https://doi.org/10.1016/0257-8972(95)02650-9

Daneshifar, M. H., Sajjadi, S. A., Zebarjad, S. M., Mohammadtaheri, M., Abbasi, M., & Mossaddegh, K. (2019). The effects of fillers on properties of automotive nanocomposite clear coats: Type, content, and surface functionalization. *Progress in Organic Coatings*, *134*, 33–39. https://doi.org/10.1016/j.porgcoat.2019.05.001

Domínguez García, S., Béjar Gómez, L., Maya Yescas, R., Lara Romero, J., Castro Cedeño, B., & Espinosa Medina, M. A. (2023). Friction coefficient dynamics of tribological coatings from engine lubricants: Analysis and interpretation. *Coatings*, *13*(10), 1753. https://doi.org/10.3390/coatings13101753

Gangwani, P., Kovac, J., Emami, N., & Kalin, M. (2024). Effect of multi-scale fillers on the tribological behavior of UHMWPE composites in water-lubricated contacts. *Tribology International*, *196*, 109669. https://doi.org/10.1016/j.triboint.2024.109669

Gao, R., Huang, Y., Zhou, X., Ma, G., Jin, G., Li, T., Wang, H., & Liu, M. (2024). Material system and tribological mechanism of plasma sprayed wear resistant coatings: Overview. *Surface and Coatings Technology, 483*, 130758. https://doi.org/10.1016/j.surfcoat.2024.130758

Gee, M., Kamps, T., Woolliams, P., Nunn, J., & Mingard, K. (2022). In situ real-time observation of tribological behaviour of coatings. *Surface and Coatings Technology, 442*, 128233. https://doi.org/10.1016/j.surfcoat.2022.128233

Ginting, J. (2022). Manufacturing of ceramic reinforced metal matrix composites (Al/Sicp) and their characterization through powder metallurgy methods. *Jurnal Matematika Dan Ilmu Pengetahuan Alam LLDikti Wilayah 1 (JUMPA), 2*(1), 52–63. https://doi.org/10.54076/jumpa.v2i1.209

Hao, Y., Zhou, X., Shao, J., & Zhu, Y. (2019). The influence of multiple fillers on friction and wear behavior of epoxy composite coatings. *Surface and Coatings Technology, 362*, 213–219. https://doi.org/10.1016/j.surfcoat.2019.01.110

Hu, K. H., Wang, J., Schraube, S., Xu, Y. F., Hu, X. G., & Stengler, R. (2009). Tribological properties of MoS_2 nano-balls as filler in polyoxymethylene-based composite layer of three-layer self-lubrication bearing materials. *Wear, 266*(11–12), 1198–1207. https://doi.org/10.1016/j.wear.2009.03.036

Jakab, B., Panaitescu, I., & Gamsjäger, N. (2021). The action of fillers in the enhancement of the tribological performance of epoxy composite coatings. *Polymer Testing, 100*, 107243. https://doi.org/10.1016/j.polymertesting.2021.107243

Jakovljevic, S., & Landek, D. (2023). Special issue: Tribological coatings-properties, mechanisms, and applications in surface engineering. *Coatings, 13*(2), 451. https://doi.org/10.3390/coatings13020451

Jansons, E., Lungevics, J., Kanders, U., Leitans, A., Civcisa, G., Linins, O., Kundzins, K., & Boiko, I. (2022). Tribological and mechanical properties of the nanostructured superlattice coatings with respect to surface texture. *Lubricants, 10*(11), 285. https://doi.org/10.3390/lubricants10110285

Joshi, S. S., Patil, V. J., & Gite, V. V. (2024). High performance coating formulation using multifunctional monomers and reinforcing functional fillers for protecting metal substrate. *Pigment & Resin Technology, 54*(5), 651–660. https://doi.org/10.1108/PRT-03-2024-0033

Keshavanarayana, G., Divakar, C., Venkatesh, K., Mohan, C. B., Bhatta, K. L., & Murali, L. (2014). Tribological studies on effect of mixture of fillers in polystyrene matrix composites. *Journal of Thermoplastic Composite Materials, 27*(9), 1192–1208. https://doi.org/10.1177/0892705712470262

Kumar, M., & Bijwe, J. (2013). Optimized selection of metallic fillers for best combination of performance properties of friction materials: A comprehensive study. *Wear, 303*(1–2), 569–583. https://doi.org/10.1016/j.wear.2013.03.053

Kumar, R., Antonov, M., Varga, M., Hussainova, I., & Rodriguez Ripoll, M. (2023). Synergistic effect of Ag and MoS2 on high-temperature tribology of self-lubricating NiCrBSi composite coatings by laser metal deposition. *Wear, 532–533*, 205114. https://doi.org/10.1016/j.wear.2023.205114

Li, R., He, X., Li, C., Zhang, R., Meng, F., Zhang, H., Cui, X., & Nong, Z. (2024a). Processes and properties of self-lubricating coatings fabricated on light alloys by using micro-arc oxidation: A review. *Crystals, 14*(10), 845. https://doi.org/10.3390/cryst14100845

Li, X., Shen, H., Xu, Z., Ye, J., Hua, P., & Liu, D. (2024b). Effects of nano-h-BN and Ag-coated Cu on the microstructure and friction-reduction properties of self-lubricating composite coatings. *Surface and Coatings Technology, 479*, 130531. https://doi.org/10.1016/j.surfcoat.2024.130531

Lin, Z., Zhang, K., Tao, P., Gao, B., Zhao, X., Zong, X., Chen, Z., & Zhang, Z. (2023). Coupled effect of filler and countersurface roughness on the tribological performance of the

PEEK-PTFE-steel hybrid wear system. *Wear, 523*, 204841. https://doi.org/10.1016/j.wear.2023.204841

Ling, X., Lin, X., Li, F., Fan, X., Li, S., Song, J., Wang, W., Zhao, X., Yang, K., & He, J. (2024). Design of solid-liquid composite lubrication coatings based on thermal sprayed ceramic template. *Ceramics International*, *50*(13, Part A), 22346–22358. https://doi.org/10.1016/j.ceramint.2024.03.335

Liu, J., Qian, Y., & Li, D., et al. (2024). Oil-soluble polymer brushes-functionalized nano-MOFs for highly efficient friction and wear reduction. *Friction*, *12*, 1499–1511. https://doi.org/10.1007/s40544-023-0823-x

Lu, X., Sui, X., Niu, D., Yan, Z., Cao, X., Hao, J., & Liu, W. (2023). Exploring the effect of V doping on the tribological mechanism of MoS_2 coatings in atomic oxygen environment. *Materials Today Communications*, *36*, 106721. https://doi.org/10.1016/j.mtcomm.2023.106721

Malaki, M., Xu, W., Kasar, A. K., Menezes, P. L., Dieringa, H., Varma, R. S., & Gupta, M. (2019). Advanced metal matrix nanocomposites. *Metals*, *9*(3), 330. https://doi.org/10.3390/met9030330

Mihoob, M. M., Albarody, T. M. B., Ahmad, F., & Alnarabiji, M. S. (2023). Optimizing the thermal spray parameters for producing high-performance Mo/ZrB_2 metal matrix composites using the Taguchi method. *Coatings*, *13*(9), 1620. https://doi.org/10.3390/coatings13091620

Moustafa, E. B., & Taha, M. A. (2023). The effect of mono and hybrid additives of ceramic nanoparticles on the tribological behavior and mechanical characteristics of an Al-based composite matrix produced by friction stir processing. *Nanomaterials*, *13*(14), 2148. https://doi.org/10.3390/nano13142148

Mukhtar, S. H., Gulzar, A., Saleem, S., Wani, M. F., Sehgal, R., Yakovenko, A. A., Goryacheva, I. G., & Sharma, M. D. (2024). Advances in development of solid lubricating MoS_2 coatings for space applications: A review of modeling and experimental approaches. *Tribology International*, *192*, 109194. https://doi.org/10.1016/j.triboint.2023.109194

Nazir, M. H., Khan, Z. A., Hussain, M. M., Rahil, A., & Zaidi, S. Z. J. (2023). A comprehensive experimental study and numerical analysis of coefficient of friction of nanocomposite coatings. *Materials Chemistry and Physics*, *301*, 127550. https://doi.org/10.1016/j.matchemphys.2023.127550

Neves, F., Duarte, N., & Trindade, B. (2007). On the formation of intermetallic matrix composites produced by mechanical alloying and sputtering. *Journal of Alloys and Compounds*, *434–435*, 378–382. https://doi.org/10.1016/j.jallcom.2006.08.281

Ning, H., Chen, F., & Su, Y., & et al. (2024). Modeling and prediction of tribological properties of copper/aluminum-graphite self-lubricating composites using machine learning algorithms. *Friction*, *12*(5), 1322–1340. https://doi.org/10.1007/s40544-023-0847-2

Pan, S., Jin, K., & Wang, T., et al. (2022). Metal matrix nanocomposites in tribology: Manufacturing, performance, and mechanisms. *Friction*, *10*, 1596–1634. https://doi.org/10.1007/s40544-021-0572-7

Raj, V. R., Praban, S. D., Jayasooriya, M., & Sairam, T. (2024). Development of aluminum-based metal matrix composite for dry bearings reinforced with titanium dioxide and silicon carbide. SAE Technical Paper 2024-01-5091. https://doi.org/10.4271/2024-01-5091

Ray, S. K., Banerjee, A., Bhangi, B. K., Pyne, D., & Dutta, B. (2023). 2 - Tribological analysis—general test standards. In S. C. George, J. T. Haponiuk, S. Thomas, R. Reghunath, & S. P. Sarath (Eds.), *Tribology of polymers, polymer composites, and polymer nanocomposites* (pp. 17–50). Elsevier. https://doi.org/10.1016/B978-0-323-90748-4.00001-7

Ren, Y., Zhang, L., Xie, G., et al. (2021). A review on tribology of polymer composite coatings. *Friction*, *9*(3), 429–470. https://doi.org/10.1007/s40544-020-0446-4

Sahoo, U. K., Jeet, S., Bagal, D. K., Sahu, M. K., Kumari, K., & Barua, A. (2022). Optimization of industrial coatings tribological parameters by studying its application on mechanical components using Taguchi coupled WASPAS method. *Materials Today: Proceedings, 50*(5), 1405–1412. https://doi.org/10.1016/j.matpr.2021.08.326

Shi, N., Li, Z., Li, X., Luo, H., Jin, J., Dong, S., & Li, H. (2024). H-BN base triple-functional filler enhances the anti-corrosion performance of epoxy coating. *Polymer, 300*, 126975. https://doi.org/10.1016/j.polymer.2024.126975

Siddiqui, T. U., & Mittal, R. (2023). Mechanical and tribological characterization of Al1070 alloy based self-lubricating metal matrix composites. *Materials Today: Proceedings*. https://doi.org/10.1016/j.matpr.2023.04.667

Singh, K., Haq, M. I. U., & Mohan, S. (2024a). Tribological behavior of novel Al_2O_3-La_2O_3 HVOF composite coatings. *Tribology International, 193*, 109427. https://doi.org/10.1016/j.triboint.2024.109427

Singh, K., Ul Haq, M. I., & Mohan, S. (2024b). Synergism of h-BN and La_2O_3 in improving the tribological performance of Al_2O_3 coatings. *Proceedings of the Institution of Mechanical Engineers, Part J: Journal of Engineering Tribology, 0*(0). https://doi.org/10.1177/13506501241272777

Singh, N., & Sinha, S. K. (2021). Tribological performances of hybrid composites of Epoxy, UHMWPE, and MoS_2 with in situ liquid lubrication against steel and itself. *Wear, 486–487*, 204072. https://doi.org/10.1016/j.wear.2021.204072

Singh Gautam, R. K., Tripathi, V. M., Gautam, J. K., Jha, P., Sahab, S., Tyagi, R., & Nautiyal, H. (2023). Investigations of high temperature synergetic tribological behavior of HVOF deposited Ni-based metallic coating with self-lubricating ceramic phase (h-BN). *Surface and Coatings Technology, 473*, 130041. https://doi.org/10.1016/j.surfcoat.2023.130041

Skarvelis, P., & Papadimitriou, G. D. (2009). Microstructural and tribological evaluation of potential self-lubricating coatings with MoS_2/MnS additions produced by the plasma transferred arc technique. *Tribology International, 42*(11–12), 1765–1770. https://doi.org/10.1016/j.triboint.2009.01.002

Smoleń, J., Olesik, P., Stępień, K., Mikuśkiewicz, M., Myalska-Głowacka, H., Kozioł, M., Gawron, A., & Godzierz, M. (2024). The influence of graphite filler on the self-lubricating properties of epoxy composites. *Materials, 17*(6), 1308. https://doi.org/10.3390/ma17061308

Somya, A., Varshney, A., Thakur, A., Kumar, A., Swarnalatha, K. S., & Kumari, P. (2024). Nano-hybrid smart coatings: Advancements in industrial efficiency and corrosion resistance. In *Advances in Industrial Coatings* (pp. 205–226). https://doi.org/10.1021/bk-2024-1469.ch009

Song, L., Jiang, T., Yu, H., Zhou, Y., Cao, X., Wang, P., Zhang, B., Yu, Z., Pan, K., & Wei, W. (2024). Research progress on the tribological properties of epoxy-based solid-lubricating composites. *Polymer Composites, 45*(18): 16452–16466.

Sukumaran, A. K., Zhang, C., Nisar, A., & Agarwal, A. (2024). Recent trends in tribological performance of aerospace materials in lunar regolith environment – A critical review. *Advances in Space Research, 73*(1), 846–869. https://doi.org/10.1016/j.asr.2023.10.039

Tian, J., Qi, X., Li, C., & Xian, G. (2023). Friction behaviour's and wear mechanisms of multi-filler reinforced epoxy composites under dry and wet conditions: Effects of loads, sliding speeds, temperatures, and water lubrication. *Tribology International, 179*, 108148. https://doi.org/10.1016/j.triboint.2022.108148

Vafaeenezhad, H., & Eslami-Farsani, R. (2024). Nano-hybrid smart coatings: Advancements in industrial efficiency and corrosion resistance. In *Advances in Industrial Coatings* (pp. 303–352). https://doi.org/10.1021/bk-2024-1469.ch014

Vo, T. D., Deng, G., Tieu, A. K., Su, L., Wu, X., Nguyen, C., Wexler, D., & Yang, J. (2023). Excellent tribological performance at elevated temperatures and associated mechanisms of novel AlCoCrFeNi-MoS_2 solid self-lubricating composite. *Tribology International, 189*, 109011. https://doi.org/10.1016/j.triboint.2023.109011

Voevodin, A. A., O'Neill, J. P., & Zabinski, J. S. (1999). Nanocomposite tribological coatings for aerospace applications. *Surface and Coatings Technology*, *116–119*, 36–45. https://doi.org/10.1016/S0257-8972(99)00228-5

Wood, R. J. K., & Lu, P. (2024). Coatings and surface modification of alloys for tribo-corrosion applications. *Coatings*, *14*(1), 99. https://doi.org/10.3390/coatings14010099

Wu, S., Tang, Y., Gu, J., Li, R., Liang, Y., Liu, P., Wang, H., An, C., Deng, Q., Zhao, L., & Hu, N. (2024). Controllable preparation of metal-based lubrication coatings in extreme environmental applications. *Materials & Design*, *241*, 112922. https://doi.org/10.1016/j.matdes.2024.112922

Xiao, J., Wu, Y., & Zhang, W., & et al. (2020). Friction of metal-matrix self-lubricating composites: Relationships among lubricant content, lubricating film coverage, and friction coefficient. *Friction*, *8*(5), 517–530. https://doi.org/10.1007/s40544-019-0270-x

Xing, Y., Liu, Z., Zhang, W., Jia, Z., Han, W., Zhen, J., & Zhang, R. (2024). The preparation of MoS_2/Metal nanocomposites functionalized with N-oleoylethanolamine: Application as lubricant additives. *Lubricants*, *12*(9), 319. https://doi.org/10.3390/lubricants12090319

Xu, X., & Zhang, Y. (2024). Effect of different chemical conversion coatings on the tribological performance of cylindrical thrust roller bearings under conditions of dry friction and solid lubrication. *Wear*, *548–549*, 205351.

Yang, D., Chen, W., Geng, Y., Chen, J., Zhu, S., Cheng, J., & La, P. (2024). High temperature tribological behaviors of the D-gun sprayed $CrFeNiAl_{0.3}Ti_{0.3}$-Ag-$(Ba,Ca)F_2$ high entropy alloy matrix self-lubricating coatings. *Tribology International*, 197, 109760. https://doi.org/10.1016/j.triboint.2024.109760

Ye, W., Shi, Y., Zhou, Q., Xie, M., Wang, H., Bou-Saïd, B., & Liu, W. (2024). Recent advances in self-lubricating metal matrix nanocomposites reinforced by carbonous materials: A review. *Nano Materials Science*, *X*, 1–15.

Ye, W., Shi, Y., Zhou, Q., Xie, M., Wang, H., Bou-Saïd, B., & Liu, W. (2024). Recent advances in self-lubricating metal matrix nanocomposites reinforced by carbonous materials: A review. *Nano Materials Science*. https://doi.org/10.1016/j.nanoms.2024.02.007

Zhang, S., Yang, X., Huang, T., Guo, F., Dai, L., Liu, Y., & Zhang, B. (2024). Tribological properties of CrN/DLC and CrN coatings under different testing conditions. *Coatings*, *14*(8), 1002. https://doi.org/10.3390/coatings14081002

Zhao, Y., Guo, J., Yan, X., Du, S., Gong, M., Sun, B., Shi, J., & Deng, W. (2024). Tribological performance of maintenance-free Cu-Sn-Ni-MoS2 composites over a wide temperature range. *Industrial Lubrication and Tribology*, *76*(1), 1–10. https://doi.org/10.1108/ILT-06-2023-0174

Zhou, M., Tong, R., & Zhang, T., et al. (2023). Application of MoS_2 in the space environment: A review. *Frontiers in Mechanical Engineering*, *18*, 39. https://doi.org/10.1007/s11465-023-0755-1

Ziyamukhamedova, U., Wasil, S., Kumar, S., Sehgal, R., Wani, M. F., Singh, C. S., Tursunov, N., & Gupta, H. S. (2024). Investigating friction and antiwear characteristics of organic and synthetic oils using h-BN nanoparticle additives: A tribological study. *Lubricants*, *12*(1), 27. https://doi.org/10.3390/lubricants12010027

11 Case Studies Related to the Tribological Performance of Reinforced Metal Matrix Composites

Comparative Analysis of Machine Learning Models for Predicting Wear Rates in AA7075/SiC Composite

Guttikonda Manohar, Vaishali, and V. S. S. Venkatesh

11.1 INTRODUCTION

Aluminum composites play a significant role among all the available matrix materials due to their unique properties and advantages. Aluminum is a lightweight metal with excellent strength-to-weight ratio, making it an ideal choice for various industrial applications (Aherwar et al., 2020). However, pure aluminum has certain limitations in terms of mechanical properties and wear resistance. This is where aluminum composites come into play. By adding reinforcement materials such as ceramic particles or fibers, the mechanical properties of aluminum can be significantly improved (Varol and Canakci, 2013). The resulting composite materials exhibit enhanced strength, stiffness, and wear resistance, making them suitable for a wide range of applications in industries such as aerospace, automotive, and manufacturing. The importance of wear rate in industrial applications cannot be overstated (Suvorova et al., 2024). Wear is a common phenomenon that occurs when two surfaces come into contact and experience frictional forces. In industries where components are subjected to constant movement and contact, wear can cause significant damage and lead to costly maintenance, repairs, and downtime. Therefore, it is crucial to minimize wear rate to ensure the longevity and reliability of industrial equipment (Sarkari Khorrami et al., 2015). Wear

DOI: 10.1201/9781003530909-14

rate refers to the amount of material loss or degradation that occurs due to wear over a specific period of time. It is typically measured as the volume or weight loss per unit distance or time. The lower the wear rate, the better the performance and durability of the materials used in industrial applications. This is especially important in high-stress environments where components are exposed to abrasive particles, high temperatures, or corrosive substances (Divakar et al., 2022).

Aluminum composites excel in terms of wear resistance, making them highly desirable for industrial applications. The addition of reinforcement materials enhances their hardness and toughness, enabling them to withstand friction and abrasive forces without significant wear. This makes aluminum composites suitable for use in components such as bearings, gears, pistons, and cutting tools that experience high levels of wear and require long-lasting performance (Karnam et al., 2018). Furthermore, reducing wear rate not only extends the life span of industrial equipment but also results in cost savings for companies. By using materials with low wear rates, businesses can minimize downtime, reduce maintenance and replacement costs, and improve overall productivity. This is particularly advantageous in industries where equipment failure or breakdowns can have severe consequences (Sharma and Mishra, 2019).

Machine learning plays a pivotal role in the present research towards predicting mechanical properties without the need for experimentation. This approach, known as supervised machine learning, utilizes previous data to train models that can accurately predict outcomes. The field of materials science and engineering greatly benefits from this innovative technique, as it eliminates the need for time-consuming and costly experimental trials. Supervised machine learning leverages existing data to develop models capable of predicting mechanical properties (Manohar et al., 2020). By feeding the algorithms with a large data set containing various material parameters and their corresponding mechanical properties, the models learn to recognize patterns and relationships. These patterns are then used to make accurate predictions for new materials based on their specific input parameters (Manohar et al., 2021). This approach has numerous advantages in the field of materials science and engineering. Firstly, it significantly reduces the time and cost associated with conducting physical experiments (Shinde et al., 2020). Traditional experimentation involves fabricating samples, subjecting them to various tests, and analyzing the results. This process can be lengthy and resource-intensive. By using supervised machine learning, researchers can bypass these steps and obtain predictions quickly and efficiently. Furthermore, supervised machine learning allows researchers to explore a wider range of materials and properties. Experimentation is often limited by factors such as budget, time constraints, and the availability of suitable equipment. However, with the use of existing data and predictive models, researchers can explore a vast database of materials to find the most suitable options for specific applications (Hasan et al., 2022).

Another advantage of using supervised machine learning in predicting mechanical properties is the ability to identify underlying trends and patterns that may not be immediately apparent through experimentation alone. The algorithms can identify complex relationships between different material parameters and their effects on mechanical properties. This deeper understanding can lead to the discovery of new materials with enhanced properties or the optimization of existing ones (Im et al., 2019). However, it is important to note that supervised machine learning is dependent on the quality and quantity of available data. The accuracy of predictions relies on the richness of the data set used for training the models. Therefore, it is necessary to ensure that the data used is representative of a wide range of materials and covers various conditions to obtain reliable predictions (Kwak et al., 2022).

In this view the present work delves into the study of prediction performance of the AA7075/SiC wear rate with respect to the affecting parameters of SiC composition, load, and sliding velocity.

11.2 METHODOLOGY

In this study, data was gathered from previous experimental values to analyze the relationship between various factors and the wear rate of a material. A total of 72 data points were collected, comprising four features: SiC composition, load, sliding velocity, and sliding distance. These features were selected as input variables for analysis, with the wear rate designated as the target variable. To visualize the relationship between these variables, a pair plot was constructed, as depicted in Figure 11.1. The pair plot facilitates the observation of scatter plots of each variable against every other variable,

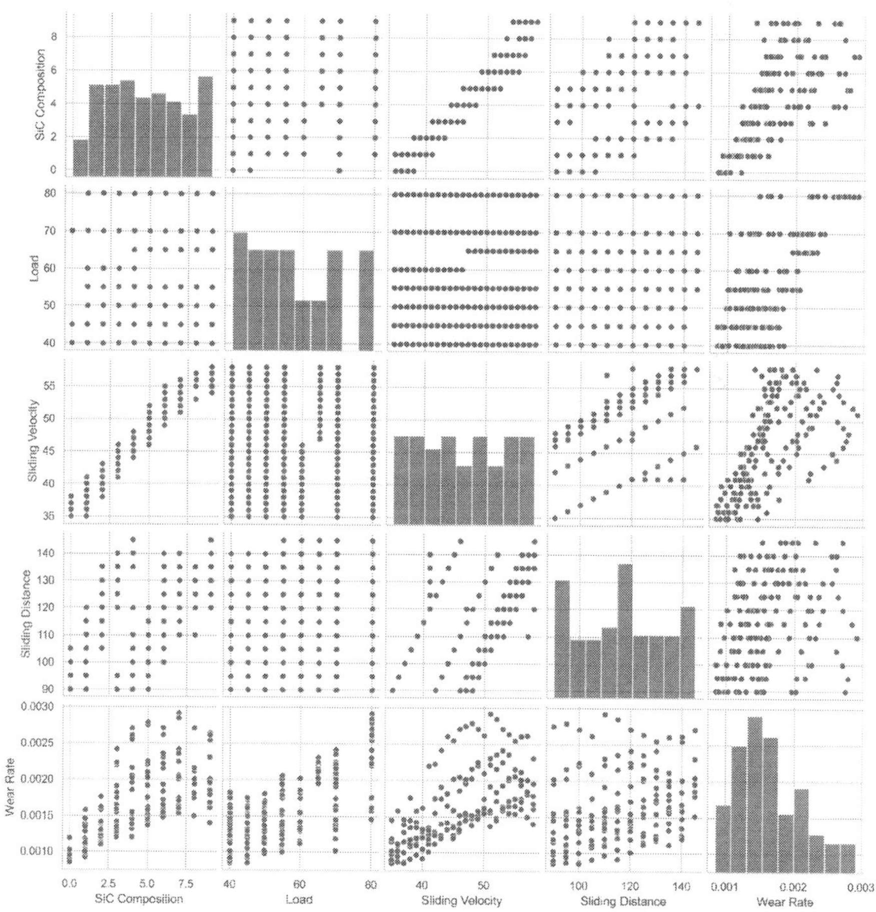

FIGURE 11.1 Pair plot between each variable.

providing insights into potential correlations. Through visual examination of the pair plot, any patterns or trends among the variables can be identified (Hasan et al., 2022).

To quantify the correlation between each variable, the Pearson correlation method was employed. Pearson correlation assesses the strength and direction of the linear relationship between two variables (Dhanesh et al., 2021). By computing the Pearson correlation coefficient for each pair of variables, the degree of their association can be determined, as shown in Figure 11.2. A coefficient nearing 1 signifies a strong positive correlation, whereas a coefficient approaching –1 indicates a strong negative correlation. By scrutinizing the pair plot and computing the Pearson correlation coefficients, a deeper understanding of how SiC composition, load, sliding velocity, and sliding distance impact the wear rate can be obtained. This insight holds significance across various applications, such as material selection and design optimization. By identifying the factors exhibiting the most pronounced correlation with wear rate, informed decisions can be made to mitigate wear and prolong material life span in diverse industries (Akinwamide et al., 2020).

In machine learning, it is standard practice to split a data set into two parts: a training set and a testing set. In this case, 80% of the data was allocated for training, while the remaining 20% was set aside for testing. This division enables the evaluation of a trained model on unseen data, offering insights into its generalization abilities. To optimize

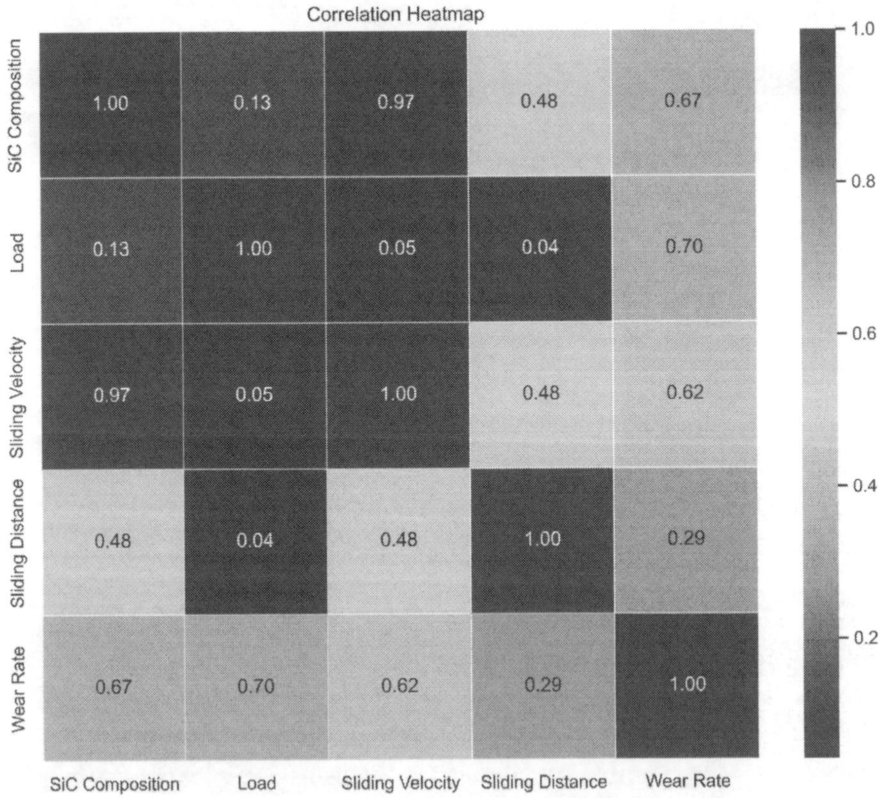

FIGURE 11.2 Correlation matrix between each variable in the data.

the model's hyperparameters, we utilized grid search cross-validation. This technique involves systematically exploring a predefined set of hyperparameter combinations and assessing their performance using cross-validation. Specifically, we employed a fivefold cross-validation scheme, dividing the data into five subsets or folds. The model underwent training and evaluation five times, with each fold serving as the testing set once and the remaining fourfold used for training. This method enhances the robustness of performance assessment by mitigating the impact of random data splitting. To aid algorithm convergence during training, we applied standard scalar scaling to the variables. Scaling is crucial when dealing with features of varying ranges or units, ensuring equal weighting and preventing any particular feature from dominating the learning process. The standard scalar method transforms variables to have zero mean and unit variance, normalizing them onto a similar scale. This normalization fosters efficient convergence and guards against numerical instability. In summary, this approach—utilizing an 80/20 split for training/testing, employing grid search cross-validation with fivefold, and implementing standard scalar scaling—is a common and effective practice in machine learning. It facilitates thorough model evaluation, hyperparameter optimization, and improved algorithm convergence (Manohar et al., 2023).

After conducting an in-depth analysis of the correlation values, it has been observed that there is a weak association between sliding distance and wear rate. To streamline the algorithm and enhance its efficiency, the decision was made to exclude sliding distance from the features. This step was taken to reduce complexity and concentrate on the variables exerting a more substantial influence on wear rate. To evaluate the performance of the machine learning models, several metrics were utilized. The first metric employed was R-squared, which gauges the proportion of the variance in the dependent variable explained by the independent variables. A higher R-squared value signifies a better fit of the model to the data (Im et al., 2019). Utilizing this metric enables the assessment of how effectively the machine learning models predict wear rate based on the remaining features. Mean absolute error (MAE) is another crucial metric utilized to evaluate machine learning model performance. MAE quantifies the average difference between the predicted wear rate and the actual wear rate (X. Liu et al., 2021). A smaller MAE value indicates a better alignment of the model with the data. This metric aids in understanding the accuracy of the machine learning models in predicting wear rate. Root mean squared error (RMSE) is a widely used metric in regression analysis for evaluating prediction accuracy. It computes the square root of the average squared differences between predicted and actual values (X. Liu et al., 2021). A lower RMSE value suggests a better fit of the model to the data. By employing this metric, the performance of the machine learning models in predicting wear rate can be determined. Through the utilization of these evaluation metrics, the performance of the machine learning models in predicting wear rate without considering sliding distance as a feature can be assessed. This enables a focus on other variables with stronger associations with wear rate, potentially enhancing prediction accuracy.

In this manuscript, a study was undertaken to forecast the wear rate of the AA7075/SiC composite employing machine learning models. The objective was to assess the efficacy of prominent models, specifically random forest and XGBoost, in accurately predicting the wear rate of this composite material. Machine learning has garnered considerable attention in recent times owing to its capacity to analyze extensive data sets and make precise predictions based on inherent data patterns and relationships (Venkatesh et al., 2024).

The AA7075/SiC composite is widely utilized across diverse industries due to its exceptional mechanical properties and wear resistance. Nonetheless, accurately predicting its wear rate poses a challenge. Conventional analytical techniques often encounter limitations in capturing the intricate relationships among various variables influencing wear rate. Machine learning models present a promising alternative by leveraging algorithms capable of identifying and learning from these complex patterns random forest stands out as a versatile and accurate machine learning model (Im et al., 2019). It operates by constructing an ensemble of decision trees, with each tree developed using a random subset of the available data. Subsequently, predictions from all trees are amalgamated to derive a final prediction. Random Forest has demonstrated successful applications across multiple domains, including finance, health care, and manufacturing, showcasing promising outcomes in predicting complex phenomena.XGBoost, an abbreviation for extreme gradient boosting, has emerged as another popular machine learning model in recent years (Ciccone et al., 2023). It represents an optimized iteration of gradient boosting, a methodology that amalgamates weak prediction models into a robust predictive model. XGBoost employs a gradient boosting framework to iteratively refine the performance of weak prediction models by minimizing errors in each iteration. This model has found extensive utility across various domains, such as natural language processing, image recognition, and time series forecasting. In this investigation, the wear rate of the AA7075/SiC composite was forecasted utilizing both random forest and XGBoost models. The data set utilized for training and testing encompassed diverse parameters, including SiC composition, sliding velocity, load, and sliding distance. The performance of both models was evaluated using metrics such as mean absolute error, mean squared error, and R-squared value (Pasha et al., 2024).

11.3 RESULTS AND DISCUSSIONS

11.3.1 Machine Learning Models Performance

The Pearson correlation heatmap (Figure 11.2) provides a systematic and quantitative visualization of the correlations between different variables involved in the study of wear rates. This analysis is pivotal for understanding the interdependencies between variables and identifying the most influential factors for further modeling and analysis.

In the heatmap, the correlation coefficients range from –1 to +1, where +1 indicates a perfect positive correlation, –1 indicates a perfect negative correlation, and values close to 0 signify no correlation (Liu et al., 2024). A striking observation from this heatmap is the weak correlation coefficients associated with the variable 'sliding distance'. Notably, 'sliding distance' shows low correlation values with all other variables, including a notably weak correlation with 'wear rate' (0.29), suggesting a minimal direct linear relationship with the wear rate compared to other variables (Bong Kim et al., 2022).

The variables 'SiC composition', 'load', and 'sliding velocity' exhibit considerably stronger correlations with 'wear rate', with coefficients of 0.67, 0.70, and 0.62, respectively. These values suggest a moderate to strong linear relationship with 'wear rate', making them more significant predictors in the context of this study. The high correlation of 'SiC composition' and 'sliding velocity' (0.97) also indicates a potential collinearity issue, which could affect the performance and interpretability of predictive models if both variables are used simultaneously without adjustments (Saride et al., 2023).

Due to the negligible correlation of 'sliding distance' with 'wear rate', it was decided to eliminate 'sliding distance' from further predictive modeling. This decision is based on the premise that including variables with weak correlations in predictive modeling can lead to reduced model accuracy and efficiency. Eliminating such variables can streamline the model, focusing on more impactful predictors and potentially improving the model's performance and interpretability (Huang et al., 2023).

The heatmap not only informs the selection of variables for modeling but also highlights the relationships that might be explored further in detailed analyses. By focusing on variables that exhibit stronger relationships with 'wear rate', the study aims to build a more robust and efficient predictive model, as suggested by the trends observed in Figure 11.2. This approach ensures that the modeling efforts are concentrated on the most influential factors, optimizing the use of computational resources and simplifying the model structure for better performance and easier understanding (Chen and Guestrin, 2016).

11.3.2 Random Forest

The performance evaluation of the random forest model, using various diagnostic plots and a high R^2 value of 0.9608, demonstrates a sophisticated ability to predict wear rates with impressive accuracy and reliability. The actual versus predicted wear rate plot (Figure 11.3) is particularly telling, showing a close correspondence

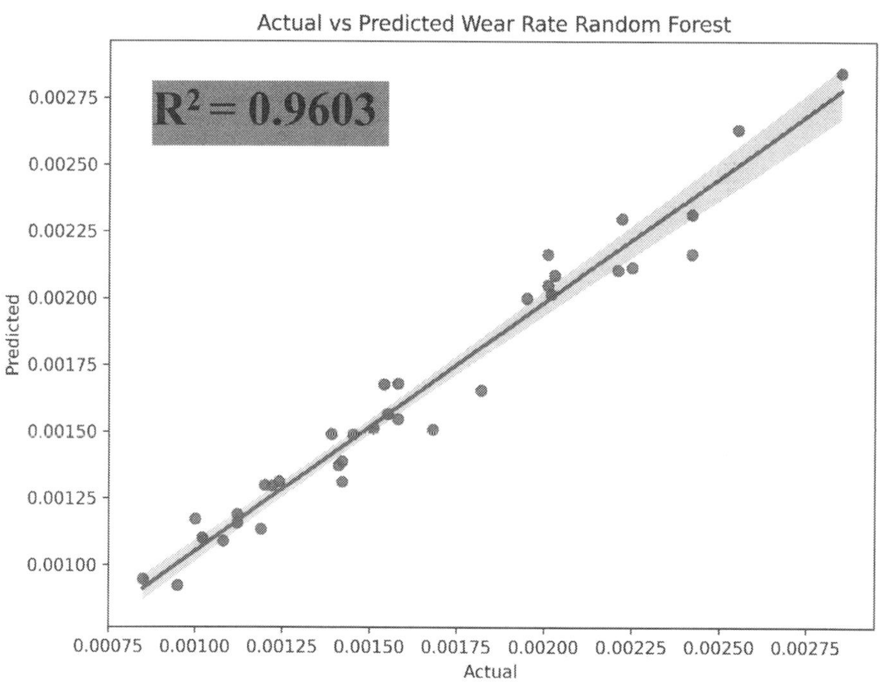

FIGURE 11.3 Regression plot between predicted versus actual wear rates in random forest model.

TABLE 11.1
Best Parameters Obtained through GridSearchCV Technique in Random Forest Model

Best Parameters	'max_depth': 10, 'min_samples_leaf': 1, 'min_samples_split': 2, 'n_estimators': 100

between predicted and actual values. The near-perfect alignment along the identity line, underscored by a narrow confidence interval, clearly indicates that the model is not only accurate but also consistent across a variety of wear conditions. The tuned hyperparameters obtained through GridSearchCV techniques are listed in Table 11.1.

The residual plot (Figure 11.4) offers further insight into the model's performance by illustrating the residuals' distribution relative to the predicted values. The clustering of residuals near zero with few outliers is indicative of the model's robustness, where the majority of predictions closely match the actual data. However, the existence of outliers, albeit few, points to potential anomalies or exceptional cases within

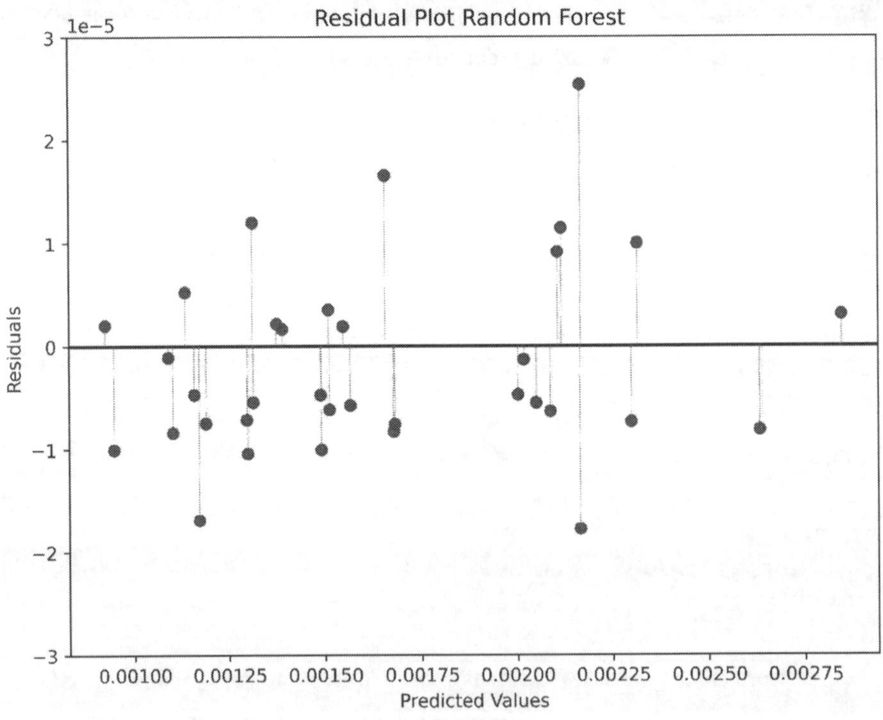

FIGURE 11.4 Residual spread versus predicted values in the random forest model.

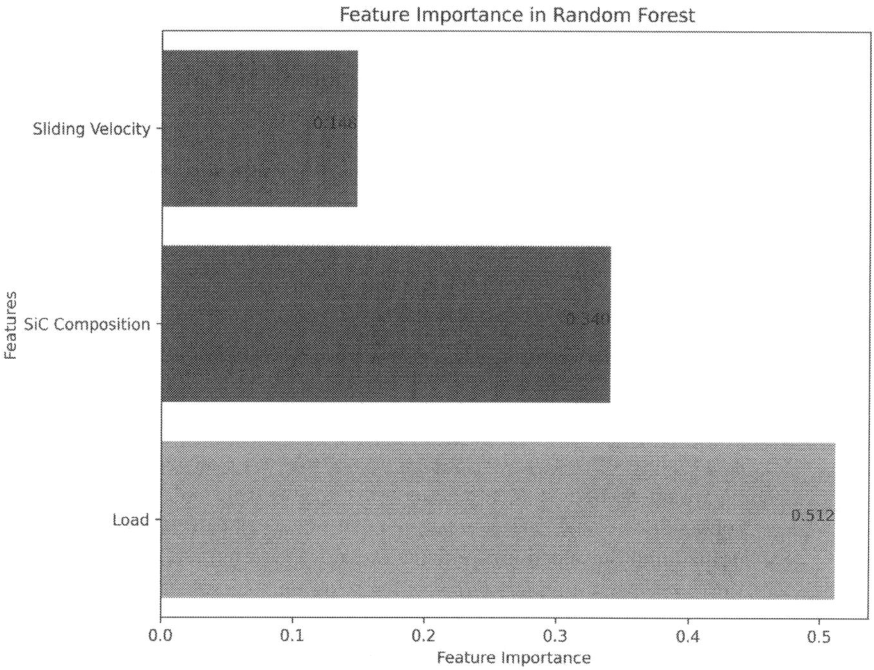

FIGURE 11.5 Feature importance in the random forest model while predicting the wear rate.

the data set that the model may not fully account for. These discrepancies could be due to extreme operating conditions, unusual wear characteristics, or noise within the data, suggesting areas where additional data processing or more complex modeling techniques might improve outcomes.

Feature importance (Figure 11.5) sheds light on the predictors that most significantly influence the model's outputs. The predominance of 'load' as a feature, contributing over 50% to the predictive performance, underscores its critical role in the wear process. The significant contributions from 'SiC composition' and 'sliding velocity' also reveal the complex interplay of mechanical and material properties in determining wear rates. This detailed breakdown not only validates the physical understanding of wear mechanisms but also directs future efforts in feature engineering and data collection strategies to refine the predictive model.

The random forest model's high R^2 value is not just a statistical measure but a practical indicator of the model's utility in industrial applications where precision in predicting wear rates can lead to better maintenance schedules, optimized operational efficiency, and prolonged machinery life. The ability to accurately predict wear rates ensures that maintenance can be conducted proactively, minimizing downtime and reducing costs associated with unexpected breakdowns.

In summary, the analysis illustrates that the random forest model, fine-tuned through GridSearchCV, is a potent tool in the predictive maintenance toolkit. The insights gained from the actual versus predicted comparisons, residuals distribution, and feature importance not only confirm the model's validity but also provide a roadmap for further enhancement. Future work could focus on refining the model by incorporating additional data, exploring more complex interactions among features, or employing advanced ensemble techniques to capture nuances that could improve prediction accuracy even further. This ongoing refinement and adaptation are key to developing a truly dynamic and responsive predictive maintenance system.

11.3.3 XGBoost

XGBoost in the assessment of the XGBoost model's predictive performance for wear rates, an in-depth analysis utilizing key evaluation metrics and plots is instrumental. The actual versus predicted wear rate plot (Figure 11.6) highlights a strong correlation, as indicated by an R^2 value of 0.9432, signifying that the model successfully captures 94.32% of the variability in the wear rate data. This plot reveals a high alignment of the predicted values with the actual ones, with the data points closely clustered along the identity line, suggesting a precise and reliable predictive ability of the model. The narrow confidence interval further enhances the credibility of the predictions across the data set. The tuned hyperparameters obtained through GridSearchCV techniques are listed in Table 11.2.

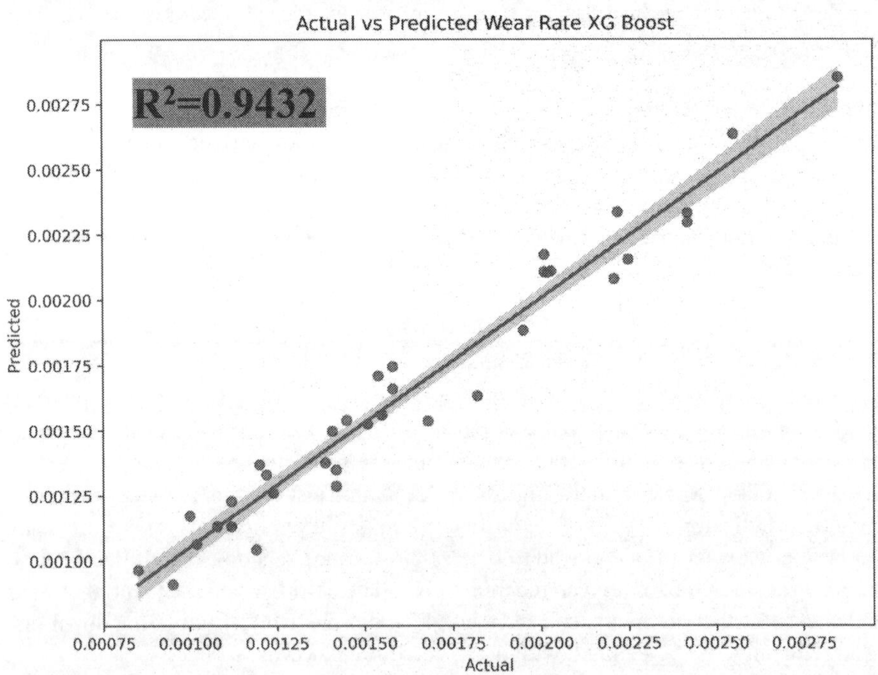

FIGURE 11.6 Regression plot between predicted versus actual wear rates in Xtreme Gradient (XGBoost) model.

TABLE 11.2
Best Parameters Obtained through GridSearchCV Technique in Xtreme Gradient (XGBoost) Model

Best Parameters	'colsample_bytree': 0.7, 'learning_rate': 0.1, 'max_depth': 3, 'n_estimators': 200, 'subsample': 0.8

The residual plot (Figure 11.7) complements the regression analysis by showing the residuals' distribution relative to the predicted values. The residuals are mostly centered around zero, but some notable outliers indicate instances where the model's predictions deviate from the actual values. The spread and pattern of these residuals can guide further refinement of the model by pinpointing the areas where prediction accuracy may be enhanced.

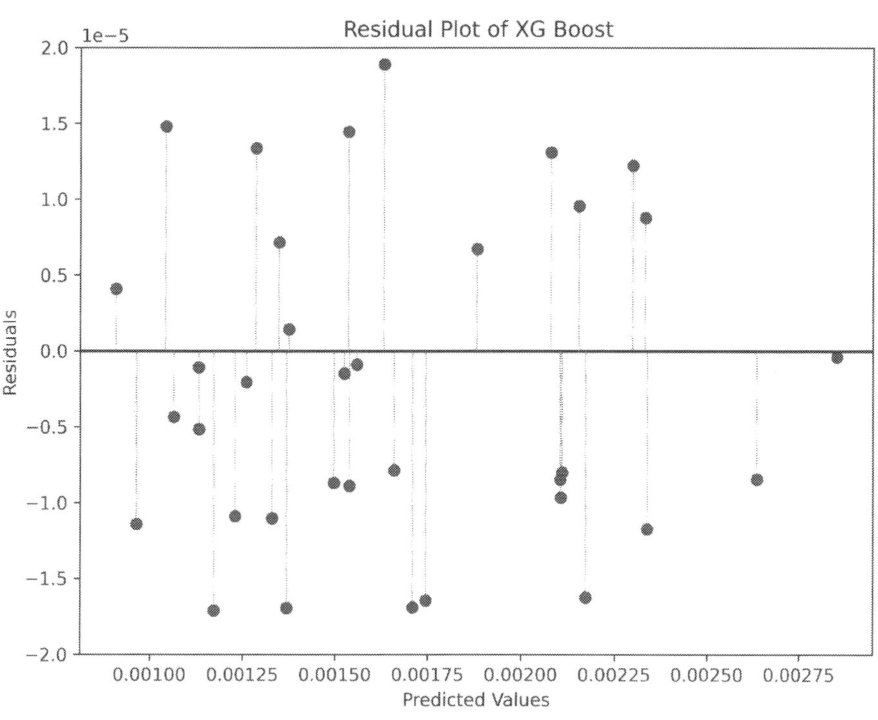

FIGURE 11.7 Residual spread versus predicted values in the Xtreme Gradient (XGBoost) model.

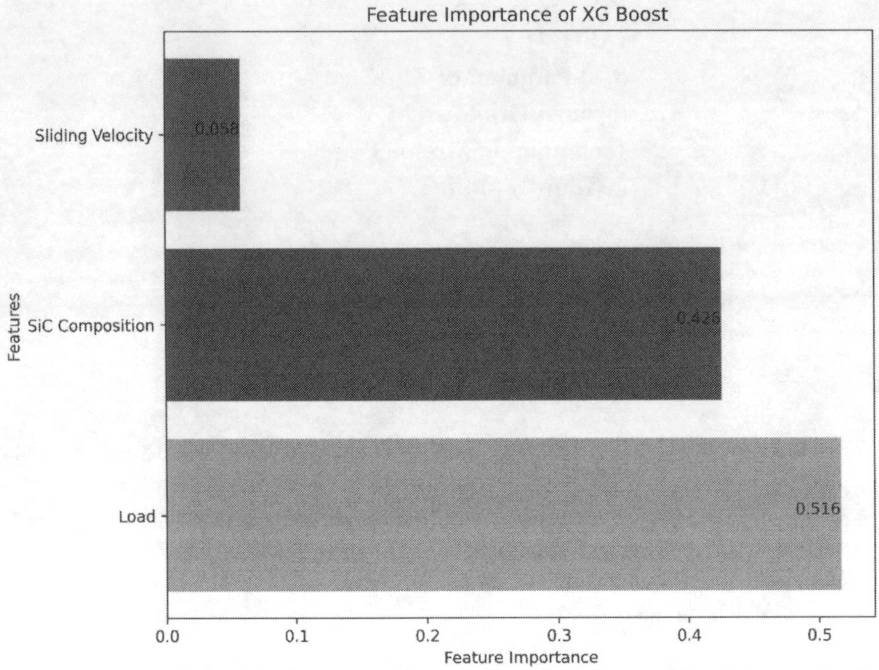

FIGURE 11.8 Feature importance in the Xtreme Gradient (XGBoost) model while predicting the wear rate.

Furthermore, the feature importance plot (Figure 11.8) provides crucial insights into which variables most significantly impact the model's predictions. 'Load' and 'SiC composition' are identified as the predominant features, with substantial importance scores of 0.516 and 0.426, respectively, affirming their critical roles in determining wear rates. In contrast, 'sliding velocity' holds a lesser importance, suggesting that its influence on wear rates is relatively minor under the conditions studied.

These analyses collectively affirm the XGBoost model's robustness, equipped with optimized parameters including a colsample_bytree of 0.7, learning_rate of 0.1, max_depth of 3, and 200 estimators with a subsample rate of 0.8. The results illustrate the model's capability to provide accurate and dependable predictions, making it an invaluable tool for predictive maintenance in industrial settings where understanding wear dynamics is crucial for operational efficiency and cost management. Further explorations and optimizations based on the residual insights and feature importances could lead to even more refined models.

11.3.4 KNN

The KNN (k-nearest neighbors) model's performance in predicting wear rates showcases substantial efficacy, as depicted by an R^2 value of 0.9322 in the actual versus predicted wear rate plot (Figure 11.9). This high correlation coefficient indicates that the model accounts for approximately 93.22% of the variability in the wear rate data, demonstrating its strong predictive capability. The scatter of data points

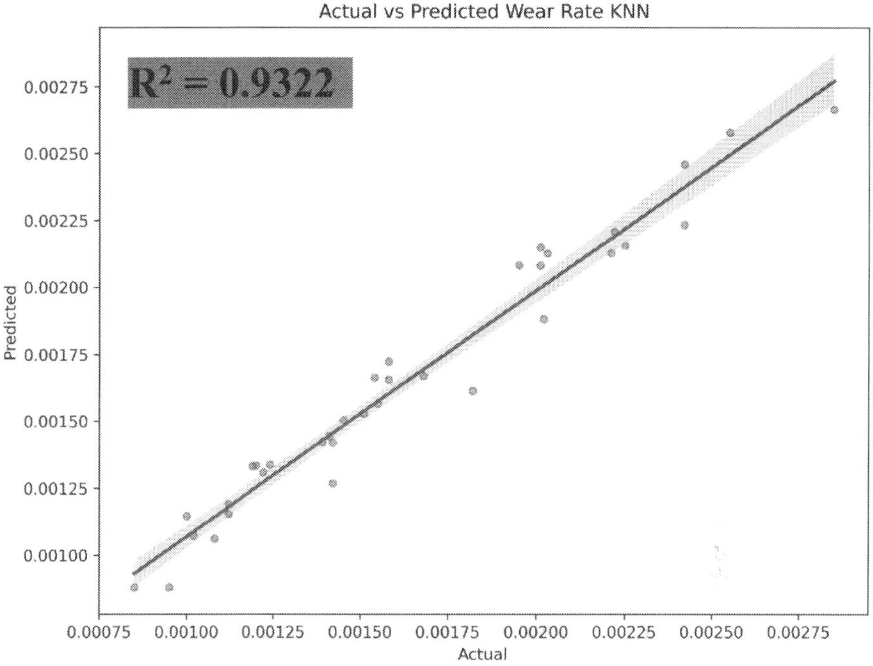

FIGURE 11.9 Regression plot between predicted versus actual wear rates in k-nearest neighbor model.

around the identity line in this plot reflects a precise alignment between the predicted and actual values, underlined by a narrow confidence interval that signifies consistent model accuracy across various conditions. The tuned hyperparameters obtained through GridSearchCV techniques are listed in Table 11.3.

The residual plot (Figure 11.10) provides further insights into the model's performance, revealing the residuals' distribution around the zero line. The spread of residuals indicates some variability in the model's accuracy, particularly noticeable in how certain predictions deviate from the actual values. This pattern suggests that while the KNN model is generally effective, there are specific areas within the range of predicted values where its performance could potentially be optimized.

A notable aspect of using KNN in this context is the absence of a feature importance plot. Unlike tree-based models such as random forest or XGBoost, which

TABLE 11.3
Best Parameters Obtained through GridSearchCV Technique in KNN Model

Best Parameters	'metric': 'manhattan', 'n_neighbors': 5, 'weights': 'distance'

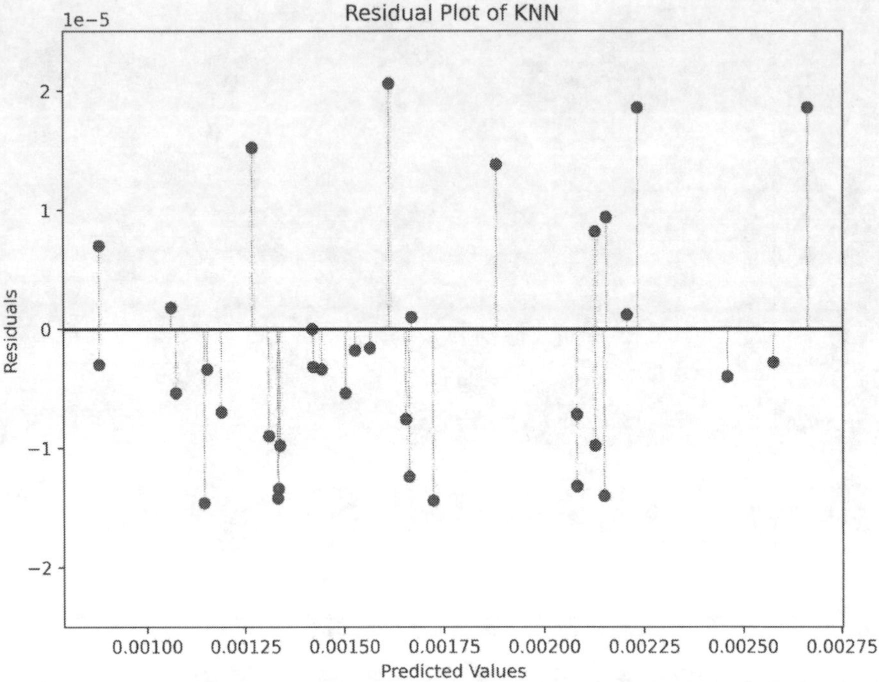

FIGURE 11.10 Residual spread versus predicted values in the k-nearest neighbor model.

intrinsically evaluate and report the importance of features, KNN operates differently. It predicts outcomes based on the proximity of the nearest labeled data points in the feature space, relying on distance metrics without differentiating the influence of individual features. This characteristic stems from KNN's method of treating all features equally in computing distances unless modifications are made to the metric for specific applications. For this analysis, the KNN model uses the 'manhattan' metric with parameters set to 'n_neighbors' of 5 and 'weights' as 'distance', enhancing the influence of nearer neighbors on the prediction.

In essence, while the KNN model confirms its robust ability to predict wear rates accurately, it does lack the capacity to provide direct insights into which features most significantly impact predictions. This absence requires users to employ other methods or external analyses to discern feature relevance, potentially integrating domain expertise or additional feature selection techniques to fully understand and utilize the model's predictions. This limitation could be viewed as a drawback when dealing with complex datasets where understanding the interplay between features is crucial for model performance and interpretability.

11.3.5 Comparative Analysis of ML Models

A detailed comparative analysis of the random forest, XGBoost, and KNN models reveals distinct performance metrics across MAE, RMSE, and R^2 values, which are essential for evaluating model accuracy, consistency, and prediction error. The

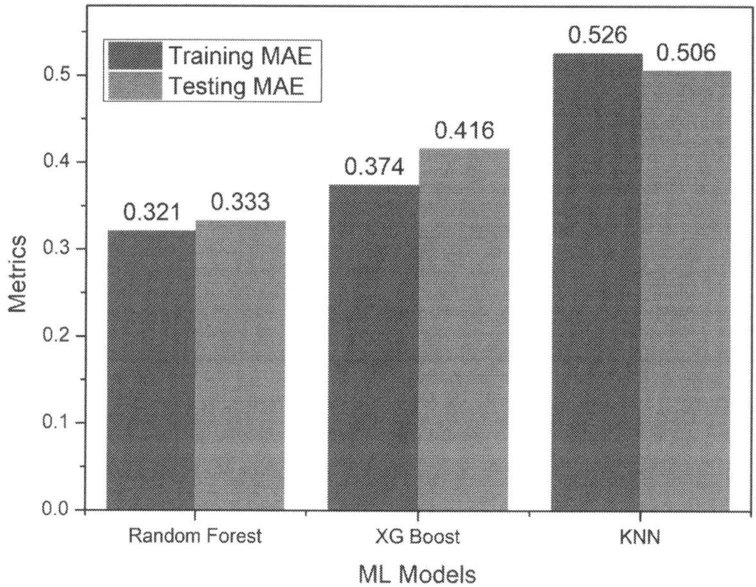

FIGURE 11.11 Mean absolute error (MAE) values of different machine learning models obtained in the present work.

random forest model exhibits the lowest MAE values (0.321 for training and 0.333 for testing), indicating minimal average prediction errors compared to the other models, as highlighted in Figure 11.11. This trend continues with the RMSE, where random forest also reports the lowest values (0.433 for training and 0.446 for testing), emphasizing its superior capability in handling both average and larger errors robustly, as seen in Figure 11.12. Furthermore, its R^2 value of 0.9608 reflects a high degree of variance explained by the model, underscoring its effectiveness in accurately predicting wear rates with consistent reliability.

XGBoost follows with slightly higher MAE values (0.374 for training and 0.416 for testing) and moderate RMSE values (0.495 for training and 0.525 for testing), suggesting a reasonable performance but with a noticeable increase in prediction errors compared to random forest. Despite these larger errors, XGBoost achieves an impressive R^2 value of 0.9432, indicating it can explain a significant portion of the variance in the data set and remains a robust choice for scenarios where speed and efficiency in model training and prediction are priorities.

KNN, on the other hand, shows the highest error metrics among the three models with MAE values of 0.526 for training and 0.506 for testing, and RMSE values of 0.650 for training and 0.614 for testing. Its R^2 value of 0.9322, while still high, is the lowest among the models, suggesting it captures less variability in the data set. This indicates that KNN may struggle with accurately predicting higher deviations and outliers, which could limit its applicability in scenarios demanding high precision (Schlenger, 2023).

In summary, the random forest model stands out as the most effective across all evaluated metrics, offering the lowest prediction errors and the highest explanatory power, making it an ideal choice for applications requiring high accuracy and

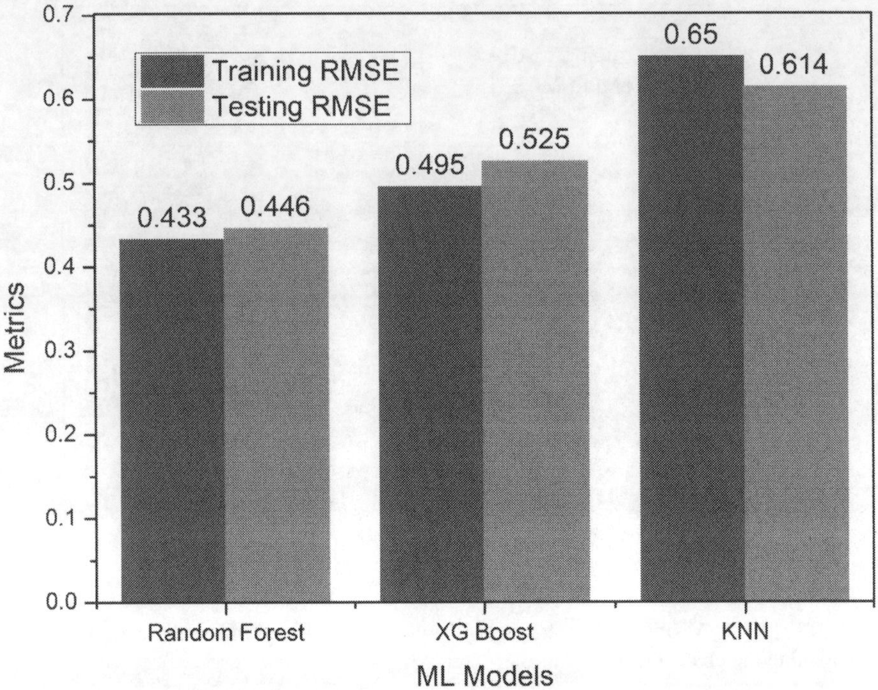

FIGURE 11.12 Root mean squared error (RMSE) values of different machine learning models obtained in the present work.

reliability. XGBoost, with its efficient performance and slightly less accuracy, serves well in situations requiring rapid model deployment and execution. Meanwhile, KNN, despite its simplicity and transparency, might be more suitable for less critical applications where ease of understanding and implementation are prioritized over achieving minimal prediction error. This analysis emphasizes the importance of aligning model choice with specific operational requirements and error tolerance levels in predictive maintenance scenarios. MAE and RMSE comparisons are illustrated in Figures 11.11 and 11.12, respectively, providing a visual representation of these findings.

11.4 CONCLUSIONS

The study effectively demonstrated the application of machine learning techniques in predicting the wear rate of AA7075/SiC composites, an essential aspect of maintenance in industrial machinery. By comparing three advanced machine learning models—random forest, XGBoost, and k-nearest neighbors (KNN)—the research provided deep insights into the predictive capabilities and effectiveness of these models under varying conditions.

1. The random forest model emerged as the most effective, offering the highest accuracy with the lowest MAE and RMSE, and achieving an R^2 value

of 0.9608. This illustrates its strong ability to capture the variability in the wear rates of the composite materials. XGBoost and KNN also showed commendable performance but were slightly outperformed by random forest in terms of both prediction accuracy and consistency.

2. A key aspect of this research was the utilization of a Pearson correlation heatmap to assess the relevance of different variables involved in wear rate prediction. The decision to eliminate 'sliding distance' from the predictive model based on its weak correlation with wear rates proved crucial in enhancing model efficiency and focus.
3. Overall, the findings affirm the substantial potential of integrating machine learning techniques into the predictive maintenance of industrial applications. Such integration not only aids in accurate wear rate prediction but also enhances operational efficiencies by reducing downtime and maintenance costs. This study underscores the importance of selecting appropriate features and machine learning models to improve the reliability and effectiveness of predictive maintenance strategies.

REFERENCES

Aherwar, A., Patnaik, A., & Pruncu, C. I. (2020). Effect of B4C and waste porcelain ceramic particulate reinforcements on mechanical and tribological characteristics of high strength AA7075 based hybrid composite. *Journal of Materials Research and Technology*, *9*(5), 9882–9894. https://doi.org/10.1016/j.jmrt.2020.07.003

Akinwamide, S. O., Lesufi, M., Akinribide, O. J., Mpolo, P., & Olubambi, P. A. (2020). Evaluation of microstructural and nanomechanical performance of spark plasma sintered TiFe-SiC reinforced aluminium matrix composites. *Journal of Materials Research and Technology*, *9*(6), 12137–12148. https://doi.org/10.1016/j.jmrt.2020.08.068

Bong Kim, D., Shao, G., & Jo, G. (2022). A digital twin implementation architecture for wire + arc additive manufacturing based on ISO 23247. *Manufacturing Letters*, *34*, 1–5. https://doi.org/10.1016/j.mfglet.2022.08.008

Chen, T., & Guestrin, C. (2016). XGBoost: A scalable tree boosting system. *Proceedings of the ACM SIGKDD International Conference on Knowledge Discovery and Data Mining, 13-17-August-2016*, 785–794. https://doi.org/10.1145/2939672.2939785

Ciccone, F., Bacciaglia, A., & Ceruti, A. (2023). Optimization with artificial intelligence in additive manufacturing: A systematic review. *Journal of the Brazilian Society of Mechanical Sciences and Engineering*, *45*(6), 1–22. https://doi.org/10.1007/s40430-023-04200-2

Dhanesh, S., Kumar, K. S., Fayiz, N. K. M., Yohannan, L., & Sujith, R. (2021). Recent developments in hybrid aluminium metal matrix composites: A review. *Materials Today: Proceedings*, *45*, 1376–1381. https://doi.org/10.1016/j.matpr.2020.06.325

Divakar, S., Chitharthan, S., Shanmugapriyan, V. G., Thalaieswaran, S., & Gopinath, K. (2022). Optimization of process parameters for mechanical properties of Al- Al_2O_3-graphite hybrid composite fabricated using stir casting process. *Materials Today: Proceedings*, *65*(xxxx), 3937–3945. https://doi.org/10.1016/j.matpr.2022.06.437

Hasan, M. S., Wong, T., Rohatgi, P. K., & Nosonovsky, M. (2022). Analysis of the friction and wear of graphene reinforced aluminum metal matrix composites using machine learning models. *Tribology International*, *170*, 107527. https://doi.org/10.1016/j.triboint.2022.107527

Huang, S.-J., Adityawardhana, Y., & Sanjaya, J. (2023). Predicting mechanical properties of magnesium matrix composites with regression models by machine learning. *Journal of Composites Science*, *7*(9), 347. https://doi.org/10.3390/jcs7090347

Im, J., Lee, S., Ko, T.-W., Kim, H. W., Hyon, Y., & Chang, H. (2019). Identifying Pb-free perovskites for solar cells by machine learning. *Npj Computational Materials*, *5*(1), 37. https://doi.org/10.1038/s41524-019-0177-0

Karnam, M., Shivaramakrishna, A., Joshi, R., Manjunatha, T. H., & Veerabhadrappa, K. (2018). Study of mechanical properties and drilling behavior of Al7075 reinforced with B4C. *Materials Today: Proceedings*, *5*(11), 25102–25111. https://doi.org/10.1016/j.matpr.2018.10.311

Kwak, S., Kim, J., Ding, H., Xu, X., Chen, R., Guo, J., & Fu, H. (2022). Machine learning prediction of the mechanical properties of γ-TiAl alloys produced using random forest regression model. *Journal of Materials Research and Technology*, *18*, 520–530. https://doi.org/10.1016/J.JMRT.2022.02.108

Liu, X., Tian, S., Tao, F., Du, H., & Yu, W. (2021). How machine learning can help the design and analysis of composite materials and structures? *AIAA Scitech 2021 Forum*, 1–21.

Liu, Y., Zheng, W., Ai, H., Zhou, H., Feng, L., Cheng, L., Guo, R., & Song, X. (2024). Application of machine learning in predicting the thermal conductivity of single-filler polymer composites. *Materials Today Communications*, *39*(March), 109116. https://doi.org/10.1016/j.mtcomm.2024.109116

Manohar, G., Kumar, A., Satyanarayana, M. V. N. V., Maity, S. R., Pandey, K. M., Reddy, V., & Pasha Mahammod, B. (2023). Effect of fabrication techniques on mechanical and microstructural behavior of AA7075/SiC/ZrC hybrid composite. *Ceramics International*, *49*(23), 37782–37792. https://doi.org/10.1016/J.CERAMINT.2023.09.106

Manohar, G., Pandey, K. M., & Maity, S. R. (2020). Aluminium (AA7075) metal matrix composite reinforced with B4C nano particles and effect of individual alloying elements in Al fabricated by powder metallurgy techniques. *Journal of Physics: Conference Series*, *1451*(1), 12024.

Manohar, G., Pandey, K. M., & Maity, S. R. (2021). Effect of processing parameters on mechanical properties of Al7175/Boron carbide (B4C) composite fabricated by powder metallurgy techniques. *Materials, Computer Engineering and Education Technology*, *105*, 8–16.

Pasha, M. B., Rao, R. N., Ismail, S., Gupta, M., & Prasad, P. S. (2024). Tribo-informatics approach to predict wear and friction coefficient of Mg/Si3N4 composites using machine learning techniques. *Tribology International*, *196*, 109696. https://doi.org/10.1016/J.TRIBOINT.2024.109696

Saride, R. K., Vajjala, S., Patel, B., Kumar, S., Kumar, R., Pappula, L., & Ginuga, J. R. (2023). Welding studies and characterisation of additively manufactured LPBF maraging steel. *International Journal of Mechanical Engineering and Applications*, *11*(5), 100–112. https://doi.org/10.11648/j.ijmea.20231105.11

Sarkari Khorrami, M., Samadi, S., Janghorban, Z., & Movahedi, M. (2015). In-situ aluminum matrix composite produced by friction stir processing using FE particles. *Materials Science and Engineering A*, *641*, 380–390. https://doi.org/10.1016/j.msea.2015.06.071

Schlenger, J. (2023). Random forest. In D. Memmert (Ed.), *Sportinformatik: Modellbildung, Simulation, Datenanalyse und Visualisierung von sportbezogenen Daten* (pp. 227–234). Springer Berlin Heidelberg. https://doi.org/10.1007/978-3-662-67026-2_24

Sharma, A., & Mishra, P. M. (2019). Effects of various reinforcements on mechanical behavior of AA7075 hybrid composites. *Materials Today: Proceedings*, *18*, 5258–5263. https://doi.org/10.1016/j.matpr.2019.07.526

Shinde, D. M., Sahoo, P., & Davim, J. P. (2020). Tribological characterization of particulate-reinforced aluminum metal matrix nanocomposites: A review. *Advanced Composites Letters*, *29*, 2633366X2092140. https://doi.org/10.1177/2633366x20921403

Suvorova, V., Volodko, S., Suvorov, D., Chernyshikhin, S., Nepapushev, A., Korol, A., Volkova, L., Sokolov, P., Khort, A., & Moskovskikh, D. (2024). Enhanced microstructure and mechanical properties of ZrN-reinforced AlSi10Mg aluminum matrix composite. *Scientific Reports*, *14*(1), 1–13. https://doi.org/10.1038/s41598-024-58614-6

Varol, T., & Canakci, A. (2013). Effect of particle size and ratio of B4C reinforcement on properties and morphology of nanocrystalline Al2024-B4C composite powders. *Powder Technology*, *246*, 462–472. https://doi.org/10.1016/j.powtec.2013.05.048

Venkatesh, V. S. S., Rao, R. N., Manohar, G., & Singh, S. N. (2024). Tribo-informative analysis of spark plasma sintered Al-SiC-HAp hybrid composite through machine learning techniques. *Arabian Journal for Science and Engineering*. https://doi.org/10.1007/s13369-024-09835-4

12 Case Studies Related to the Tribological Performance of Reinforced Metal Matrix Composites

Optimization of Tribological Properties in AA7075-WS2-TiC Hybrid Composites Using the Taguchi-GRA Method

Shilpa B. Sahare, P. Satish Kumar, Rajasekaran Saminathan, Nagoor Meeran M., Arun Prakash J., Prashant D. Kamble, and Abhijit Bhowmik

12.1 INTRODUCTION

Extensive research has been conducted on the mechanical, thermal, and corrosion properties of metal matrix composites (MMCs). The matrix of common metal matrix composites are metal alloys such as Ti, Al, and Mg, while the reinforcement materials are ceramics such as SiC, B_4C, and Al_2O_3. In comparison to monolithic materials, composites made of metal matrices including aluminum exhibit a number of desirable properties, such as enhanced thermal conductivity, decreased coefficient of thermal expansion, improved wear resistance, and higher specific strength. Using WS_2 can make MMCs lighter since WS_2 particles are lighter than matrices [1]. One type of aluminum-based MMC, known as an alloy-WS_2 composite has recently gained popularity for usage in a wide range of commercial products, including satellites, power semiconductors, heat sinks for microchips, heat exchange devices, precision electronics, ultra-precision metrology, and thermal stability [2].

DOI: 10.1201/9781003530909-15

The tribological behavior of aluminum matrix composites reinforced with silicon carbide particles was investigated by researchers in a study that examined the effects of WS_2 particles as an additional reinforcement [3]. Dry sliding wear experiments with different loads, sliding speeds, and distances have been conducted for reinforcement percentages of 2.5%, 5%, 7.5%, and 10% of WS_2 and SiC particles, respectively. Wear characteristics of hybrid composites reinforced with WS_2 are superior than those of composites reinforced with SiC alone, according to the study [4]. The wear behavior of AA6061 and AA7075 reinforced with boron carbide and WS_2 at 10 and 5 wt.%, respectively, has been examined by the authors [5]. It was discovered that the composites' wear resistance had enhanced. Researchers have studied the tribological properties of self-lubricated aluminum/silicon carbide hybrid composites produced by semi-solid powder densification, with varying concentrations of WS_2 added [6]. Research has shown that up to 5% WS_2 content causes wear to stabilize, wear debris particles to shrink, and fracture toughness to decline monotonically. Hybrid composites composed of alumina reinforcing an Al-Mg-Si alloy matrix and Rice Husk Ash (RHA) were studied for their mechanical and physical properties [7]. Authors used Al_2O_3 particles mixed with 0, 2, 3, and 4 wt.% RHA to make a reinforcing phase that was 10 wt.%. These three properties' hardness, yield strength (YS), and TS rise in direct proportion to the Al_2O_3 content.

Chemical inertness, low friction, lack of intrinsic abrasiveness, film-forming ability on metal surfaces, and relatively inoffensive to nature are some of the encouraging qualities of WS_2, making it a very popular solid lubricant. The addition of WS_2 particles in the appropriate quantity can effectively decrease the wear rate (WR) of aluminum matrix composites. When combined, the matrix and reinforcement phases give the material characteristics that no other material can match [8]. On the other hand, WS_2 in an aluminum matrix decreases hardness and flexural strength. The incorporation of ceramics into hybrid composites enhances their mechanical and tribological properties, which helps to circumvent this issue. The chemical composition of TiC is listed in Table 12.1. This material has been explored as a secondary reinforcing alternative to WS_2 in this study due to its high SiO_2 and Al_2O_3 content (up to 81% by weight), which suggests it could act as a ceramic addition.

TABLE 12.1
Chemical Composition of TiC

Property	Value
Chemical Formula	TiC
Crystal Structure	Face-centered cubic (FCC)
Molar Mass	59.89 g/mol
Density	4.93 g/cm^3
Hardness	9-9.5 on Mohs scale
Melting Point	3,160° C (5,720° F)

TABLE 12.2
Comparison of AA7075, WS_2 and TiC Properties

Materials	Units	AA7075-T6	WS2	TiC
UTS	MPa	310	14–68	138–206
CS	MPa	188	60–85	100–250
Young's Modulus	MPa	71,000	9.6	35
Coefficient of Thermal Extension	m/(m°C)	25.4	3–7	7.8
Bulk Density	kg/m^3	2700	1750–1900	1720–1810
TC	W/m°C	170	72.4–124.2	1.9–5
Brinell Hardness	BHN	108	70	855

Table 12.2 shows the mechanical property comparisons of AA7075, WS_2, and TiC. For appropriate mixing at stir casting and to give the finished composite coherency hardening, the density of the aluminum must be greater than that of the particles. When it comes to making aluminum matrix composites, stir casting is clearly the way to go. In order to further enhance the wear resistance of the synthetic composites, our study utilized the precipitation hardening technique, which improves tensile strength and hardness more rapidly than natural aging. ANOVA was used to identify the main factors impacting the wear behavior of composites. The grey relational grade (GRG), a quality parameter, was improved by conducting confirmation tests with the optimal values of design variables.

12.2 MATERIALS AND PROCEDURES

12.2.1 Preparation Phases for Reinforcement

TiC particles were automatically dried to eliminate water content and volatile compounds after being rinsed with water to remove impurities. In their studies, authors used WS_2 particles with a size of 150 m, while successfully used WS_2 grains with a size of 80 m [9, 10]. To avoid agglomeration and ensure even distribution inside the AA7075 during stir casting, the average size of the reinforcing materials used in this study, WS_2 and TiC, must be correspondingly bigger than that of other common reinforcements like B_4C, SiC, and TiC, which have much lower bulk densities. With their nearly identical densities, WS_2 and TiC were considered to have an optimal particle size of 100 m for our investigation. Particulates with an average size of 100 μm or less were obtained by passing the fines through an ASTM140 sieve. Additionally, WS_2 particles purchased from the neighborhood market were screened to extract particles with an average size of 100 μm or smaller.

12.2.2 Stir Casting of AA7075-WS2-TiC Composites

Specifications and composition identification of stir-cast AA7075 composites are displayed in Table 12.3. Melted in a furnace at 780°C were commercially available AA6061 rods with a 1.2″ diameter that were cut to particular specifications according

TABLE 12.3
Specifications and Composition Identification of Stir-Cast AA7075 Composites

Structure Description	Structure Details
3W	3 wt.% WS_2, Remaining AA7075
3W 3T	3 wt.% WS_2, 3 wt.% TiC, Rest AA7075
3W 5T	3 wt.% WS_2, 5 wt.% TiC, Rest AA7075
5T	5 wt.% TiC, Rest AA7075
7T	7 wt.% TiC, Rest AA7075

to the weight that was needed. Once the melting process was finished, the nitrogen gas probe was used to degasify the mixture for approximately 5 minutes. Before being added to the melt, the cast-iron reinforcements and permanent billet molds were warmed in a separate furnace to 500°C for 2 hours. The AA7075-T6 composites, made with WS_2 and TiC particles, were cast using five different compositions.

The induction furnace was then equipped with the mechanical stirrer mechanism and funnel for adding reinforcement was positioned. The reinforcements were added to the melt in two phases with the help of a stirrer to create a vortex, which prevented heat loss and kept the furnace temperature high. It took an hour for the furnace to reach 780°C after two-step stirring and reinforcement addition was complete. Then, the molten mixture was put into the permanent molds that had been preheated and left to harden for another 10 hours.

In order to achieve composites with consistent grain size, the as-cast specimens were first homogenized at 560°C for 8–10 hours, cooled slowly in the furnace, and then treated with solution heat for 2 hours before being quenched in water. The hardness of hybrid composites is enhanced with aging treatment at various temperatures following solutionizing [11, 12]. For the purpose of artificial aging, the cast components were heated to 180°C for 4–5 hours, or until they reached their peak hardness. Table 12.4 displays the mechanical properties that were acquired for the composites that were manufactured; 3W5T has the highest tensile strength (according to ASTM-E8 criteria) of the four compositions tested, although 7T is very brittle

TABLE 12.4
Mechanical Behaviors of the Composites Created with 180°C as the Peak Aging Temperature

Mechanical Behaviors	Units	AA7075-T6	3W	3 W 3T	3 W 5T	5T	7T
Peak Hardness	BHN	107	93.12	108.62	115.12	112.12	125.62
Density	kg/m^3	2700	2663.22	2586.82	2561.62	2577.42	2564.22
TS	MPa	312	275.37	307.74	328.55	303.89	298.18

and so has a lower tensile strength. After 5T composition, 3W5T achieves the second-best hardness.

The results display that the mechanical properties are improved when TiC is added to WS_2. Polishing and buffing the specimens, etching them with Keller's solution, and finally analyzing them with a METJI Inverted metallurgical microscope allowed us to investigate their microstructure.

Cylindrical specimens were cut to dimensions of 8 mm in diameter and 30 mm in height (in accordance with ASTM-G99 standards) for the wear test on two bodies. Following their respective peak aging periods, the WT samples were treated with solutioning/precipitation hardening for the duration of the test. For each of the five compositions, specimens were subjected to two body wear tests at three different sliding speeds (1.0, 1.3, 1.6 m s^{-1}) over 550 m of distance and with a 60 mm of track radius. At every speed, specimens were tested with three various loads (15, 25, 35 N) to examine the variation in coefficient of friction (COF) and wear rates. A pin-on-disk apparatus is depicted in Figure 12.1.

The current research utilizes the volumetric rate of wear because, as shown in Table 12.4, the composition densities vary. Normally, the weight loss technique is used to specify the wear rates, which are expressed in milligrams. For every experiment, the apparatus recorded the COF values at certain time intervals, which were then used to create the appropriate graphs. Equation (12.1) uses the trial weight loss 'W,'

FIGURE 12.1 Pin-on-disk apparatus.

the composite sample mass density, the distance traveled 'S,' and the applied load 'F,' to provide the particular wear rate [13, 14].

$$\dot{W} = \frac{\Delta W}{\rho F S} \tag{12.1}$$

12.2.3 Experimental Process

The wear performance of the composite specimens was evaluated utilizing three control parameters selected from the available literature [15]. The factors consist of the applied load (X), sliding velocity (Y), and distance (Z). Each element was examined at three different levels, as tabulated in Table 12.5.

An orthogonal array (OA) known as Taguchi L9 was utilized to conduct tests, consisting of nine runs [16].

The wear performance was evaluated utilizing two response parameters: the COF and the specific wear rate (SWR). The COF was calculated by dividing the F_f (kinetic frictional force) by F_n (applied normal load) using Equation (12.2). The value of specific energy requirement (SER) was determined using Equation (12.3) [17].

$$\text{COF} = \frac{F_f}{F_n} \tag{12.2}$$

$$SWR = \frac{\Delta W}{\rho L D} (\text{mm}^3 / \text{Nm}) \tag{12.3}$$

The variables in the equation are: ΔW denotes the pin's weight loss (g), ρ for density of the sample (g/mm^3), X (N), and Z (m).

12.2.4 Ratios of Signal to Noise in the Taguchi Technique

The Taguchi techniques utilizes OA to reduce volatility and improve the process variables. The signal-to-noise ratio is a performance metric that the Taguchi technique utilizes to assess process resilience and deviations from target values [18]. The S/N ratio is calculated by logarithmically comparing the average S/N standard deviation [19]. An increase in the S/N ratio is selected to lessen the influence of noise and other uncontrollable variables [20]. Improved product quality is indicated by

TABLE 12.5
Input Variables and Their Levels

Parameters	Symbol	Units	Levels 1	Levels 2	Levels 3
Applied Load	X	N	15	30	45
Sliding Velocity	Y	m.s^{-1}	1.5	2.5	3.5
Sliding Distance	Z	m		550	

maximum S/N ratios. Equations (12.4) through (12.6) represent the three types of S/N ratios: smaller-the-better (STB), nominal-the-best (NTB), and higher-the-better (HTB).

$$\left(\frac{S}{N}\right)_{STB} = -10 \cdot \log_{10}\left(\frac{1}{n}\Sigma_{i=1}^{n} y_i^2\right) \quad (12.4)$$

$$\left(\frac{S}{N}\right)_{NTB} = 10 \cdot \log_{10}\left(\frac{\bar{y}^2}{s^2}\right) \quad (12.5)$$

$$\left(\frac{S}{N}\right)_{HTB} = -10 \cdot \log_{10}\left(\frac{1}{n}\Sigma_{i=1}^{n}\frac{1}{y_i^2}\right) \quad (12.6)$$

where

n number of tests
y_i response value of ith test in the orthogonal array
y^2 average
s^2 variance of the data that was observed.

12.2.5 Optimizing Multiple Responses Using GRA

Taguchi's experimental technique is well suited for identifying the best process parameter settings for specific response characteristics. When dealing with several responses that have different quality characteristics, the preferred solution is multi-response optimization utilizing GRA. GRA can evaluate the similarity of apparently irregular and restricted data sets [21]. The study conducts multi-response optimization of wear characteristics utilizing the subsequent methods in GRA.

12.2.5.1 GRG Analysis

In GRA, the importance of the components is disregarded when the reference sequence range and standard value are substantially enhanced. Different objectives and paths of components might lead to inaccurate outcomes in the GRA. The pre-processing of the data involves normalizing the initial reference sequences to a range of 0 to 1 in order to facilitate a more accurate comparison [22]. The process of normalizing data into a series of sequences is known as GRG. There are two parameters that govern the response of the changed sequences following GRA preprocessing: smaller-the-better and larger-the-better. Use Equation (12.7) to normalize the sequence in order to obtain a STB feature.

$$x_i^*(k) = \frac{\max y_i(k) - y_i(k)}{\max y_i(k) - \min y_i(k)} \quad (12.7)$$

where

$x_i^*(k)$ modified reference sequence for the ith trial
$y_i(k)$ initial average response sequence.

12.2.5.2 Calculation of GRC and GRG

Using Equation (12.8), the deviation sequence of the standard sequence is then calculated following the standardization of the sequence.

$$\xi_i\left(k\right)=\frac{\Delta_{\text{min}}+\zeta\Delta_{\text{max}}}{\Delta_{0\text{i}}\left(k\right)+\zeta\Delta_{\text{max}}} \tag{12.8}$$

$\xi_i(k)$ indicates the comparability sequence, $\Delta_{0i}(k)$ denotes the deviation sequence, and larger-the-better. The GRC is calculated using Equation (12.9).

$$\xi_i\left(k\right)=\frac{\Delta_{\text{min}}+\zeta\Delta_{\text{max}}}{\Delta_{0\text{i}}\left(k\right)+\zeta\Delta_{\text{max}}} \tag{12.9}$$

$\xi_i(k)$ denotes the GRC of individual response factors calculated based on Δ_{max} and Δ_{min}, which are the largest and smallest variations in each response parameter. The identification coefficient, denoted by ζ and ranging from 0 to 1, is often assigned a value of 0.5 to evenly distribute weights among all parameters. Equation (12.10) illustrates the computation of a composite GRG by summing the GRC of each input factor [23].

$$\gamma_i=\frac{1}{n}\sum\nolimits_{i=1}^{n}\xi_i\left(k\right) \tag{12.10}$$

where

γ_i the GRG value calculated for the ith test.

n the total number of performance parameters.

After establishing the best amount of the elements with the GRG method, the next stage involves predicting and confirming the quality features using Equation (12.11) [24].

$$\gamma_{\text{predicted}}=\gamma_{\text{m}}+\sum\nolimits_{\text{i}=1}^{\text{q}}\gamma_{\text{o}}-\gamma_{\text{m}} \tag{12.11}$$

where

γ_o the highest average GRG at the optimum factor level.

γ_m the average GRG.

The quantity of variables impacting answer values is represented by the variable q.

12.2.6 ANOVA

ANOVA is commonly utilized to determine if the factors in an experimental design significantly affect the responses. Parameter interactions and their effects on dependent variables are often investigated using the ANOVA table [25]. The F-test is frequently employed to evaluate the influence of factors on the test results. If the 'Prob > F' value, sometimes referred to as the p-value, is < 0.05, at a level of confidence of 95%, the variables and relations are measured important [26].

The F-value indicates the extent to which a process variable affects the performance parameter. The ANOVA utilizes the R^2adj (adjusted correlation coefficient) to evaluate the accuracy of the model. R^2adj quantifies the proportion of variance in the response variables that is specifically accounted for by the independent factors and interactions. It is important for the models to have high and similar R^2 and R^2adj values to indicate a good fit with the experiments.

12.3 RESULTS AND DISCUSSIONS

Figure 12.2 shows the time-dependent changes in friction coefficients for three various loads 15, 25, and 35 N at a speed of 1.0 m s $^{-1}$ and distance of 550 m.

As the load increases, the friction coefficients decrease. Compositions with 6S and 4S have recorded the highest friction coefficient values. The tougher TiC particles that are released at the interface cause the friction coefficients to be higher because they get in the way of the worn specimen's contact with the disk. The dynamic character of friction will inevitably come into play as loads grow, resulting in reduced friction coefficients across all compositions. As the number of asperities subjected to elastic deformation increases, these asperities act as linear springs, reducing friction, which in turn reduces friction at greater loads [27, 28]. The dry lubricant property of WS_2 particles is crucial in lowering friction for those WS_2-containing formulations.

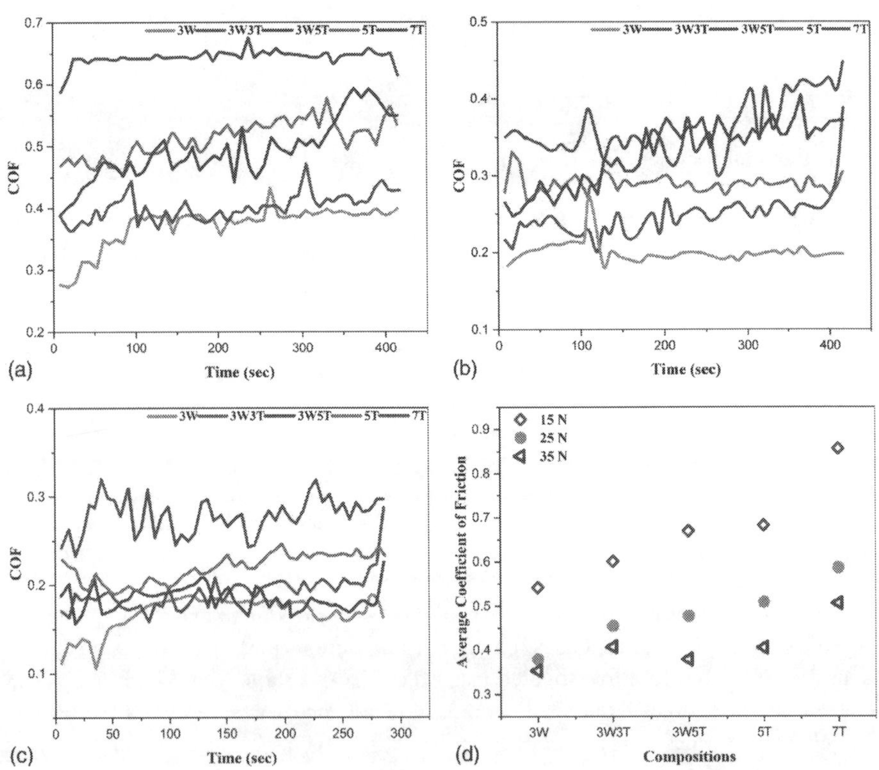

FIGURE 12.2 Change in coefficient of friction with time (in seconds): (a) 15 N, 550 m, 1.0 m s^{-1}; (b) 25 N, 550 m, 1.0 m s^{-1}; (c) 35 N, 550 m, 1.0 m s^{-1} (d) Average COF for every composition.

Wear specimens and plates slide smoothly together due to the dispersion of fine WS_2 particles at the contact surface, which improves with increasing loads.

The friction coefficients change with different weights, speeds (1.3 m/s), and distances (550 m), as shown in Figure 12.3. The COF is shown to rise as the weight percentage of TiC increases. In general, the friction coefficients are lower at 1.3 m s^{-1} compared to 1.0 m s^{-1}. Despite the lack of lubricant for the 5T and 7T specimens, the Stribeck effect is supported by the decreasing friction coefficients as speeds increase. By reducing the strain hardening rate and lowering the friction coefficient, the sample subsurface below the wear surface can be detected at 1.3 m s^{-1}, this intermediate speed. As weights are increased, the sliding disc and wear specimen make more consistent contact, which helps reduce friction coefficients. Specimens 3W, 3W3T, and 3W5T benefit from WS_2 since their friction coefficients are noticeably lower than those of specimens made with TiC alone.

Figure 12.4 displays that COF decreases more rapidly the speed is improved to 1.6 m s^{-1}. Here, too, the decrease in friction with increasing weights is preserved, and all compositions record the minimum of the friction values relative to the first two speeds. This means that the COF drops as one's speed rises. At slower speeds,

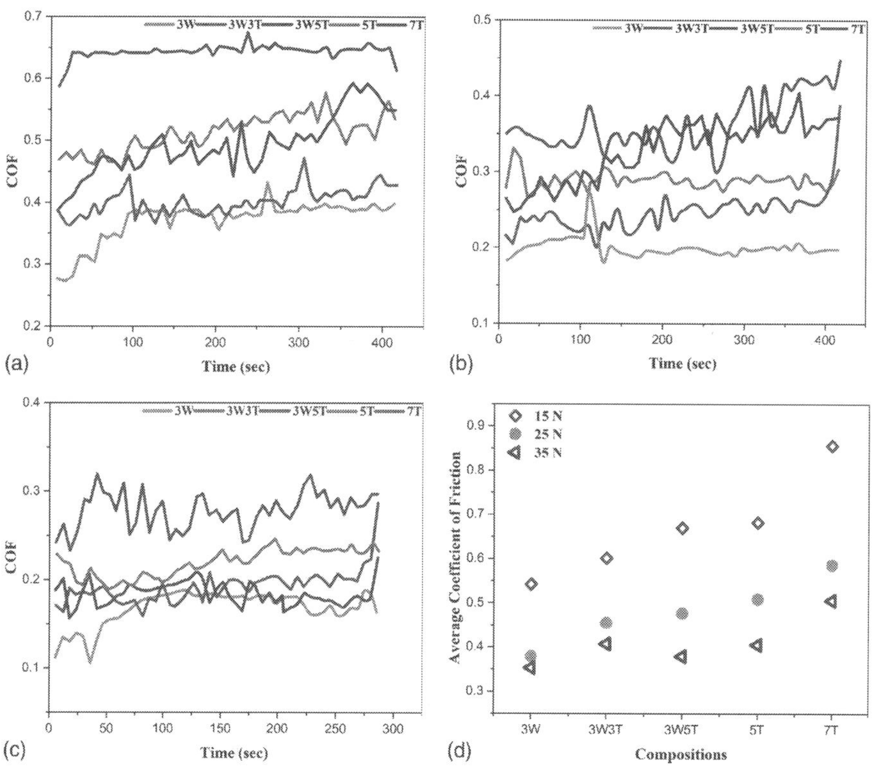

FIGURE 12.3 Change in COF over time (in seconds): (a) 15 N at 550 m and 1.3 m/s^{-1}; (b) 25 N at 550 m and 1.3 m/s^{-1}; (c) 35 N at 550 m and 1.3 m/s^{-1}; (d) average COF for every composition.

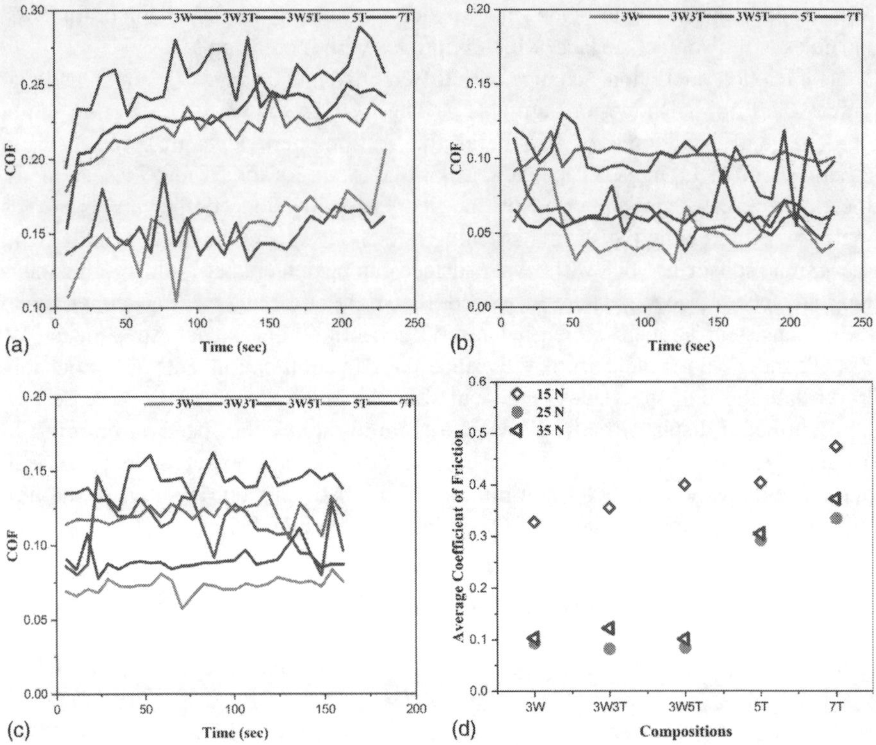

FIGURE 12.4 Change in COF over time (in seconds): (a) 15 N at 550 m and 1.6 m/s^{-1}; (b) 25 N at 550 m and 1.6 m/s^{-1}; (c) 35 N at 550 m and 1.6 m/s^{-1}; (d) average COF for every composition.

viscous and Coulomb's friction coexist, shifting the system from main stiction. When it comes to wear performance, the 3W, 3W3T, and 3W5T specimens with WS_2 particles continue to improve the lubricity. However, our study only allowed a maximum of 3 wt.% of WS_2 because adding more of it tends to lower the final composite's hardness and strength.

3W5T has the lowest COF of all the compositions tested under all weights and speeds. Due of the potential for brittleness, the upper limit for the quantity of TiC that can be incorporated into hybrids is 5 wt.%. The inherent induced brittleness of TiC makes it an inappropriate microfiller for use in composites containing an aluminum alloy matrix. In addition to increasing the friction coefficient, the TiC particles generated during wear contaminate the surfaces as well as provide extra forces at micro-asperities. Therefore, WS_2 is an essential secondary filler.

Table 12.6 shows the mean values and standard deviations of COF to compare. Across all five materials, the variation in friction coefficients at low and high speeds (1.0 m s^{-1} and 1.6 m s^{-1}) is negligible. At the intermediate speed (1.3 m s^{-1}), the friction coefficients exhibit substantially greater dispersion, whereas 5T and 7T exhibit the largest dispersion in comparison to the other three compositions.

TABLE 12.6
Standard Deviations (σ) and Average COF (μ) for the Various Wear Parameter Levels

Load (N)	Speed (m/s)	Sliding Distance (m)	3W		3W3T		3W5T		5T		7T	
			μ	σ	μ	σ	μ	σ	μ	σ	μ	σ
15	1.0	550	0.331	0.032	0.451	0.025	0.421	0.036	0.584	0.022	0.695	0.026
25	1.0	550	0.26	0.038	0.249	0.017	0.358	0.031	0.397	0.036	0.464	0.019
35	1.0	550	0.135	0.032	0.158	0.016	0.216	0.032	0.23	0.033	0.289	0.017
15	1.3	550	0.397	0.045	0.425	0.034	0.516	0.059	0.532	0.041	0.659	0.024
25	1.3	550	0.233	0.025	0.28	0.04	0.321	0.024	0.36	0.053	0.392	0.047
35	1.3	550	0.208	0.033	0.23	0.031	0.223	0.026	0.254	0.029	0.312	0.033
15	1.6	550	0.182	0.033	0.18	0.026	0.246	0.026	0.254	0.031	0.278	0.032
25	1.6	550	0.099	0.027	0.106	0.025	0.108	0.019	0.141	0.017	0.139	0.027
35	1.6	550	0.108	0.018	0.146	0.03	0.125	0.02	0.155	0.02	0.177	0.022

In order to examine the effect of WS_2 and TiC reinforcements in the AA7075 matrix, a sliding wear study was performed on a plain AA7075 specimen (without reinforcements) that had been artificially aged for 3.5 hours at 180°C.

WR for various levels of wear parameters for each of the five manufactured composite compositions are shown in Table 12.7. Since the wear rates are significantly lowered after adding WS_2 and TiC, reinforcing base alloy is definitely beneficial. Reducing wear is a major function of WS_2 addition. It is possible that the wear resistance of base alloy will not be improved by utilizing TiC as a lone reinforcement, as the WRs are maximum and TiC content exceeds 6 wt.%.

The exact WRs for all samples at varied speeds and weights are displayed in Figure 12.5. Despite showing minimum WR at all speeds and loads, 3W and

TABLE 12.7
Wear Rates According to the Various Wear Parameter Levels

Load (N)	Speed ($m\ s^{-1}$)	Sliding Distance (m)	Specific WR ($\times 10^{-5}$ mm^3/Nm)					
			AA7075	3W	3W3T	3W5T	5T	7T
15	1.0	550	29.03	15.93	15.96	15.96	21.90	21.05
25	1.0	550	33.63	16.34	16.37	16.39	22.33	23.14
35	1.0	550	29.92	16.48	17.89	17.54	22.18	23.94
15	1.3	550	32.71	18.41	20.11	18.98	23.18	21.34
25	1.3	550	37.75	18.83	19.90	20.50	22.76	23.29
35	1.3	550	35.22	19.38	20.52	21.00	24.47	24.44
15	1.6	550	36.96	19.66	22.19	20.28	28.68	25.24
25	1.6	550	40.88	19.04	21.98	22.87	26.23	32.28
35	1.6	550	39.33	20.07	23.15	23.45	25.19	33.13

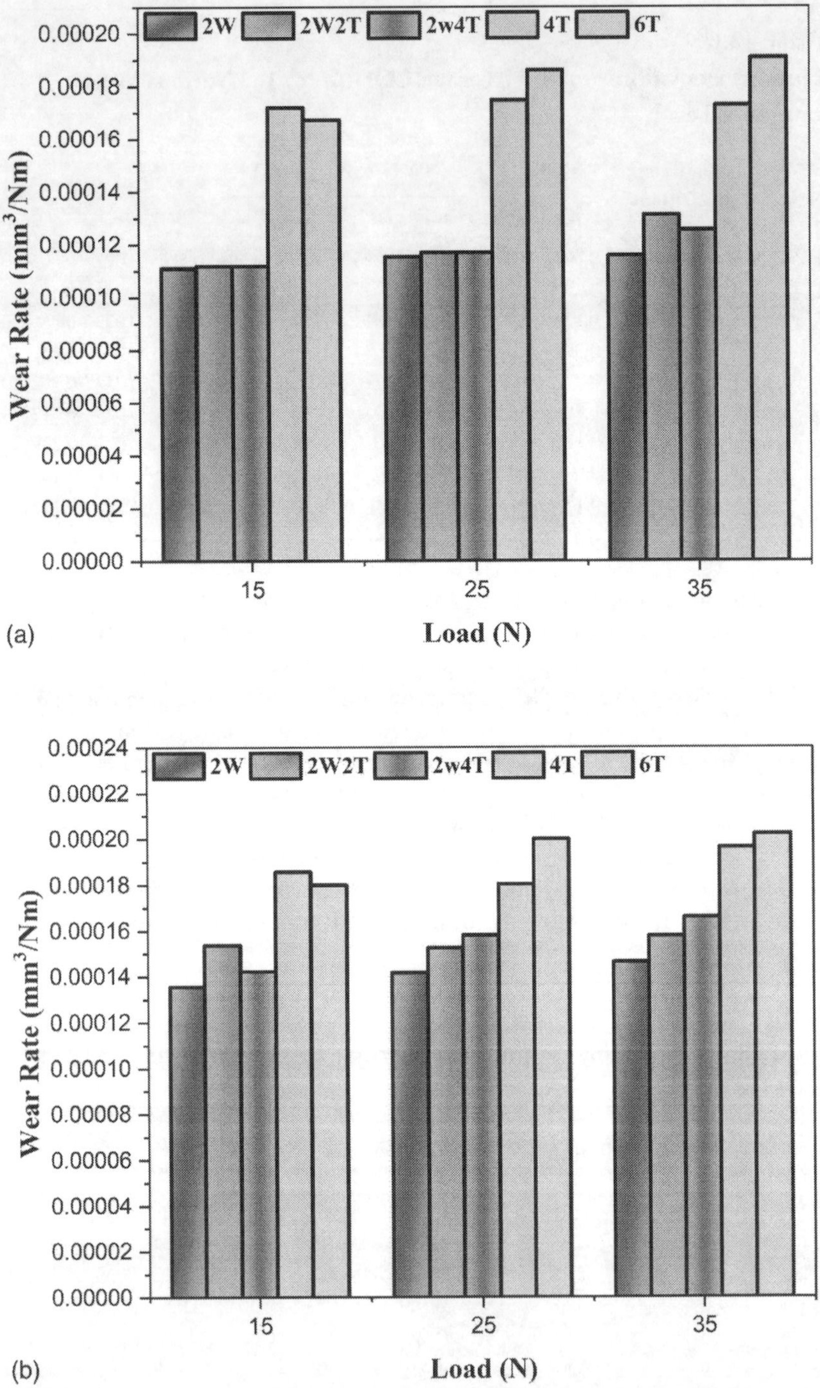

FIGURE 12.5 Specific volumetric WR (a) 1.0 m s^{-1}, 550 m, (b) 1.3 m s^{-1}, 550 m, (c) 1.6 m s^{-1}, 550 m. *(Continued)*

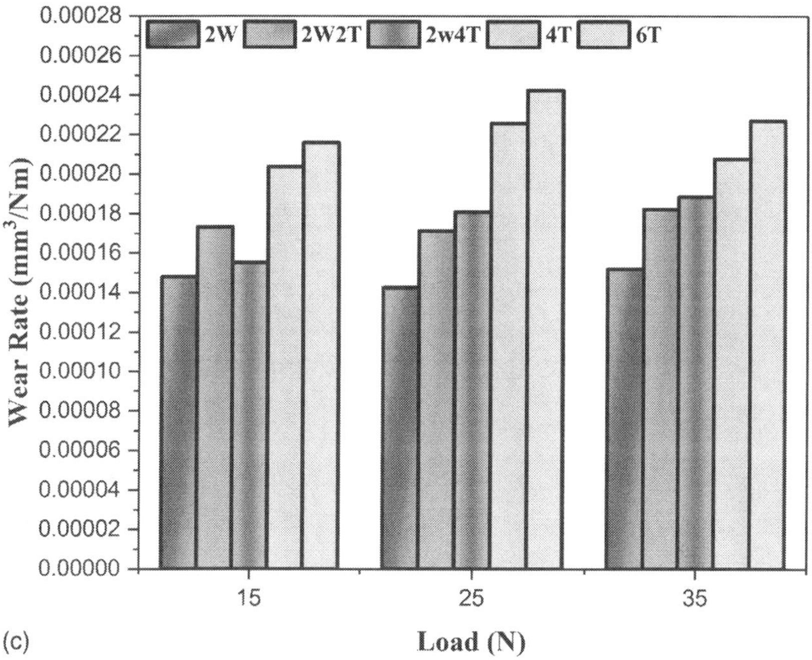

FIGURE 12.5 *(Continued)*

3W3T are not suitable for high-strength applications such as IC engine pistons and clutch cylinders because to their reduced hardness and strength. Wear rates are higher for 7T and 5T because they have higher coefficients of friction. Because its WR rises with increasing loads at all speeds, 7T is clearly the least desirable of the five compositions. The wear rate in 5T and 7T is directly proportional to the speed. The separation of brittle TiC particles as large debris, caused by rubbing action against the sample's strain-hardened surface. The 3W5T composition, which excels in hardness and TS, shows a WR that is similar to 3W composition for the majority of wear parameters, with the exception of extreme loads and speeds.

As the speed increases, strain hardening accelerates and the interface releases brittle debris more quickly, leading to a higher wear rate. 3W5T is the most optimal composition since it exhibits moderate WR across all loads and speeds. As a result of WS_2 particles' ability to mitigate TiC particles' brittleness, the COF and WR are both decreased.

12.3.1 Results on Taguchi-Grey Relational Analysis

The data from Table 12.8 was converted into S/N ratios in order to analyze Taguchi's L9 OA utilizing the grey relational analysis (GRA) method. This study examines the influence of modifying the control parameters X, Y, and Z on the different reactions of COF and SWR. Here, the sliding distance is constant. Because of that it is

TABLE 12.8
Taguchi L9 OA in Combination with S/N Ratio Multi-Response Analysis

	Parameters			Output		S/N Ratio	
Run	**X**	**Y**	**Z**	**COF (μ)**	**SWR × 10^{-3} (mm^3/Nm)**	**COF**	**SWR**
1	15	1.0	550	0.421	15.96	7.51	−24.06
2	25	1.0	550	0.358	16.39	8.92	−24.29
3	35	1.0	550	0.216	17.54	13.31	−24.88
4	15	1.3	550	0.516	18.98	5.75	−25.57
5	25	1.3	550	0.321	20.50	9.87	−26.24
6	35	1.3	550	0.223	21.00	13.03	−26.44
7	15	1.6	550	0.246	20.28	12.18	−26.14
8	25	1.6	550	0.108	22.87	19.33	−27.19
9	35	1.6	550	0.125	23.45	18.06	−27.40

not considered for Taguchi analysis. An analysis was conducted to assess the impact of factors on different outcomes, using a "smaller-the-better" superiority parameter, as lower COF and SWR values are favored.

12.3.2 Impact of the Control Parameters on COF

The S/N ratios were calculated using Minitab 20, a statistical program, utilizing the COF experimental data found in Table 12.8. After analyzing the delta data, we were able to determine the primary governing variables, and the results for the signal to noise ratios are displayed in Table 12.9. The delta values were calculated by taking the maximum and lowest average values of each variable and subtracting them. The differential value was used to establish the rankings. One of the main factors affecting COF is the most highly ranked delta value. There are three significant variables in Table 12.9, with Y being the most prominent at 6.321 and X following at 6.975.

TABLE 12.9
Signal-to-Noise Ratio Response Table for COF

Level	X	Y
1	9.916	8.481
2	9.550	12.708
3	16.525	14.802
Delta	6.975	6.321
Rank	1	2

TABLE 12.10
ANOVA for COF

Source	DF	Seq SS	Contribution	Adj MS	F-Value	P-Value	
X	2	0.06756	46.46%	0.033780	10.94	0.024	Significant
Y	2	0.06552	45.05%	0.032761	10.61	0.025	Significant
Error	4	0.01235	8.49%	0.003087			
Total	8	0.14543	100.00%				
		S	0.05556	R^2	91.51%	Adj.R^2	83.02%

Table 12.10 displays the outcomes of an analysis of variance (ANOVA) that was conducted to determine the relative importance of each element in determining the friction coefficient after factor values were specified. At 46.46%, X has the greatest effect on COF, followed by Y at 45.05%. A 95% confidence level is used to establish statistical significance when the p-value is less than 0.05. The fact that A and B both have p-values less than 0.05 indicates that they play a significant role in impacting the change in COF. The values of R^2 and adjust R^2 at their maximum and comparable ranges are displayed in Table 12.5. The model's ability to accurately reflect the data is demonstrated by this. A main effects plot displaying signal-to-noise ratios of COF is shown in Figure 12.6.

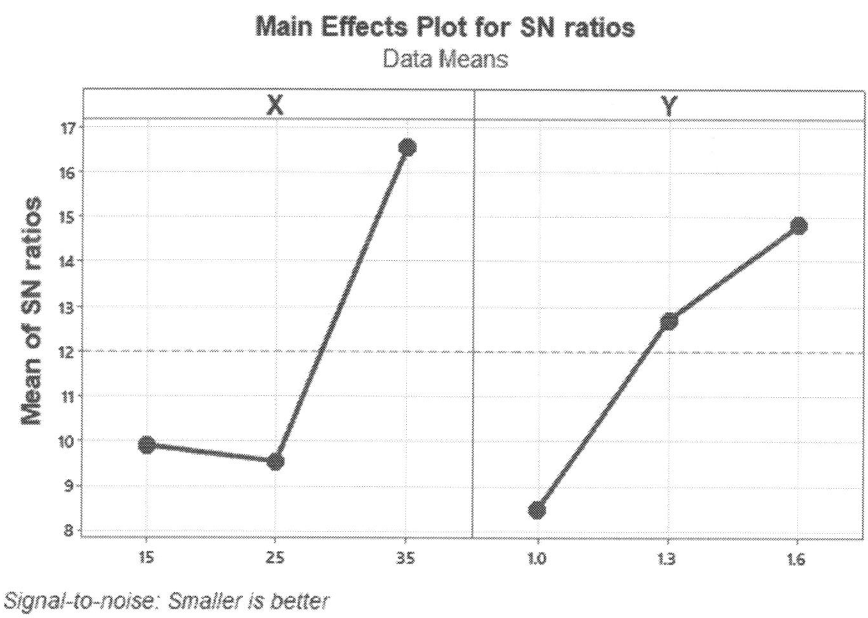

FIGURE 12.6 Main effects plot displaying signal-to-noise ratios of COF.

TABLE 12.11
S/N Ratio Response Table for SWR

Level	X	Y
1	–24.41	–25.26
2	–26.08	–25.90
3	–26.91	–26.24
Delta	2.50	0.99
Rank	1	2

12.3.3 Influence of the Control Variables on SWR

The parameters that influenced the SWR were identified by computing the S/N ratios of the experimental data. The Taguchi technique, which is comparable to COF analysis in that it emphasizes the importance of small samples, was selected for the purpose of studying variable effects. Table 12.11 shows the outputs of the statistical analysis of S/N ratios (SWR). Delta values of 2.50 for X indicate the greatest impact on SWR, followed by 0.99 for Y, respectively, according to the results.

The response table for S/N ratios was utilized to construct the primary impacts plot for SWR, as shown in Figure 12.7. The ratio exhibits no change with sliding velocity and declines with increasing applied weight. According to Figure 12.7, the

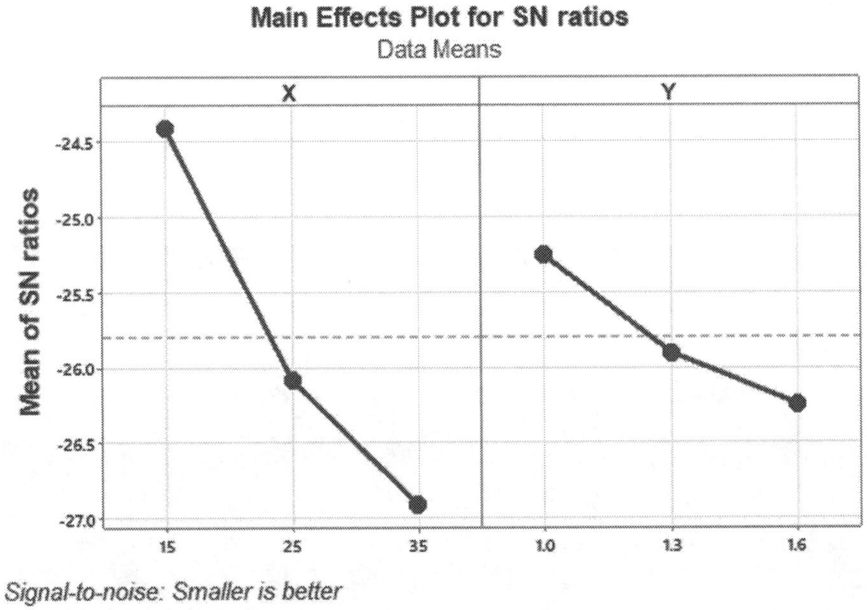

FIGURE 12.7 S/N ratios of SWR: main effects plot.

TABLE 12.12
ANOVA for SWR

Source	DF	Seq SS	Contribution	Adj MS	F-Value	P-Value	
X	2	47.647	83.75%	23.8237	72.70	0.001	Significance
Y	2	7.935	13.95%	3.9676	12.11	0.020	Significance
Error	4	1.311	2.30%	0.3277			
Total	8	56.893	100.00%				
		S	0.5724	R^2	97.70%	Adj.R^2	95.39%

best S/N ratios for SWR are achieved when X1 and Y1 are set to their initial levels. The S/N ratios show consistent levels for the components, as seen in Table 12.11. Table 12.11 shows that among the three variables influencing SWR, X, Y, the delta data shows that X is the most important.

The components that were determined to be significant and their percentage contributions to the response variable SWR were determined by an ANOVA, as shown in Table 12.12. The maximum percentage involvement of the applied load is 83.44% at a 95% conferencing level, as shown in Table 12.12, with sliding speed following at 13.95%. It is worth noting that the X component is the only one with a statistically significant p-value (< 0.05). A highest R-value of 97.70% indicates that the model is reliable since it accounts for the majority of the variation.

By applying the previously mentioned S/N ratio optimization to two sets of input variables each, the two outputs were fine-tuned. When dealing with practical applications, finding the optimal mix of parameters is of the utmost importance. This study makes use of the GRA to optimize three parameters and two responses all at once.

12.3.4 Optimizing Multiple Responses Using GRA

The primary use of GRA is to tackle real-world problems with sparse data. A common application of this strategy is to estimate the behavior of uncertain systems that do not have a clear-cut answer. In the grayscale scheme, white signifies full data and black denotes incomplete data.

According to the sources cited, the GRA has seen substantial application in optimizing scenarios involving multiple inputs and replies. The normalized reference series of results, which fluctuate between 0 and 1, was generated using Equation (12.7). To determine the sequences of deviations, Equation (12.8) was utilized. Table 12.13 displays the results of the normalization and the sequence deviation.

Once the sequences of deviations were found, the GRC for each response value was computed using Equation (12.9). By averaging the GRCs, the GRG (γ_i) was determined. In order to determine the appropriate S/N ratios, the predicted values from GRGs Table 12.14 were utilized. The testing data is very near to the best GRG value when the S/N ratio is large, which is useful. Authors can see the relationship between S/N ratio and GRG in Figure 12.8. According to the results, the first experimental run had the optimal S/N ratio. As predicted, the first experimental trial was

TABLE 12.13
Normalization and Sequence Deviation

	Normalization x^*_i		Sequence Deviation, Δ_{0i}	
Run	**COF**	**SWR**	**COF**	**SWR**
1	0.233	1.000	0.767	0.000
2	0.387	0.943	0.613	0.057
3	0.735	0.789	0.265	0.211
4	0.000	0.597	1.000	0.403
5	0.478	0.394	0.522	0.606
6	0.718	0.327	0.282	0.673
7	0.662	0.423	0.338	0.577
8	1.000	0.077	0.000	0.923
9	0.958	0.000	0.042	1.000

given the top spot. Figure 12.8 shows the S/N ratios, and the previous explanation is complemented by the GRG's trailing feature.

A response table for the GRG was subsequently generated using the ranks. After choosing the GRG for each variable at the necessary level, we averaged it to find the overall GRG. From iteration 4 through 6, the "applied load" parameter was set to level 2. Equation (12.12) was used for computation using the GRG technique values taken from Table 12.14.

As shown in the previous section, Table 12.15 shows the outcomes of our computations for the average of the GRGs. The connections between the reference and corresponding GRA series are shown by the grades in the table. A stronger correlation is shown by a higher average value of GRGs. The GRG response table is located in Table 12.15, where

TABLE 12.14
GRG with S/N Ratios

	GRC (ξi (k))				
Run	**COF**	**SWR**	**GRG (γ_i)**	**S/N Ratio**	**Rank**
1	0.395	1.000	0.6973	−3.132	1
2	0.449	0.897	0.6732	−3.437	4
3	0.654	0.703	0.6786	−3.368	2
4	0.333	0.554	0.4435	−7.063	9
5	0.489	0.452	0.4706	−6.547	8
6	0.639	0.426	0.5329	−5.467	6
7	0.596	0.464	0.5304	−5.508	7
8	1.000	0.351	0.6757	−3.404	3
9	0.923	0.333	0.6282	−4.038	5

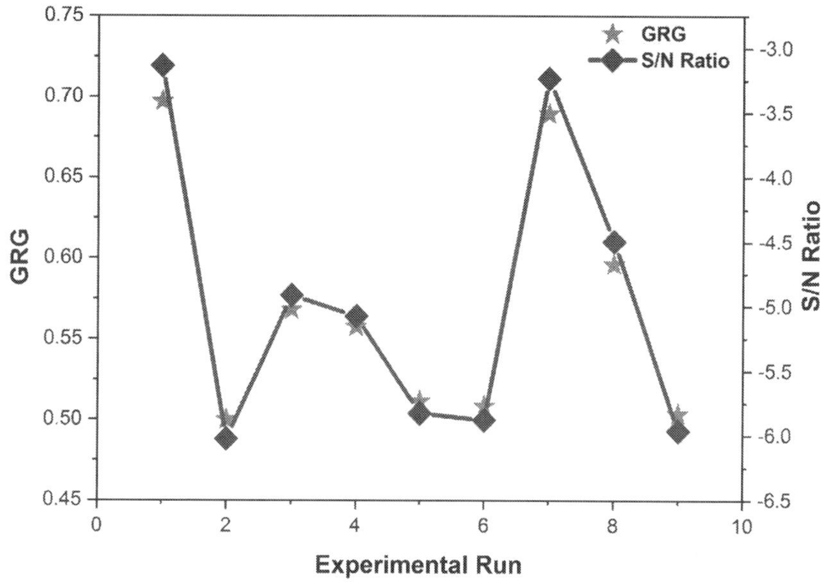

FIGURE 12.8 Graph of S/N ratios of GRG and GRG.

one can discover the optimal combination of parameters that enhances the total response. According to Table 12.15, the locations with the highest GRG are X3 and Y3, and Z. Parameters applied load (X) = 45 N, sliding velocity (Y) = 1.6 m/s, and sliding distance (Z) = 550 m are optimal for increasing the wear resistance of composites.

$$L(level2) = \frac{0.44346 + 0.47062 + 0.5329}{3} = 0.48232 \quad (12.12)$$

12.3.5 Anova for GRG

To determine the relative importance of each component's effect on composite's wear performance metrics, we ran an analysis of variance (ANOVA) using a 95% confidence interval. According to Table 12.16, the biggest impact on the GRG is 80.01%

TABLE 12.15
Grey Relational Grade Response Table

Level	X	Y
1	0.6830	0.5571
2	0.4823	0.6065
3	0.6115	0.6132
Delta	0.2007	0.0562
Rank	1	2
Average of GRG = 0.5923		

TABLE 12.16
ANOVA for GRG

Source	DF	Seq SS	Contribution	Adj MS	F-Value	P-Value	
X	2	0.062070	80.01%	0.031035	12.59	0.019	Significant
Y	2	0.005645	7.28%	0.002823	1.15	0.404	Non- significant
Error	4	0.009860	12.71%	0.002465			
Total	8	0.077575	100.00%				

for the X, followed by 7.28% for the Y. It is also clear that all parameters significantly affect the composite's wear characteristics because the p-values for all components are less than 0.05. The generated model is trustworthy if the R-values are high.

12.3.6 Confirmation Testing

The final step of GRA, following optimal setting determination, is to forecast and confirm the solutions' improved performance. The analytical findings were double-checked with confirmation studies, and the average GRG from all three trials was calculated. Under ideal circumstances, the COF and SWR were found to be 0.2381 and 17.63, respectively. Table 12.17 shows that the results of the confirmation experiment agree with the expected values quite well. We achieved a 38.21% increase in GRG. Using the Taguchi technique in conjunction with GRA to increase composite wear performance was validated by the better testing findings.

12.4 CONCLUSIONS

In order to determine the optimal composition that demonstrates exceptional tribological performance, the present study synthesized five distinct hybrid AA7075-T6 composites reinforced with TiC and WS_2 small particulates and subjected them to two-body wear tests. Application of heat treatment to a solution and liquid stir casting at 560°C were successful methods for producing AA7075 hybrid composites. Subsequently, the created composites were artificially aged at 180°C for 45 hours to enhance their hardness and strength.

TABLE 12.17
Confirmation Test Results

		Optimal Factors	
Initial Factors		**Prediction**	**Experiment**
Setting Level	X_1Y_1	$A_1B_1C_1$	$A_1B_1C_1$
GRG	0.6973	0.647805	0.6973
Increase in GRG		39.11%	38.21%

The addition of WS_2 and TiC to AA7075 alloy, followed by heat treatment by precipitation hardening, greatly enhanced the wear resistance. Based on the results of the wear testing, TiC holds great potential as a ceramic filler and a cost-effective alternative to more expensive fillers such as B_4C and SiC. Hybrid composites of Al alloy (specifically AA7075-T6) with WS_2 and TiC particles are superior for tribological applications because WS_2 inclusion improves lubricity at the wear interface. In order to manage the sliding wear phenomenon, it is essential to examine the mechanically mixed layers at the exhausted hybrids surfaces (3W3T, 3W5T). Wearing away occurs in the non-peeling zones. The optimal composition of 3% WS_2 and 5% TiC displays the best potential for tribological efficiency when compared to the base alloy and other mixes.

This is supported by its higher hardness, TS, and wear resistance. A study was conducted utilizing a Taguchi L9 OA and GRA technique to analyze the effects of applying load, sliding velocity, and sliding distance on the COF and SWR. The optimal conditions for enhancing the composite's wear resistance were determined to be an X of 15 N, a Y of 1.0 $m.s^{-1}$, and a Z of 550 m according to the GRG's response table. All the parameters in the GRG analysis of variance had p-values that were less than 0.05, indicating significance. Confirmation experiments were carried out to verify a 38.21% enhancement in GRG, 0.6973 with the ideal parameters ($X_1Y_1Z_1$).

REFERENCES

1. C. T. Scaria, R. Pugazhenthi, A. A. Daniel, and K. Santhosh, "Mechanical and wear studies on AA7075/nano TiC/graphite hybrid composites for tribological applications," Advances in Materials Science and Engineering, vol. 2022, pp. 5320363, 2022.
2. E. Salur, M. Acarer, and İ. Şavkliyildiz, "Improving mechanical properties of nano-sized TiC particle reinforced AA7075 Al alloy composites produced by ball milling and hot pressing," Materials Today Communications, vol. 27, pp. 102202, 2021.
3. R. Saravanakumar, T. Rajasekaran, and C. Pandey, "Optimisation of underwater friction stir welding parameters of aluminum alloy AA5083 using RSM and GRA," Proceedings of the Institution of Mechanical Engineers, Part E: Journal of Process Mechanical Engineering, vol. 237, no. 6, pp. 2334–2348, 2023.
4. N. A. Patil, S. R. Pedapati, O. Mamat, and A. M. H. S. Lubis, "Morphological characterization, statistical modeling and wear behavior of AA7075-Titanium Carbide-Graphite surface composites via Friction stir processing," Journal of Materials Research and Technology, vol. 11, pp. 2160–2180, 2021.
5. T. Zheng, S. Pan, N. Murali, B. Li, and X. Li, "Selective laser melting of novel 7075 aluminum powders with internally dispersed TiC nanoparticles," Materials Letters, vol. 319, 132268, 2022.
6. A. Dhulipalla et al., "Synthesis and machining characteristics of novel TiC ceramic and MoS2 soft particulate reinforced aluminium alloy 7075 matrix composites," Manufacturing Letters, vol. 24, pp. 82–86, 2020.
7. D. Dinesh Kumar et al., "Study of microstructure and wear resistance of AA5052/B4C nanocomposites as a function of volume fraction reinforcement to particle size ratio by ANN," Journal of Chemistry, vol. 2023, pp. 2554098, 2023.
8. L. Mihlyuzova, H. M. Enginsoy, S. Slavov, D. Dontchev, and E. Bayraktar, "Experimental and finite element study of recycled aluminium (AA7075) matrix composites reinforced of TiC/MoS2/γ-Al_2O_3Fibre/Nb2Al," in Conference Proceedings of the Society for Experimental Mechanics Series, 2022, pp. 41–47.

9. M V a Ramakrishna and K. Srinivas, “Grey relational analysis of friction stir welding parameters for the development of dissimilar joints between AA6082 and AA7075,” Advances in Materials and Processing Technologies, vol. 8, no. sup2, pp. 954–965, 2022.
10. S. Sundaramoorthy, R. Gopalan, and R. Thulasiram, “Microstructural characterization, tribological and corrosion behavior of AA7075 -TiC composites,” China Foundry, vol. 21, no. 4, pp. 334–342, 2024.
11. R. Dinesh Kumar, M. S. Ilhar Ul Hassan, S. Muthukumaran, T. Venkateswaran, and D. Sivakumar, “Single and multi-response optimization and validation of mechanical properties in dissimilar friction stir welded AA2219-T87 and AA7075-T73 alloys using T-GRA,” Experimental Techniques, vol. 43, no. 3, pp. 245–259, 2019.
12. D. Vijay Praveen, D. Ranga Raju, and M. V Jagannadha Raju, “Assessment of optimal parameters of wire EDM on Ni-coated Al_2O_3p/AA7075 MMCs using PCA coupled GRA,” Arabian Journal for Science and Engineering, vol. 46, no. 8, pp. 7953–7966, 2021.
13. G. Katundi, D. Katundi, E. Bayraktar, and I. Miskioglu, “Alternative composite design from recycled aluminium (AA7075) chips for knuckle applications-II,” in Conference Proceedings of the Society for Experimental Mechanics Series, 2020, pp. 9–14.
14. P. Rajendran, V. Duraisamy, A. R. Rajendran, and R. V Loganathan, “Optimization on the electrical discharge machining (EDM) process parameters of aged AA7075/TiC metal matrix composites,” Revista de Metalurgia, vol. 59, no. 3, pp. e245–e245, 2023.
15. M. M. Quazi, M. A. Fazal, A. S. M. A. Haseeb, F. Yusof, H. H. Masjuki, and A. Arslan, “Laser composite surfacing of Ni-WC coating on AA5083 for enhancing tribomechanical properties,” Tribology Transactions, vol. 60, no. 2, pp. 249–259, 2017.
16. B. Davoodi, M. R. Eslami, and G. H. Payganeh, “Determination of mechanical behavior of AA5083 alloy under machining conditions, applicable in metal cutting simulation,” Advanced Materials Research, Vol. 189, pp. 1507–1512, 2011.
17. A. Pareek, J. K. Jain, V. Raghavender, N. Gupta, V. Bhojak, and A. Saxena, “Investigation on mechanical and wear performance of friction stir processed on GMAW cladded AA5083/ER4043,” International Journal on Interactive Design and Manufacturing, Vol. 18, pp. 1401–1414, 2024.
18. X. G. Yang, D. L. Duan, X. Zhang, S. L. Jiang, S. Li, and H. C. Zhang, “Impact behavior of polyetheretherketone/nickel foam co-continuous composites,” Journal of Materials Engineering and Performance, vol. 28, no. 10, pp. 6380–6390, 2019.
19. A. Hojjati Najafabadi, R. Shoja Razavi, R. Mozaffarinia, and H. Rahimi, “A new approach of improving rain erosion resistance of nanocomposite sol-gel coatings by optimization process factors,” Metallurgical and Materials Transactions A Physical Metallurgy and Materials Science, vol. 45, no. 5, pp. 2522–2531, 2014.
20. K. Shunmugesh and P. Kavan, “Investigation and optimization of machining parameters in drilling of carbon fiber reinforced polymer (CFRP) composites,” Pigment and Resin Technology, vol. 46, no. 1, pp. 21–30, 2017.
21. P. Achuthamenon Sylajakumari, R. Ramakrishnasamy, G. Palaniappan, and R. Murugan, “Multi-response optimization of end milling parameters for Al-Zn-Mg/SiC co-continuous composite using response surface methodology,” Medziagotyra, vol. 25, no. 4, pp. 471–477, 2019.
22. A. Viswanath, M. Modrek, K. A. Khan, and R. K. Abu Al-Rub, “Deep learning for topology optimization of triply periodic minimal surface based Gyroid-like structures,” in 36th Technical Conference of the American Society for Composites 2021: Composites Ingenuity Taking on Challenges in Environment-Energy-Economy, ASC 2021, 2021, pp. 1088–1100.
23. S. Liao, L. Zhou, K. Yin, J. Wang, and C. Jiang, “Research status of β-Si3N4 ceramics based on high thermal conductivity,” Cailiao Daobao/Materials Reports, vol. 34, no. 21, pp. 21105–21114, 2020.

24. N. Santhosh, B. A. Praveena, A. Chandrashekar, V. Mohanavel, S. Raghavendra, and D. Basheer, "Wear behaviour of aluminium alloy 5083/SiC/fly ash inoculants based functional composites - optimization studies," Materials Research Express, vol. 9, no. 7, p. 076513, 2022.
25. S. H. Tomadi, J. A. Ghani, C. H. Che Haron, M. S. Kasim, and A. R. Daud, "Optimization of cutting parameters of multiple performance characteristics in end milling of ALSi/AIN MMC-taguchi method and grey relational analysis," Journal Teknologi (Sciences and Engineering), Vol 68, pp. 33–40, 2014.
26. M. J. Umar, Pl. K. Palaniappan, P. Maran, and R. Pandiyarajan, "Investigation and optimization of friction stir welding process parameters of stir cast AA6082/ZrO2/B4C composites," Materials Science- Poland, vol. 38, no. 4, pp. 715–730, 2020.
27. R. Gopalakrishnan, B. P. R. Dennis, N. R. R. Ramakrishnan, and A. R. Rajendran, "Influence of MoS2 with TiC on the tribological and wear properties of hybrid aluminum composites," in Smart Innovation, Systems and Technologies, Springer, 2023, pp. 275–283.
28. E. Bayraktar, I. Miskioglu, D. Katundi, and F. Gatamorta, "Manufacturing of recycled aluminum matrix composites reinforced of TiC/MoS2/Al_2O_3 fiber through combined method: Sintered + Forging," in Conference Proceedings of the Society for Experimental Mechanics Series, 2020, pp. 15–25.

Section IV

Tribological Performance of Ceramic Matrix Composites

13 Influence of Micro-Sized Filler Reinforcement in Tribological Performance of Ceramic Matrix Composites

Shrishail B. Sollapur, Sharath P. C., Bharathkumar B. R., Hemanth K., Vishwanath B. R., and Anand K. Hosamani

13.1 INTRODUCTION TO CERAMIC MATRIX COMPOSITES (CMCs)

Ceramic matrix composites (CMCs) have emerged as a revolutionary class of materials within the field of materials engineering, offering a blend of properties that are indispensable for high-performance applications. CMCs achieve enhanced mechanical, thermal, and tribological properties by incorporating ceramic fibers or particles into a ceramic matrix. These composites were developed to meet the demand for materials that can withstand extreme conditions like high temperatures, corrosive environments, and high wear rates, where traditional materials such as metals and polymers often fail. Ceramics are naturally valued for their high hardness, excellent thermal stability, and strong resistance to corrosion (Ran et al., 2024). However, their widespread application has historically been hindered by their inherent brittleness and low fracture toughness, making them unsuitable for many demanding engineering applications. CMCs effectively address these limitations by introducing reinforcements that enhance the toughness and ductility of the ceramic matrix, all while retaining the desirable properties of ceramics. This beneficial combination of properties has made CMCs highly sought after in industries such as aerospace, automotive, defense, and energy, where materials must perform reliably under the most extreme conditions (Tulshibagwale et al., 2024).

The development of CMCs began in the 1960s and 1970s when researchers started exploring the use of ceramic materials for high-temperature applications. Initial efforts focused on developing oxide-based ceramics such as alumina and zirconia, which were subsequently reinforced with fibers like silicon carbide (SiC) and alumina to enhance their mechanical performance (Rashid et al., 2024a). Over the decades, advancements in material science and manufacturing techniques have resulted in a wide array of CMCs, each specifically tailored for different

DOI: 10.1201/9781003530909-17

applications grounded on the selection of matrix and reinforcement materials. The structure of CMCs typically consists of a ceramic matrix, which serves as the primary load-bearing component, and a reinforcement phase, which can take the form of continuous or discontinuous fibers, whiskers, or particulates. The ceramic matrix is usually composed of materials such as alumina, silicon carbide, zirconia, or silicon nitride, while the reinforcement phase might include materials like carbon fibers, silicon carbide fibers, or alumina fibers. The choice of matrix and reinforcement materials is vital to the overall performance of CMCs. For example, silicon carbide matrices are preferred for their high thermal conductivity and stability, making them suitable for high-temperature environments (Morales-Florez & Dominguez-Rodriguez, 2022). Conversely, alumina matrices are renowned for their excellent chemical resistance, rendering them suitable for applications in corrosive environments. The reinforcement phase, whether fibers or particles, enhances the composite's toughness and fracture resistance by bridging and deflecting cracks, thus preventing catastrophic failure (Shrivastava et al., 2024). The global market for CMCs is projected to reach USD$22.31 billion by 2030, growing from USD$7.92 billion in 2021 at a compound annual growth rate (CAGR) of 12.2%, as represented in Figure 13.1.

In recent years, the addition of micro-sized fillers into ceramic matrix composites has attracted significant attention as a method of further improving their tribological performance. These micro-sized fillers, which can include materials such as micro-alumina, silicon carbide, or carbon nanotubes, are incorporated into the ceramic matrix to enhance properties such as wear resistance, friction coefficient, and mechanical strength (Ogunbiyi et al., 2024). The introduction of micro-sized

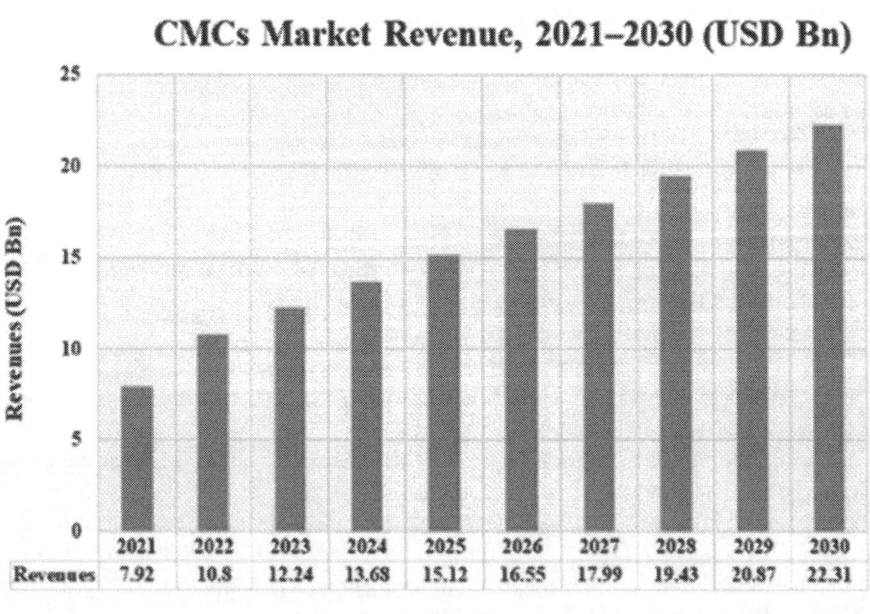

FIGURE 13.1 The global market of ceramic matrix composites. (Reproduced with permission from MDPI, Shrivastava et al., 2024.)

fillers into CMCs marks a significant advancement in materials engineering, especially for applications where tribological performance is crucial. While this is a relatively new research area, early studies show that these fillers can substantially improve CMC performance in various ways. For instance, micro-sized fillers can reduce the friction coefficient of CMCs, leading to decreased wear and improved material life span (Kovalčíková et al., 2020). Micro-sized fillers can also enhance the mechanical strength of CMCs by providing additional reinforcement to the ceramic matrix. This is particularly important in applications where the material faces high stress or impact, as the added reinforcement can help prevent cracking or breaking (Zhou et al., 2024). Another important benefit of adding micro-sized fillers to CMCs is their ability to influence the microstructure of the material. The addition of micro-sized fillers can help to create a more uniform distribution of stress within the composite, which can lead to improved crack resistance and a reduction in the likelihood of catastrophic failure (Chandran et al., 2024). Micro-sized fillers can also enhance the mechanical strength of CMCs by providing additional reinforcement to the ceramic matrix. This is particularly important in applications where the material faces high stress or impact, as the added reinforcement can help prevent cracking or breaking. The influence of micro-sized fillers on the tribological performance of CMCs is a topic of ongoing research, and there is still much to learn about how these fillers interact with the ceramic matrix (Li et al., 2018). Early studies have demonstrated that adding micro-sized fillers can significantly improve the wear resistance, friction coefficient, and mechanical strength of CMCs. This has sparked increased interest in using micro-sized fillers in various applications, particularly in industries such as aerospace, automotive, defense, and energy, where materials must perform reliably under extreme conditions (Wyatt et al., 2021). The addition of micro-sized fillers to ceramic matrix composites represents a significant advancement in materials engineering. These fillers have the potential to greatly enhance the tribological performance of CMCs, making them suitable for a wider range of applications (Markandan et al., 2017). As research in this area progresses, we are likely to see further advancements in the use of micro-sized fillers in CMCs, leading to the development of new materials with even greater performance capabilities (Nassar et al., 2021). This ongoing research will undoubtedly contribute to the continued evolution of CMCs as a critical material in modern engineering, particularly in applications where high performance and reliability are of utmost importance.

13.1.1 Manufacturing Techniques for CMCs

The fabrication of CMCs involves a variety of manufacturing techniques, each suited to producing composites with specific properties and microstructures. Some of the most common methods include:

1. Infiltration Techniques: These methods involve infiltrating a preform, typically made of fibers or particles, with a ceramic matrix material. One example is the polymer infiltration and pyrolysis (PIP) process, where a ceramic precursor is infiltrated into a fiber preform and then pyrolyzed to form a ceramic matrix.

2. Hot Pressing and Sintering: In hot pressing, the ceramic matrix and reinforcement materials are mixed and then subjected to high temperature and pressure to densify the composite. This method is particularly useful for producing dense CMCs with high mechanical properties.
3. Chemical Vapor Infiltration (CVI): It involves depositing a ceramic matrix onto a fiber preform from a gaseous precursor. This process is commonly used to produce SiC/SiC composites, known for their high-temperature stability.
4. Sol-Gel Processing: Sol-gel processing transforms a colloidal suspension of ceramic particles (sol) into a gel, which is then dried and heat-treated to form a ceramic matrix. This technique allows for precise control over the composition and microstructure of the CMC.
5. Liquid Metal Infiltration (LMI): It involves infiltrating a fiber preform with molten metal, which reacts to form a ceramic matrix upon cooling. This method is used to produce metal-ceramic composites with improved thermal and mechanical properties.

13.1.2 Properties of CMCs

The unique properties of CMCs result from the synergistic interaction between the ceramic matrix and the reinforcement phase. Key properties that make CMCs attractive for high-performance applications include:

1. High-Temperature Stability: CMCs can withstand temperatures exceeding 1000°C without significant degradation of their mechanical properties, making them ideal for use in high-temperature environments such as gas turbines, rocket nozzles, and heat shields.
2. Resistance to Oxidation and Corrosion: The ceramic matrix in CMCs provides excellent resistance to oxidation and corrosion, making them suitable for use in harsh environments where metals would corrode or degrade over time.
3. Enhanced Fracture Toughness: The incorporation of fibers or particles within the ceramic matrix significantly improves the fracture toughness of CMCs, reducing their susceptibility to brittle failure. This is achieved through mechanisms such as crack deflection, crack bridging, and fiber pull-out.
4. Low Density: CMCs have a lower density compared to metals, making them ideal for applications where weight reduction is critical, such as in aerospace and automotive components.
5. Wear Resistance: The hardness and wear resistance of the ceramic matrix make CMCs highly resistant to abrasive and erosive wear, extending the life span of components in tribological applications.

13.1.3 Influence of Micro-Sized Filler Reinforcement

In recent years, the incorporation of micro-sized fillers into CMCs has gained attention as a means of further enhancing their tribological performance. These fillers, which may include materials such as micro-alumina, silicon carbide, or carbon nanotubes, are added to the ceramic matrix to enhance properties like wear resistance, friction coefficient, and mechanical strength. The addition of micro-sized fillers can also influence the microstructure of the CMC, leading to improved crack resistance and a more uniform distribution of stress within the composite (Nie et al., 2020). The impact of micro-sized filler reinforcement on the tribological performance of CMCs is an ongoing research topic. Studies focus on optimizing the size, shape, and distribution of fillers to achieve desired properties (Chan et al., 2021). Adding fillers can also affect CMC processing and manufacturing, necessitating adjustments to parameters like temperature, pressure, and processing time for optimal results.

13.2 IMPORTANCE OF TRIBOLOGICAL PERFORMANCE

Tribological performance, which refers to the behavior of materials under conditions of friction, lubrication, and wear, is a critical factor in the design and application of engineering materials, particularly for CMCs. The importance of optimizing tribological performance in CMCs cannot be overstated, as these materials are often employed in environments where they are subjected to extreme stresses, high temperatures, and abrasive conditions. In such applications, even minor improvements in tribological properties can significantly enhance the overall performance, durability, and reliability of the composite, thereby extending the life span of components and reducing maintenance costs (Manu et al., 2021). Ceramic matrix composites while inherently resistant to wear and high temperatures, are still prone to issues such as brittleness and low fracture toughness. These challenges can lead to material degradation and eventual failure when exposed to severe tribological conditions. Therefore, the introduction of micro-sized fillers into CMCs has been a significant advancement in addressing these challenges. By reinforcing the ceramic matrix with micro-sized particles, researchers and engineers can tailor the tribological properties of the composite to better resist wear, reduce friction, and improve load-bearing capacity.

The role of micro-sized fillers in enhancing the tribological performance of CMCs is multifaceted. For instance, these fillers can create a more homogeneous microstructure within the composite, leading to a more uniform distribution of stress and reducing the initiation and propagation of cracks. This is particularly important in high-stress applications where uneven stress distribution could lead to catastrophic failure (Rashid et al., 2024b). Additionally, micro-sized fillers can act as barriers to dislocation movement within the ceramic matrix, further enhancing the material's resistance to wear and improving its overall toughness. Moreover, the specific nature of the micro-sized fillers used can also influence the frictional properties of the CMCs. Fillers such as micro-alumina, silicon carbide, and carbon nanotubes can reduce the coefficient of friction, leading to lower wear rates and improved material efficiency. This is especially crucial in high-temperature environments where traditional lubricants may not be effective or where the wear and tear on components can lead to significant operational challenges (Chan et al., 2021). By reducing friction, these fillers not only enhance the wear resistance of the composite but also contribute to more efficient energy usage in mechanical systems, ultimately improving the performance and longevity of the equipment.

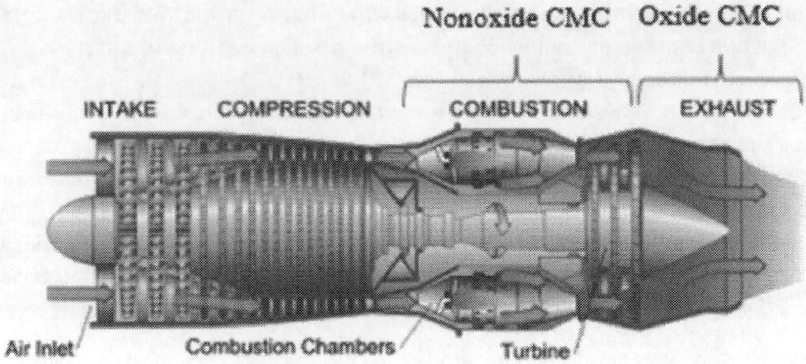

FIGURE 13.2 CMC component parts in the engine. (Reproduced with permission from MDPI, Karadimas and Salonitis, 2023.)

In high-performance industries like aerospace, automotive, and defense, where CMCs are frequently utilized, the importance of superior tribological performance cannot be underestimated. Components made from CMCs in these sectors must perform reliably over extended periods under extreme conditions, where failure is not an option. Therefore, the incorporation of micro-sized fillers to enhance the tribological performance of CMCs represents a vital advancement that meets the demanding requirements of these applications (Karadimas & Salonitis, 2023). Figure 13.2 illustrates the use of CMCs with oxide and non-oxide fillers in aero-engine applications. Tribological performance is a key consideration in the development and application of ceramic matrix composites. Using micro-sized fillers to enhance these properties has created new opportunities for improving the durability, reliability, and efficiency of CMCs in critical engineering applications (Dhanasekar et al., 2022). As research continues in this area, further advancements in the understanding and application of micro-sized fillers will likely lead to even greater improvements in the tribological performance of CMCs, solidifying their role as essential materials in modern engineering.

13.3 ROLE OF MICRO-SIZED FILLER REINFORCEMENTS IN CERAMIC MATRIX COMPOSITES

CMCs have attracted significant attention recently due to their superior properties, including high temperature resistance, excellent mechanical strength, and low density. These materials are used in various industries, including aerospace, automotive, and energy sectors. Incorporating micro-sized filler reinforcements plays a crucial role in enhancing CMC performance and properties. This section will discuss the various aspects of micro-sized filler reinforcements and their impact on CMCs.

13.3.1 Types of Micro-Sized Filler Reinforcements

Micro-sized fillers have gained considerable attention in ceramic matrix composites due to their ability to enhance the mechanical properties of these materials (Kong et al., 2023). Micro-sized filler reinforcements used in CMCs can be classified into several categories based on their composition and morphology:

1. Ceramic particles: These include oxide ceramics (e.g., alumina, zirconia, silica), non-oxide ceramics (e.g., silicon carbide, boron carbide, titanium carbide), and nitrides (e.g., silicon nitride, aluminum nitride).
2. Metallic particles: Common metallic fillers include titanium, aluminum, and nickel-based alloys.
3. Carbon-based fillers: These include carbon fibers, graphite particles, and carbon nanotubes.
4. Whiskers: Single-crystal, high-aspect-ratio reinforcements such as silicon carbide whiskers.

The choice of filler material depends on the desired properties and the intended application of the CMC.

13.3.2 Strengthening Mechanisms

The selection of filler material depends on the desired properties and intended application of the CMC:

a. Load transfer: Micro-sized fillers act as load-bearing elements, transferring stress from the ceramic matrix to the reinforcement phase, thereby improving tensile strength and overall mechanical performance (Falodun et al., 2019; Makarov et al., 2019; Yin et al. 2020).
b. Crack deflection: The presence of micro-sized particles can deflect propagating cracks, increasing the energy required for crack growth and improving fracture toughness. This mechanism is crucial for enhancing the durability and reliability of CMCs (Wang et al., 2021).
c. Grain refinement: Micro-sized fillers restrict grain growth during sintering, leading to a finer grain structure. This refinement results in improved mechanical properties such as increased hardness and strength (Shirvanimoghaddam et al., 2016).
d. Differences in thermal expansion coefficients between the matrix and filler generate residual stresses. These stresses contribute to strengthening and toughening mechanisms, enhancing the composite's resistance to thermal and mechanical loads (Li et al., 2019a).

Micro-sized filler reinforcements significantly enhance the mechanical properties of CMCs through multiple mechanisms. These include effective load transfer, crack deflection, grain refinement, and the beneficial effects of thermal expansion mismatch. Collectively, these mechanisms contribute to improved strength, toughness, and overall performance of CMCs, making them suitable for a wide range of advanced applications.

13.3.3 Influence on Mechanical Properties

The incorporation of micro-sized filler reinforcements in CMCs can significantly impact their mechanical properties. This synthesis explores the various effects of micro-sized fillers on the mechanical behavior of CMCs, drawing insights from multiple research studies.

The use of micro-sized powders such as Al and Ti as active fillers in CMCs can lead to a denser ceramic matrix, which improves mechanical properties by reducing interbundle pores (Stantschev et al., 2005). Micro-fillers like coir powder in polyester matrices enhance tensile and flexural properties, with the highest mechanical properties observed at maximum filler reinforcement (Babu et al., 2018). The addition of micro-fillers such as MoS_2, SiC, and Al_2O_3 in glass-carbon hybrid composites can impair tensile and flexural behavior but improve impact properties and thermal stability (Rudresh et al., 2019). Hybrid fillers in PA66/PP blends show enhanced tensile and flexural behavior, although micro-fillers slightly decrease strength due to loss of ductility (Lingesh et al., 2018). Combining micro- and nano-fillers in composites can lead to improved wear resistance and stiffness, although impact strength may suffer. The synergistic effect between different filler sizes plays a crucial role in enhancing overall performance. Incorporating micro-fillers such as talc and calcium carbonate in high-density polyethylene (HDPE) composites significantly improves tensile, flexural, and impact strengths, highlighting the importance of filler type, size, and dispersion (Alshammari et al., 2022). Microcrystalline cellulose (MC) as a filler in thermoplastic pea starch composites increases thermal stability and tensile strength, demonstrating good adhesion and improved mechanical properties. Generally, micro-sized fillers in CMCs enhance mechanical properties such as tensile and flexural strength, impact resistance, and thermal stability. The effectiveness of these improvements depends on the type, size, and dispersion of the fillers. Combining micro- and nanofillers can yield synergistic effects, further enhancing composite performance. Overall, micro-sized fillers play a crucial role in optimizing the mechanical properties of CMCs.

13.3.4 Challenges and Future Directions of Micro-Sized Fillers in CMCs

The use of micro-sized filler reinforcements in CMCs offers numerous benefits with enhanced physical properties. However, several challenges and areas for future research remain to be addressed to fully exploit these materials. While micro-sized filler reinforcements offer numerous benefits for CMCs, there are several challenges and areas for future research:

- Dispersion and Porosity Issues: Effective dispersion of fillers like graphene nanoplatelets (GNPs) and carbon nanotubes (CNTs) within the ceramic matrix is crucial to avoid agglomeration and ensure uniform properties. High porosity in the final product can negatively impact mechanical properties (Groetsch et al., 2022).
- Processing Techniques: Advanced processing techniques such as electrophoretic deposition (EPD) and resin transfer molding (RTM) are being explored to improve the infiltration of fillers and decrease the number of polymer impregnation and pyrolysis (PIP) cycles needed for densification (Zapata-Solvas et al., 2012).
- Mechanical Property Enhancement: The incorporation of micro-sized fillers like GNPs and CNTs can significantly enhance the mechanical properties of CMCs, including stiffness and toughness. However, achieving consistent improvements across different composite systems remains a challenge.

- Functional Property Improvements: Fillers such as GNPs and CNTs also provide CMCs with improved electrical and thermal conductivity, making them suitable for multifunctional applications. However, the control over these properties through precise filler dispersion is still an area needing further research.
- Microstructural Challenges: The presence of large interbundle pores and the need for multiple PIP cycles to achieve a dense matrix are significant challenges. Novel approaches using active fillers like Al and Ti powders are being investigated to address these issues.
- Sustainable and Biocompatible Fillers: The use of sustainable fillers like cellulose nanocrystals (CNCs) is being explored for their non-cytotoxic properties and potential in bio-electronics and tissue scaffolds. However, the fabrication of these composites at the microscale remains challenging (Zhou et al., 2019).

Micro-sized filler reinforcements are crucial in improving the properties and performance of ceramic matrix composites. By carefully selecting and optimizing filler type, size, and content, CMCs can be customized to meet the demanding requirements of various advanced applications. Ongoing research and development in this field will likely lead to further enhancements in CMC performance and expand their potential uses in high-technology industries.

13.4 THE TRIBOLOGICAL BEHAVIOR OF CERAMIC MATRIX COMPOSITES (CMCs) IN METALS AND POLYMERS

CMCs are advanced materials known for their high-temperature stability, wear resistance, and mechanical strength. Understanding their tribological behavior, especially when reinforced with materials like CNTs and graphene, is crucial for their application in metal and polymer matrices.

Ceramic matrix composites have attracted significant interest in various industrial applications, especially in aerospace and automotive sectors, due to their exceptional mechanical properties and resistance to wear and corrosion (Bai et al., 2021). These materials offer a unique combination of properties, including high strength, stiffness, and thermal stability, making them an attractive alternative to traditional metal alloys.

A key aspect of ceramic matrix composites is their tribological behavior, which refers to the study of friction, wear, and lubrication. CMCs generally exhibit lower friction coefficients compared to monolithic ceramics when sliding against metals. This is often due to the presence of reinforcing phases that can act as solid lubricants. At elevated temperatures, some CMCs may form tribo-films on the metal counterface, which can alter friction characteristics. The type and amount of reinforcement in CMCs can significantly influence friction. For example, carbon fiber–reinforced CMCs often show lower friction against metals due to the lubricating nature of carbon (Yin et al., 2020).

Researchers have investigated the tribological performance of these materials, particularly in comparison to metal systems, to understand their suitability for various applications. Aluminum-based metal matrix composites reinforced with ceramic particles, such as titanium diboride (TiB^-), have been extensively researched for their improved tribological characteristics (Pradhan et al., 2023). The addition of ceramic reinforcements improves the wear resistance and hardness of the composite material, making it more suitable for applications in harsh environments (Mohapatra et al., 2018). In contrast, metal systems, such as traditional metal alloys, often exhibit different tribological behavior. Metal systems tend to have lower wear resistance and are more susceptible to adhesive wear and galling, which can lead to premature failure in certain applications.

In contrast to metal systems, CMCs generally exhibit reduced friction coefficients. CMCs frequently demonstrate lower friction coefficients than metals, particularly at elevated temperatures. This phenomenon is attributed to the formation of lubricating tribofilms on the contact surfaces.

- Enhanced wear resistance: The hardness and thermal stability inherent to ceramic materials contribute to superior wear resistance compared to most metals, especially in high-temperature applications.
- Superior high-temperature performance: CMCs maintain their tribological properties at elevated temperatures, whereas many metal systems experience rapid degradation due to oxidation and softening. Reduced thermal expansion in CMC, the lower coefficient of thermal expansion in CMCs results in improved dimensional stability and diminished thermal stresses during operation.
- Increased chemical inertness: CMCs generally exhibit greater resistance to chemical attack and corrosion compared to metal systems, which is advantageous in aggressive environments. Factors influencing tribological behavior are load, sliding speed, temperature, and environment. Generally, higher loads generally increase both friction and wear rates in CMC-metal systems. The effect of sliding speed can vary depending on the specific CMC-metal pair and environmental conditions. Elevated temperatures can significantly alter tribological behavior, often leading to increased wear rates and changes in friction coefficients. Humidity, presence of lubricants, and atmospheric composition can all influence the tribological behavior of CMC-metal contacts.

It is important to recognize that CMCs also have certain limitations compared to metal systems:

- Increased brittleness: CMCs are generally more brittle than metals, which may lead to catastrophic failure under specific loading conditions.
- Complex manufacturing processes: The production of CMCs often involves more intricate and costly manufacturing processes compared to metal systems.
- Limited ductility: CMCs typically exhibit lower ductility than metals, which may affect their capacity to accommodate localized stresses and strains (Veeman et al., 2023).

The tribological behavior of CMCs in contact with metals presents several challenges and opportunities for future research (Rashid et al., 2024b).

- Interfacial reactions: At high temperatures, chemical reactions between the CMC and metal counterface can occur, leading to the formation of complex interfacial layers. These reactions can significantly alter the wear behavior and friction characteristics of the tribological system. Future research should focus on understanding and controlling these interfacial reactions to optimize wear resistance.

- Thermal mismatch: The difference in thermal expansion coefficients between CMCs and metals can lead to thermal stresses and potential damage at the interface. Investigating methods to mitigate these thermal stresses, such as the development of functionally graded materials or novel interface designs, is a promising area for future research.
- Transfer film formation: The formation and stability of transfer films on metal counterfaces when sliding against CMCs are not well understood. Future studies should explore the mechanisms of transfer film formation, their composition, and their role in reducing friction and wear.
- Environmental effects: The tribological behavior of CMC-metal contacts in extreme environments, such as high temperatures, corrosive atmospheres, or radiation, requires further investigation. Understanding the synergistic effects of these environmental factors on wear mechanisms is crucial for developing CMCs for advanced applications.
- Modeling and simulation: Developing accurate models to predict the tribological behavior of CMC-metal contacts remains a challenge. Future research should focus on multi-scale modeling approaches that incorporate microstructural features, interfacial phenomena, and environmental effects to improve predictive capabilities.
- Novel CMC compositions: Exploring new CMC compositions and reinforcement architectures tailored for specific metal counterfaces could lead to improved tribological performance. This includes investigating the potential of hybrid composites or nanocomposites for enhanced wear resistance.
- In-situ characterization: Developing advanced techniques for in-situ characterization of CMC-metal contacts during sliding would provide valuable insights into wear mechanisms and tribochemical reactions. This could involve the use of high-resolution microscopy, spectroscopy, or synchrotron-based techniques.
- Lubricant interactions: Understanding the interactions between lubricants and CMC-metal tribological systems is essential for optimizing performance in lubricated conditions. Future research should explore novel lubricant formulations and surface treatments to enhance the compatibility of CMCs with existing lubrication systems.
- Wear debris characterization: Detailed analysis of wear debris generated during CMC-metal contact can provide valuable information about wear mechanisms and material degradation. Developing standardized methods for collecting and analyzing wear debris from CMC-metal tribological systems is an important area for future research.
- Long-term performance: Investigating the long-term tribological behavior of CMC-metal contacts, including the effects of aging, fatigue, and environmental exposure, is crucial for predicting the service life of components in real-world applications.
- Surface texturing: Investigating the potential of surface texturing techniques to enhance the tribological performance of CMC-metal contacts

could lead to improved wear resistance and reduced friction. This includes investigating the effects of different texture patterns, sizes, and distributions on tribological behavior.

- Tribo-corrosion: The combined effects of wear and corrosion in CMC-metal tribological systems, particularly in aggressive environments, require further study. Understanding the mechanisms of tribo-corrosion and developing strategies to mitigate its effects is an important challenge for future research.

The tribological behavior of ceramic matrix in polymer composites is a crucial area of study due to the potential for enhanced wear resistance, reduced friction, and improved mechanical properties. The study of the tribological behavior of CMCs in polymer environments is crucial for applications where these materials come into contact under conditions of relative motion. This interaction is particularly relevant in fields such as automotive, aerospace, and industrial machinery where lightweight, high-performance materials are increasingly in demand.

Adding ceramic particles such as boron carbide (B_4C), tungsten carbide (WC), and zirconia (ZrO_2) to polymer matrices significantly improves wear resistance and hardness. For example, AZ31/ZrO_2 nanocomposites showed a 120% increase in hardness compared to the base metal, and hybrid composites with B_4C and ZrO_2 exhibited improved wear resistance and friction performance (Jalilvand & Mazaheri, 2020). Epoxy composites reinforced with ceramic carbides (e.g., silicon and boron carbides) demonstrated improved wear behavior, with enhancements increasing with the weight fraction of fillers (Kadhim et al., 2020). Titanium-based ultra-high-temperature ceramic particulates (e.g., titanium carbide, titanium boride, titanium nitride) in AZ91D alloy composites demonstrated superior tribological performance at both room and elevated temperatures, with titanium boride (ATB) composites performing the best (Surendran & Gnanavelbabu, 2021). The addition of nanoparticles such as alumina (Al_2O_3) and calcium silicate ($CaSiO_3$) to epoxy resins significantly enhances mechanical and tribological properties. Nanoparticles at low volume fractions (1–2 vol.% Al_2O_3) effectively improve the epoxy resin's performance, and further addition of microparticles yields synergistic effects (Wetzel et al., 2003). Composites containing CNTs exhibit self-lubricating characteristics, reducing the coefficient of friction and wear volume. CNTs improve the mechanical, tribological, and thermal properties of the composites (Ujah et al., 2023). The presence of an autolubricating layer in 3Y-TZP/Ta ceramic-metal composites against ultra-high molecular weight polyethylene (UHMWPE) results in lower volume loss and friction coefficient under dry conditions (Smirnov et al., 2018). Adding graphene-based fillers (e.g., graphene nanoplatelets, reduced graphene oxide sheets) to Si_3N_4/SiC composites significantly reduces the friction coefficient and enhances wear resistance. The best performance is observed with reduced graphene oxide sheets due to the nature of the lubricating tribofilm (Llorente et al., 2020). Carbon fiber–reinforced ceramic composites, particularly carbon fiber–reinforced carbon and silicon carbide dual-matrix composites (C/C-SiC), are recognized for their high-performance frictional properties. These composites are used in applications requiring high thermal stability and low chemical reactivity, such as brake discs and friction clutches (Duan et al., 2021). The use of nanoparticles and graphene-based fillers further optimizes these properties, while self-lubricating characteristics of certain composites offer additional benefits.

Generally, the friction coefficient between CMCs and polymers is lower compared to CMC-metal pairs. This is often due to the softer nature of polymers and their ability to form transfer films. During sliding contact, polymers tend to form a thin transfer film on the surface of the CMC. This film can act as a solid lubricant, reducing friction and wear. The type of ceramic matrix and reinforcement in the CMC can significantly influence the friction behavior. For example, carbon fiber–reinforced CMCs might exhibit lower friction against polymers due to the lubricating nature of carbon (Chikkegouda et al., 2022; Kumar & Sharath, 2022).

13.5 WEAR MECHANISM AND INFLUENCE OF FILLER SIZE

CMCs exhibit complex wear mechanisms due to their unique microstructure and properties. The primary wear mechanisms in CMCs include (Diaz et al., 2019):

1. Abrasive wear: This occurs when hard particles or asperities on the counterface material plough or cut into the CMC surface. In CMCs, abrasive wear can lead to fiber pullout, matrix cracking, and debris formation.
2. Adhesive wear: Adhesive wear involves material transfer between the contacting surfaces due to strong adhesive forces. In CMCs, this mechanism is less prevalent compared to metals but can occur under certain conditions, particularly at high temperatures.
3. Tribo-oxidation: At elevated temperatures, oxidation of the CMC surface can occur, leading to the formation of a tribofilm. While this film can act as a lubricant, reducing friction and wear, it may also contribute to material loss through spallation.
4. Fatigue wear: Cyclic loading during sliding contact can induce fatigue-related damage in CMCs, including matrix cracking and fiber-matrix debonding.
5. Erosive wear: In applications involving particle-laden flows, CMCs may experience erosive wear due to the impact of solid particles on their surface.

The wear behavior of CMCs is influenced by several factors (Khaliq, 2021):

1. Fiber-matrix interface: The strength and nature of this interface are crucial in determining the wear resistance of CMCs. A weak interface can lead to fiber pullout and increased wear, while a strong interface may result in brittle fracture.
2. Matrix composition and porosity: These factors affect the matrix's hardness, toughness, and wear resistance. Generally, higher porosity leads to increased wear rates.
3. Fiber properties: The mechanical and thermal properties of the reinforcing fibers influence the overall wear behavior of the composite.
4. Operating conditions: Temperature, load, sliding speed, and environmental factors significantly impact the wear mechanisms and rates in CMCs.
5. Counterface material: The properties of the counterface material, such as hardness and surface roughness, affect the dominant wear mechanisms and overall wear behavior of the CMC.

TABLE 13.1
Comparison of Different Ceramic Matrix Composites with Micro-Fillers

CMC Type	Micro-Filler	Tribological Properties	References
SiC-Matrix	SiC particles	High hardness, wear resistance, low friction coefficient	Chen et al. 2017
Al_2O_3-Matrix	Al_2O_3 particles	Good wear resistance, high temperature stability	Li et al. 2019a
ZrO_2-Matrix	ZrO_2 particles	High fracture toughness, thermal shock resistance	Liu et al. 2018
B_4C-Matrix	B_4C particles	High hardness, low density, good wear resistance	Wang et al. 2018
Si_3N_4-Matrix	Si_3N_4 particles	High strength, good thermal shock resistance	Wang et al., 2021
SiC-Matrix	Carbon nanotubes	Enhanced wear resistance, reduced friction	Zhang et al. 2020
Al_2O_3-Matrix	Graphite particles	Improved thermal conductivity, reduced friction	Zhao et al. 2020
ZrO_2-Matrix	Diamond carbide particles	Exceptional hardness, wear resistance	Zheng et al. 2016
Polyamide 6 (PA6)-Matrix	Ceramic microspheres	Significantly enhances wear resistance	Savaş et al. 2023
Epoxy Resin	Silicon carbide (SiC) nanoparticles	Improve wear resistance and provide optimal tribological performance	Venugopal et al. 2023
Polyetherimide (PEI) and Polyimide (PI)	Solid lubricant fillers	A very low coefficient of friction (CoF) of approximately 0.05 at room temperature, while PI-based composites have higher CoF and less effective tribological performance at elevated temperatures	Panin et al. 2022
Low-Density Polyethylene (LDPE)	Hybrid fillers (0.5 wt% Al_2O_3 + 0.5 wt% SiC)	LDPE improve tensile strength by 5.38% and hardness by 8.15%, with a 12.5% reduction in CoF under normal load	Alnaser et al. 2023
Polyimide (PI) and Polyetherimide (PEI)	Chopped carbon fibers (CFs) and solid lubricant fillers	Forms protective tribological layers, improving wear resistance and reducing friction	Panin et al. 2022

The comparison of different ceramic matrix composites with micro-fillers is given in Table 13.1.

Filler sizes also affect the wear mechanisms like micro-sized fillers, ranging from 1 to 100 μm, often enhance wear resistance by elevating their hardness and toughness. They can distribute stresses more evenly by serving as load-bearing components. Also, they could lessen abrasive wear by producing a smoother wear surface and increase fracture toughness by causing crack deflection. Nano-sized fillers (less than 100 nm) noticeably improve the matrix-filler interaction by increasing the surface-area-to-volume ratio. This can result in refining the grain, enhancing its mechanical qualities overall. It may lessen porosity, producing composites that are denser and more resistant to wear. It can improve the tribo-films' ability to protect. Mixing fillers of micro- and nano-sizes can have synergistic effects. While smaller particles occupy interstitial areas, larger particles offer load-bearing capacity.

Filler size significantly influences specific wear mechanisms in CMCs. For abrasive wear, larger fillers can obstruct abrasive particles, while nanofillers can enhance overall hardness. In adhesive wear, nanofillers may reduce available surface area for adhesion, whereas micro-fillers can create heterogeneous surfaces affecting adhesion patterns. Regarding fatigue wear, nanofillers strengthen grain boundaries and micro-fillers can deflect cracks, both improving wear resistance. Tribo-chemical wear may be accelerated by nanofillers due to increased reactivity, while larger fillers might provide a more stable surface. Several factors influence the effect of filler size, including volume fraction, distribution uniformity, interface properties, and matrix material type. Higher volume fractions of nanofillers can lead to agglomeration, and their even distribution is more challenging compared to micro-fillers. The strength of the filler-matrix interface is crucial, with nanofillers generally providing a larger interfacial area (Muraliraja et al., 2021). Lastly, the impact of filler size can vary depending on the ceramic matrix type, such as oxide versus nonoxide matrices.

13.6 CONCLUSION AND FUTURE PERSPECTIVES

In conclusion, the incorporation of micro-sized filler reinforcements has demonstrated significant potential in enhancing the tribological performance of CMCs. These fillers, when carefully selected and optimally dispersed, can effectively improve wear resistance, reduce friction coefficients, and increase the overall durability of CMCs under various operating conditions. The synergistic effects between the ceramic matrix and micro-sized reinforcements have been shown to create a more robust and resilient material structure, capable of withstanding higher loads and more severe environmental factors than traditional monolithic ceramics. Moreover, recent studies have highlighted the ability of micro-sized fillers to tailor specific wear mechanisms, such as abrasive, adhesive, and fatigue wear, offering unprecedented control over the tribological behavior of CMCs.

Looking towards the future, research in this field should focus on further optimizing the selection, sizing, and distribution of micro-sized fillers to maximize their beneficial impact on CMCs' tribological properties. This includes exploring novel filler materials, such as advanced ceramics or intermetallics, that could offer unique combinations of hardness, toughness, and thermal stability. Additionally, investigating the potential of hybrid reinforcement systems, combining micro- and nano-sized fillers, could lead to even more advanced CMCs with superior wear characteristics and multifunctional properties. The development of in-situ synthesis techniques for filler generation within the ceramic matrix presents another promising avenue for achieving more uniform distribution and stronger interfacial bonding.

Another critical area for future studies is the development of predictive models and simulation techniques to better understand and forecast the tribological behavior of these composite materials under diverse applications. This includes multiscale modeling approaches that can account for the complex interactions between fillers, matrix, and counterface materials across different length scales. Advanced characterization techniques, such as in-situ TEM and high-resolution 3D tomography, will be essential in validating these models and providing deeper insights into wear mechanisms at the micro- and nano-scale.

Furthermore, the integration of smart materials and self-healing mechanisms into micro-filler-reinforced CMCs could revolutionize their performance in extreme environments. This might involve the incorporation of phase-change materials for thermal management or the development of composites that can autonomously repair wear damage during operation.

As industries continue to demand materials with enhanced performance and longevity, particularly in aerospace, energy, and advanced manufacturing sectors, the ongoing refinement of micro-sized filler-reinforced CMCs will likely play a crucial role in meeting these evolving needs. The potential for these materials to enable new technologies, such as hypersonic vehicles or next-generation turbines, highlights the importance of continued research and development in this field. By addressing these challenges and opportunities, future advancements in micro-filler-reinforced CMCs promise to push the boundaries of material performance in tribological applications, opening new possibilities for engineering design and innovation.

REFERENCES

Alnaser, I., Fouly, A., Aijaz, M., Mohammed, J., Elsheniti, M., Ragab, S., & Abdo, H. (2023). Enhancing the tribo-mechanical performance of LDPE nanocomposites utilizing low loading fraction Al2O3/SiC hybrid nanostructured oxide fillers. *Inorganics*, 11(9), 354. https://doi.org/10.3390/inorganics11090354

Alshammari, B., Alenad, A., Al-Mubaddel, F., Alharbi, A., Al-shehri, A., Albalwi, H., Alsuabie, F., Fouad, H., & Mourad, A. (2022). Impact of hybrid fillers on the properties of high density polyethylene based composites. *Polymers*, 14(16), 3427. https://doi.org/10.3390/polym14163427

Babu, N., Muthukumaran, S., Arokiasamy, S., & Ramesh, T. (2018). Thermal and mechanical behavior of the coir powder filled polyester micro-composites. *Journal of Natural Fibers*, 17, 1058–1068. https://doi.org/10.1080/15440478.2018.1555503

Bai, Y., Wei, J., Lei, N., Li, J., Guo, Y., & Liu, M. (2021). Effect of VN and TiB2-TiCx reinforcement on wear behavior of Al 7075-based composites. *Materials*, 14(12), 3389.

Chan, J. X., Wong, J. F., Petrů, M., Hassan, A., Nirmal, U., Othman, N., & Ilyas, R. A. (2021). Effect of nanofillers on tribological properties of polymer nanocomposites: A review on recent development. *Polymers*, 13(17), 2867.

Chan, K. F., Zaid, M. H. M., Mamat, M. S., Liza, S., Tanemura, M., & Yaakob, Y. (2021). Recent developments in carbon nanotubes-reinforced ceramic matrix composites: A review on dispersion and densification techniques. *Crystals*, 11(5), 457.

Chandran, A. J., Rangappa, S. M., Suyambulingam, I., & Siengchin, S. (2024). Micro/nano fillers for value-added polymer composites: A comprehensive review. *Journal of Vinyl and Additive Technology*, 30(5), 1083–1123.

Chen, Y., Zhang, J., Liu, G., & Wang, Y. (2017). Tribological behavior of Si3N4 particle-reinforced Si3N4 matrix composites. *Composites Part B: Engineering*, 114, 171–178.

Chikkegouda, S. P., Gurudath, B., Sharath, B. N., Karthik, S., & Mahale, R. S. (2022). Mechanical and tribological characteristics of aluminium 2618 matrix composite reinforced with boron carbide. *Biointerface Research in Applied Chemistry*, 12(4), 4544–4556.

Dhanasekar, S., Ganesan, A. T., Rani, T. L., Vinjamuri, V. K., Nageswara Rao, M., Shankar, E., & Misganaw Golie, W. (2022). A comprehensive study of ceramic matrix composites for space applications. *Advances in Materials Science and Engineering*, 2022(1), 6160591.

Diaz, O. G., Luna, G. G., Liao, Z., & Axinte, D. (2019). The new challenges of machining Ceramic Matrix Composites (CMCs): Review of surface integrity. *International Journal of Machine Tools and Manufacture*, 139, 24–36.

Duan, J., Zhang, M., Chen, P., Li, Z., Pang, L., Xiao, P., & Li, Y. (2021). Tribological behavior and applications of carbon fiber reinforced ceramic composites as high-performance frictional materials. *Ceramics International*, 47(14), 19271–19281. https://doi.org/10.1016/J.CERAMINT.2021.02.187

Falodun, O., Obadele, B., Oke, S., Okoro, A., & Olubambi, P. (2019). Titanium-based matrix composites reinforced with particulate, microstructure, and mechanical properties using spark plasma sintering technique: a review. *The International Journal of Advanced Manufacturing Technology*, 102, 1689–1701. https://doi.org/10.1007/S00170-018-03281-X

Groetsch, A., Stelzl, S., Nagel, Y., Kochetkova, T., Scherrer, N., Ovsianikov, A., Michler, J., Pethö, L., Siqueira, G., Nyström, G., & Schwiedrzik, J. (2022). Microscale 3D Printing and Tuning of Cellulose Nanocrystals Reinforced Polymer Nanocomposites. *Small*, 19(3), e2202470. https://doi.org/10.1002/smll.202202470

Jalilvand, M., & Mazaheri, Y. (2020). Effect of mono and hybrid ceramic reinforcement particles on the tribological behavior of the AZ31 matrix surface composites developed by friction stir processing. *Ceramics International*, 46, 20345–20356. https://doi.org/10.1016/j.ceramint.2020.05.123

Kadhim, M., Kamal, H., & Majeed, A. (2020). The Tribological Behaviour of Epoxy Matrix Composite Reinforced by Ceramic Carbides. *Journal of Advanced Research in Fluid Mechanics and Thermal Sciences*, 70(2), 76–88. https://doi.org/10.37934/arfmts.70.2.7688

Karadimas, G., & Salonitis, K. (2023). Ceramic matrix composites for aero engine applications—A review. *Applied Sciences*, 13(5), 3017.

Khaliq, J. (2021). Ceramic matrix composites (CMCs). In S. M. Rangappa, J. Parameswaranpillai, S. Siengchin, & T. Ozbakkaloglu (Eds.), *Advances in Machining of Composite Materials: Conventional and Non-Conventional Processes* (pp. 285–309). Cham: Springer International Publishing.

Kong, U., Mohammad Rawi, N. F., & Tay, G. S. (2023). The potential applications of reinforced bioplastics in various industries: A review. *Polymers*, 15(10), 2399.

Kovalčíková, A., Tatarko, P., Sedlak, R., Medveď, D., Chlup, Z., Múdra, E., & Dusza, J. (2020). Mechanical and tribological properties of TiB2-SiC and TiB2-SiC-GNPs ceramic composites. *Journal of the European Ceramic Society*, 40(14), 4860–4871.

Kumar, S. V., & Sharath, P. C. (2022). Elastomer-based filler composites. In S. M. Rangappa, J. Parameswaranpillai, S. Siengchin, & T. Ozbakkaloglu (Eds.), *Elastomer Blends and Composites* (pp. 45–55). Elsevier.

Li, B., Qian, B., Xu, Y., Liu, Z., Zhang, J., & Xuan, F. (2019a). Additive manufacturing of ultrafine-grained austenitic stainless steel matrix composite via vanadium carbide reinforcement addition and selective laser melting: Formation mechanism and strengthening effect. *Materials Science and Engineering: A*, 745, 495–508. https://doi.org/10.1016/J.MSEA.2019.01.008

Li, X., Gao, Y., Pan, W., Zhong, Z., & Song, L. (2019b). Tribological properties of B4C particle-reinforced B4C matrix composites. *Journal of Materials Science*, 54(1), 74–83.

Li, Z., Cao, Y., He, J., & Wang, Y. (2018). Mechanical and tribological performances of C-SiC nanocomposites synthetized from polymer-derived ceramics sintered by spark plasma sintering. *Ceramics International*, 44(12), 14335–14341.

Lingesh, B., Ravikumar, B., Rudresh, B., & Madhu, D. (2018). Hybridization Effect of Micro Fillers on Mechanical, Thermal and Morphological Behavior of PA66/PP Blend Based Hybrid Thermoplastic Composites. *Transactions of the Indian Institute of Metals*, 71, 763–773. https://doi.org/10.1007/s12666-017-1209-4

Liu, Y., Zhu, J., Chen, J., & Li, Y. (2018). Tribological behavior of graphite particle-reinforced A12O3 matrix composites. *Wear*, 409, 178–186.

Llorente, J., Ramírez, C., & Belmonte, M. (2020). Two-step strategy for improving the tribological performance of Si3N4 ceramics: Controlled addition of SiC nanoparticles and graphene-based nanostructures. *Journal of The European Ceramic Society*, 40, 5298–5304. https://doi.org/10.1016/j.jeurceramsoc.2020.06.053

Makarov, P., Bakeev, R., Peryshkin, A., Zhukov, A., Ziatdinov, M., & Promakhov, V. (2019). Modelling of the deformation and destruction of a TiNi-TiB2 metal-ceramic composite fabricated by direct laser deposition. *Engineering Fracture Mechanics*, 222, 106712. https://doi.org/10.1016/j.engfracmech.2019.106712

Manu, B. R., Gupta, A., & Jayatissa, A. H. (2021). Tribological properties of 2D materials and composites—a review of recent advances. *Materials*, 14(7), 1630.

Markandan, K., Chin, J. K., & Tan, M. T. (2017). Recent progress in graphene based ceramic composites: a review. *Journal of Materials Research*, 32(1), 84–106.

Mohapatra, S. K., Maity, K., Bhuyan, S. K., & Satpathy, M. P. (2018, March). Mechanical and tribological characteristics of the AMC, prepared by P/M route along with thermo-mechanical treatment. *IOP Conference Series: Materials Science and Engineering*, 338(1), 012023.

Morales-Florez, V., & Dominguez-Rodriguez, A. (2022). Mechanical properties of ceramics reinforced with allotropic forms of carbon. *Progress in Materials Science*, 128, 100966.

Muraliraja, R., Tamilarasan, T. R., Udayakumar, S., & Arvinda Pandian, C. K. (2020). The effect of fillers on the tribological properties of composites. In *Tribological Applications of Composite Materials* (pp. 243–266). Singapore: Springer Singapore.

Nassar, A., Salem, M., El-Batanony, I., & Nassar, E. (2021). Improving wear resistance of epoxy/SiC composite using a modified apparatus. *Polymers and Polymer Composites*, 29(9_suppl), S389–S399.

Nie, C., Wang, H., & He, J. (2020). Evaluation of the effect of adding carbon nanotubes on the effective mechanical properties of ceramic particulate aluminum matrix composites. *Mechanics of Materials*, 142, 103276.

Ogunbiyi, O., Jamiru, T., Adekoya, G. J., & Rominiyi, A. L. (2024). Structure-property relationship and emerging applications of nano-and micro-sized fillers reinforced sialon composites: a review. *Journal of the Australian Ceramic Society*, 60, 1–32.

Panin, S., Luo, J., Buslovich, D., Alexenko, V., Berto, F., & Kornienko, L. (2022). Effect of Transfer Film on Tribological Properties of Anti-Friction PEI- and PI-Based Composites at Elevated Temperatures. *Polymers*, 14, 1215. https://doi.org/10.3390/polym14061215

Panin, S., Luo, J., Buslovich, D., Alexenko, V., Kornienko, L., Byakov, A., Paimushin, V., & Shugurov, A. (2022). Role of testing conditions in formation of tribological layers at line contacts of antifriction CF-reinforced PI- and PEI-based composites. *Molecules*, 27, 6376. https://doi.org/10.3390/molecules27196376

Pradhan, R. C., Das, D., Sahoo, B. P., Rout, C., Panda, A., & Barla, E. (2023). Machining investigations of squeeze cast TiB 2/Al 7075 composites through EDM: Regression modelling and weighted principal component analysis. *Archives of Metallurgy and Materials*, 68, 551–562.

Ran, Y., Sun, J., Kang, R., Dong, Z., & Bao, Y. (2024). Towards understanding the machining process in grinding of ceramic matrix composites: A review. *Composites Part B: Engineering*, 284, 11165.

Rashid, A. B., Haque, M., Islam, S. M., & Labib, K. R. U. (2024a). Nanotechnology-enhanced fiber-reinforced polymer composites: Recent advancements on processing techniques and applications. *Heliyon,* 10(2), e24692.

Rashid, A. B., Haque, M., Islam, S. M., Uddin Labib, K. R., & Chowdhury, P. (2024b). Breaking boundaries with ceramic matrix composites: A comprehensive overview of materials, manufacturing techniques, transformative applications, recent advancements, and future prospects. *Advances in Materials Science and Engineering*, 2024(1), 2112358.

Rudresh, B., Kumar, B., & Madhu, D. (2019). Combined effect of micro- and nano-fillers on mechanical, thermal, and morphological behavior of glass–carbon PA66/PTFE hybrid nano-composites. *Advanced Composites and Hybrid Materials*, 2, 176–188. https://doi.org/10.1007/s42114-019-00089-5

Savaş, S., Pecenek, H., Savaş, L., & Dogan, M. (2023). Mechanical and tribological performances of ceramic microsphere reinforced polyamide 6 composites. *Journal of Reinforced Plastics and Composites*, 43(15–16), 889–896. https://doi.org/10.1177/07316844231197257

Shirvanimoghaddam, K., Khayyam, H., Abdizadeh, H., Akbari, M., Pakseresht, A., Abdi, F., Abbasi, A., & Naebe, M. (2016). Effect of B4C, TiB2 and ZrSiO4 ceramic particles on mechanical properties of aluminium matrix composites: Experimental investigation and predictive modelling. *Ceramics International*, 42, 6206–6220. https://doi.org/10.1016/J.CERAMINT.2015.12.181

Shrivastava, S., Rajak, D. K., Joshi, T., Singh, D. K., & Mondal, D. P. (2024). Ceramic matrix composites: Classifications, manufacturing, properties, and applications. *Ceramics*, 7(2), 652–679.

Smirnov, A., Smirnov, A., Volosova, M., Peretyagin, P., & Bartolomé, J. (2018). Tribological behaviour of a 3Y-TZP/Ta ceramic-metal biocomposite against ultrahigh molecular weight polyethylene (UHMWPE). *Ceramics International*, 44, 1404–1410. https://doi.org/10.1016/J.CERAMINT.2017.09.186

Stantschev, G., Friess, M., Kochendörfer, R., & Krenkel, W. (2005). Long fibre reinforced ceramics with active fillers and a modified intra-matrix bond based on the LPI process. *Journal of The European Ceramic Society*, 25, 205–209. https://doi.org/10.1016/J.JEURCERAMSOC.2004.08.014

Surendran, K., & Gnanavelbabu, A. (2021). Tribological behaviour of AZ91D/ultra-high-temperature ceramic composites at room and elevated temperatures. *Proceedings of the Institution of Mechanical Engineers, Part J: Journal of Engineering Tribology*, 236, 1855–1870. https://doi.org/10.1177/13506501211010035

Tulshibagwale, N., Brodnik, N., Muir, C., Hilmas, A., Kiser, J. D., Smith, C. E., … Daly, S. (2024). Acoustic emission in ceramic matrix composites. *Applied Mechanics Reviews*, 76(4), 040801.

Ujah, C., Kallon, D., & Aigbodion, V. (2023). Tribological properties of CNTs-reinforced nano composite materials. *Lubricants*, 11(3), 95. https://doi.org/10.3390/lubricants11030095

Veeman, D., Katiyar, J. K., & Ruggiero, A. (2023). Tribo-mechanical performance of brake composite material: a comprehensive review. *Tribology-Materials, Surfaces & Interfaces*, 17(4), 271–294.

Venugopal, C., Prasad, K., & Reddy, Y. (2023). Effect of nanoparticles on tribological behavior of functionality graded FRP nanocomposite. *Journal of Advancements in Material Engineering*, 11(3), 95. https://doi.org/10.46610/joame.2023.v08i03.002

Wang, N., Wang, S., Liu, G., Zhang, Y., Zhao, K., Ren, B., & Wang, Y. (2021). In situ high-entropy solid solution and ceramic particles co-reinforced Ni-based composites with outstanding strength-ductility synergy and good pitting resistance. *Materials Science and Engineering: A*, 806, 140842. https://doi.org/10.1016/J.MSEA.2021.140842

Wang, Y., Li, J., Deng, X., & Wang, X. (2018). Tribological properties of Al2O3 particle-reinforced Al2O3 matrix composites. *Ceramics International*, 44(15), 18956–18961.

Wang, Z., Yang, Y., Liu, X., & Zhang, G. (2021). Tribological properties of carbon nanotube-reinforced SiC matrix composites. *Carbon*, 174, 401–410.

Wetzel, B., Haupert, F., & Zhang, M. (2003). Epoxy nanocomposites with high mechanical and tribological performance. *Composites Science and Technology*, 63, 2055–2067. https://doi.org/10.1016/S0266-3538(03)00115-5

Wyatt, B. C., Nemani, S. K., & Anasori, B. (2021). 2D transition metal carbides (MXenes) in metal and ceramic matrix composites. *Nano Convergence*, 8(1), 16.

Yin, J., Zhou, P., Liang, H., Yao, D., Xia, Y., Zuo, K., & Zeng, Y. (2020). Microstructure and mechanical properties of Cu matrix composites reinforced by TiB2/TiN ceramic reinforcements. *Acta Metallurgica Sinica (English Letters)*, 33, 1609–1617. https://doi.org/10.1007/s40195-020-01100-5

Zapata-Solvas, E., Gómez-García, D., & Domínguez-Rodríguez, A. (2012). Towards physical properties tailoring of carbon nanotubes-reinforced ceramic matrix composites. *Journal of The European Ceramic Society*, 32, 3001–3020. https://doi.org/10.1016/J.JEURCERAMSOC.2012.04.018

Zhang, H., Liu, Y., Chen, K., & Wang, J. (2020). Tribological behavior of ZrO2 particle-reinforced ZrO2 matrix composites. *Materials Science and Engineering: A*, 782, 139315.

Zhao, X., Liu, Y., Wen, Q., & Wang, Y. (2020). Tribological properties of diamond particle-reinforced ZrO2 matrix composites. *Diamond and Related Materials*, 103, 107635.

Zheng, L., Zhou, X., Yan, Y., & Chen, W. (2016). Tribological behavior of SiC particle-reinforced SiC matrix composites. *Wear*, 352–353, 669–677.

Zhou, B., Fan, S., Fan, Y., Zheng, Q., Zhang, X., Jiang, W., & Wang, L. (2019). Recent progress in ceramic matrix composites reinforced with graphene nanoplatelets. *Rare Metals*, 39, 513–528. https://doi.org/10.1007/s12598-019-01306-2

Zhou, K., Xiao, G., & Huang, Y. (2024). Understanding machinability improvements and removal mechanism of ceramic matrix composites during laser-ablating assisted grinding. *Wear*, 538, 205199.

14 Influence of Nano-Sized Filler Reinforcement in Tribological Performance of Ceramic Matrix Composites

Md. Farid Hossain, Manoj Kumar, Abhijit Bhowmik, Raman Kumar, Sehijpal Singh, and Abayomi Adewale Akinwande

14.1 INTRODUCTION

Ceramic matrix composites (CMCs) have gained widespread recognition for their outstanding mechanical properties, including high strength, excellent thermal stability, and resistance to wear and corrosion [Chatterjee et al., 2023]. These attributes make them highly suitable for applications in extreme environments such as aerospace, automotive, and energy sectors. Despite these advantages, improving the tribological performance of CMCs—encompassing friction, wear, and lubrication characteristics—remains crucial for ensuring their long-term reliability and efficiency [Erappa Rajj et al., 2024]. Addressing challenges related to wear resistance and friction reduction is essential to expand the practical utility of CMCs. Recent advances in material science have highlighted the potential of nano-sized fillers to significantly enhance the tribological properties of CMCs. Fillers such as carbon nanotubes (CNTs), graphene, silicon carbide, and other nanoparticles exhibit unique characteristics, including high surface area, superior hardness, and excellent load-bearing capabilities [Dass et al., 2017]. These features enable synergistic interactions with the ceramic matrix, resulting in improvements in mechanical strength, thermal conductivity, and tribological performance. Incorporating nano-sized fillers into CMCs has demonstrated promising results, including reduced wear rates and lower friction coefficients [Faisal and Kumar, 2018]. These enhancements are attributed to the uniform dispersion of fillers, their ability to form protective tribo-films, and their role in enhancing load transfer at the filler-matrix interface. Additionally, the nanoscale dimensions of these fillers help mitigate defects within the matrix, thereby improving structural integrity under dynamic loading conditions [Gangwar and Pathak, 2021]. This research investigates the impact of nano-sized fillers on the tribological performance of CMCs, focusing on factors such as filler type, size,

DOI: 10.1201/9781003530909-18

concentration, and distribution. By examining the resulting microstructural and mechanical modifications, this study aims to deepen the understanding of the mechanisms driving tribological improvements [Jayanthi et al., 2022]. The insights gained will contribute to the development of next-generation CMCs with tailored properties for demanding applications.

14.2 OVERVIEW OF CERAMIC MATRIX COMPOSITES

CMCs are advanced materials developed to overcome the brittleness and low fracture toughness of traditional ceramics. By incorporating reinforcing elements like fibers or particles into the ceramic matrix, these composites achieve superior mechanical strength, thermal stability, and wear resistance, making them suitable for challenging applications in aerospace, automotive, and industrial fields [Kundie et al., 2018]. Despite these advantages, the tribological performance of CMCs—defined by friction, wear, and lubrication characteristics—remains a crucial factor affecting their durability and operational efficiency [Mohammadi et al., 2020]. The use of nano-sized fillers as reinforcements has emerged as an effective strategy to enhance the tribological properties of CMCs. Nanofillers such as carbon nanotubes, graphene, and silicon carbide nanoparticles possess exceptional properties, including high aspect ratios, remarkable hardness, and excellent thermal conductivity [Naskar et al., 2009]. When uniformly distributed within the ceramic matrix, these fillers improve load transfer, minimize microstructural defects, and facilitate the formation of protective tribo-films during sliding contact. Studies have shown that incorporating nanofillers significantly reduces the coefficient of friction and wear rates, thereby improving the reliability of CMCs under dynamic and high-temperature conditions [Padhan et al., 2020]. Additionally, the interaction between the matrix and the fillers plays a vital role in determining key factors such as interfacial bonding, filler dispersion, and matrix densification, all of which directly influence tribological performance. These advancements underline the potential of nano-sized fillers to transform the functionality and longevity of ceramic matrix composites in demanding environments [Richard et al., 2017].

14.3 NANO-SIZED FILLERS: CHARACTERISTICS AND SELECTION

Nano-sized fillers are materials with at least one dimension ranging from 1 to 100 nanometers. Their nanoscale dimensions and high aspect ratios enable substantial alterations to the microstructure and properties of the base material. In CMCs, commonly utilized nanofillers include the following.

14.3.1 Carbon-Based Materials

Carbon-based nanofillers, such as CNTs and graphene, are extensively utilized to reinforce CMCs due to their outstanding mechanical, thermal, and tribological characteristics. CNTs are known for their exceptional strength-to-weight ratio, high flexibility, and excellent electrical and thermal conductivity. Graphene, with its two-dimensional structure and large surface area, offers superior mechanical strength and self-lubricating properties. Both materials contribute to reducing friction and wear by forming protective tribo-films during sliding contact [Arunachalam and Saravanan, 2023]. Furthermore, their nanoscale dimensions facilitate uniform dispersion within the matrix, enhancing load transfer and minimizing microstructural defects. The combination of lightweight properties and multifunctional capabilities makes carbon-based fillers particularly effective for improving the tribological performance of composites in demanding applications.

14.3.2 Oxides

Oxide-based nano-fillers, including aluminum oxide (Al_2O_3), zirconium oxide (ZrO_2), and titanium dioxide (TiO_2), are commonly used in CMCs due to their remarkable hardness, chemical stability, and thermal resistance. These fillers enhance the wear resistance and high-temperature performance of CMCs, making them ideal for use in extreme conditions. Aluminum oxide is known for its outstanding hardness and abrasion resistance, while zirconium oxide enhances toughness through its transformation toughening mechanism. Al_2O_3 and TiO_2 particles' scanning electron microscopy–energy-dispersive X-ray spectroscopy (SEM–EDS) data in Figure 14.1 demonstrate that the particle diameters vary. The uniform distribution of Al_2O_3 particles, each with an average size of greater than 50 mm, is shown as individual particles in Figure 14.1(a). Compared to TiO_2 particles, they are substantially bigger [Sosiati et al., 2024].

TiO_2 stands out for its high refractive index and photocatalytic properties. The strong compatibility of oxide fillers with various ceramic matrices ensures uniform dispersion and robust interfacial bonding, resulting in improved mechanical strength and superior tribological performance in demanding applications.

14.3.3 Nitrides and Carbides

Nitrides and carbides are widely recognized as effective nanofillers for enhancing the performance of CMCs due to their superior hardness, thermal stability, and wear resistance. Silicon carbide (SiC) and titanium carbide (TiC) are commonly employed

FIGURE 14.1 SEM micrographs-EDS spectrum of (a) Al_2O_3, and (b) TiO_2 (Sosiati et al., 2014).

for their ability to reinforce the matrix, reducing wear rates and boosting tribological performance in abrasive and high-temperature environments [Xiao et al., 2024]. Similarly, nitrides such as silicon nitride (Si_3N_4) and boron nitride (BN) provide excellent thermal conductivity, chemical stability, and low friction coefficients [Dash et al., 2018]. These materials establish strong interfacial bonds with the ceramic matrix, facilitating efficient load transfer and minimizing structural defects. The choice of filler is typically guided by specific application requirements, including temperature resilience, mechanical performance, and cost considerations.

14.3.4 Metallic Nanoparticles

Metallic nanoparticles are highly regarded for their distinctive properties, including a high surface-area-to-volume ratio; customizable optical, electrical, and magnetic features; and exceptional catalytic activity. Their size, shape, and surface chemistry play a critical role in determining their behavior and suitability for various applications, such as medicine, electronics, and environmental remediation. For example, gold nanoparticles are valued for their biocompatibility and unique optical properties, while silver nanoparticles are known for their strong antimicrobial effects [Friedrich et al., 2005].

The choice of metallic nanoparticles is guided by the specific application, taking into account factors such as particle size, shape, surface charge, stability, and compatibility with the surrounding medium. Synthesis techniques, including chemical reduction, green synthesis, and physical methods, influence these attributes and impact cost-effectiveness and scalability [Ghahremani et al., 2022]. By fine-tuning these parameters, metallic nanoparticles can be tailored to meet the demands of specific applications, offering significant potential for advancements in diverse scientific and industrial fields.

14.4 TRIBOLOGICAL PERFORMANCE ENHANCEMENT

Tribology, which focuses on the study of friction, wear, and lubrication, plays a vital role in determining the performance of CMCs in engineering applications. Nano-sized fillers enhance tribological properties through several mechanisms.

14.4.1 Reduction in Friction

The addition of nano-sized fillers substantially improves the tribological performance of CMCs by reducing friction. Fillers such as CNTs, graphene, and boron nitride possess exceptional self-lubricating properties, enabling the formation of protective tribo-films at the contact interface during sliding [Kala et al., 2014]. These films minimize direct surface-to-surface interactions, thereby decreasing frictional forces. Furthermore, the uniform distribution of nanofillers aids in even load distribution and prevents localized stress concentrations, further lowering the coefficient of friction. The nanoscale size and high aspect ratios of these fillers facilitate smoother surface interactions, reducing wear and energy loss [Thirupathy et al., 2024]. These enhancements contribute to extended service life and improved reliability of CMCs in demanding and extreme operating environments.

14.4.2 Improved Wear Resistance

Enhancing wear resistance is a critical factor in improving the tribological performance of CMCs. Nano-sized fillers, such as silicon carbide (SiC) and graphene, significantly reduce wear rates by reinforcing the matrix and increasing its load-bearing capacity [Sharma et al., 2022]. These fillers form protective tribo-films on the surface during sliding contact, minimizing direct matrix wear and material loss. Moreover, their uniform dispersion throughout the matrix helps prevent crack propagation and improves structural integrity under dynamic loads. With their exceptional hardness and thermal stability, nanofillers ensure superior performance in abrasive and high-temperature conditions, enhancing the durability and reliability of CMCs in demanding tribological applications.

14.4.3 Enhanced Load-Bearing Capacity

Incorporating nano-sized fillers into CMCs significantly enhances their load-bearing capacity, a critical factor in improving tribological performance. Fillers such as SiC and graphene increase the composite's resilience to high contact pressures and dynamic loads during sliding or abrasive interactions [Singh and Yadav, 2017]. By evenly distributing stress, these fillers reduce localized strain and inhibit crack propagation, thereby reinforcing the matrix. Their high hardness and stiffness also aid in forming protective tribo-films, which minimize surface wear and friction. This improved load-bearing capacity enhances the durability and reliability of CMCs, making them well suited for demanding applications in aerospace, automotive, and industrial environments.

14.4.4 Thermal Stability

Thermal stability is essential for enhancing the tribological performance of CMCs in high-temperature environments. Nano-sized fillers like SiC, BN, and Al_2O_3 play a pivotal role in preserving mechanical integrity and minimizing wear at elevated temperatures. These fillers offer excellent thermal conductivity, aiding in the efficient dissipation of heat generated during frictional contact and preventing thermal degradation of the matrix [Singhal et al., 2024]. Additionally, their ability to form stable tribo-films under high-temperature conditions helps reduce friction and wear. The combined benefits of thermal stability and wear resistance provided by nanofillers ensure the reliable performance of CMCs in demanding applications such as aerospace, automotive, and energy sectors.

14.4.5 Self-Lubricating Behavior

Improving the tribological performance of self-lubricating materials involves optimizing their ability to minimize friction and wear across diverse operating conditions. This is often achieved by integrating solid lubricants such as graphite, molybdenum disulfide (MoS_2), or polytetrafluoroethylene (PTFE) into the material matrix. These lubricants form a protective film at the contact interface, reducing direct surface-to-surface interaction and enhancing durability [Babic et al., 2020].

Advanced approaches, including nanotechnology, incorporate nanoparticles like graphene or CNTs to further boost lubrication by decreasing particle size and increasing surface area. Additionally, refining the material's microstructure, hardness, and thermal stability enhances its load-bearing capacity and wear resistance [Nilagiri Balasubramanian and Ramesh, 2018]. Optimizing operational parameters such as load, speed, and temperature ensures consistent self-lubrication, enabling superior performance in demanding applications like aerospace, automotive, and industrial machinery.

14.5 PROCESSING TECHNIQUES

Effectively incorporating nano-sized fillers into ceramic matrices necessitates advanced processing techniques to achieve uniform dispersion and robust interfacial bonding. Key methods include the following.

14.5.1 Spark Plasma Sintering (SPS)

Spark plasma sintering (SPS) is a cutting-edge processing technique commonly utilized for fabricating CMCs reinforced with nano-sized fillers. This method combines uniaxial pressure and pulsed direct current to achieve rapid sintering, enabling densification at lower temperatures and shorter durations compared to traditional sintering processes [Azadi et al., 2018]. The localized heating mechanism in SPS enhances diffusion, limits grain growth, and maintains the nanostructure of the fillers, making it particularly suitable for high-performance applications.

In the context of CMCs, SPS ensures the uniform distribution of nano-sized fillers such as graphene, carbon nanotubes, and silicon carbide within the ceramic matrix. The technique fosters robust interfacial bonding between the fillers and matrix, leading to superior mechanical strength, thermal stability, and wear resistance. Moreover, SPS enables the production of dense, defect-free composites with customizable properties [Sosiati et al., 2024]. Its efficiency and versatility in processing various materials make SPS a preferred method for developing advanced composites for aerospace, automotive, and industrial applications.

14.5.2 Sol-Gel Method

The sol-gel method is a versatile and commonly employed technique for fabricating CMCs with nano-sized filler reinforcements. This process involves the hydrolysis and condensation of metal alkoxides or other precursors to create a colloidal sol, which subsequently forms a gel-like network. Nano-sized fillers such as SiC, titanium nitride (TiN), or graphene can be effectively dispersed within the sol, ensuring uniform distribution throughout the ceramic matrix.

Using a specially created setup, the above prepared sols, ZY, AZ, AS, and A, were infiltrated into a high alumina fiber preform (M/s Thermal Ceramics, USA) with a fiber content of 30%, length of 200–250 mm, and diameter of 3–7 mm. The preform's dimensions were 125 mm in diameter by 25 mm in thickness (Figure 14.2) [Naskar et al., 2009].

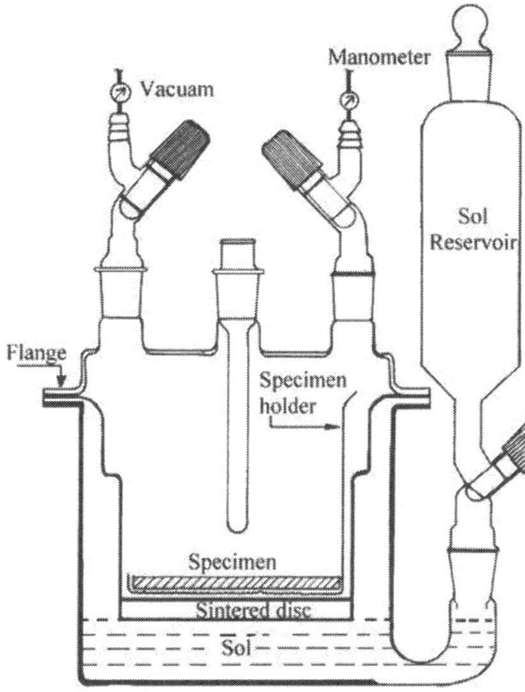

FIGURE 14.2 Schematic of the sol infiltration set up for the fabrication of near-netshape ceramic fiber–reinforced ceramic matrix composite (Naskar et al., 2009).

This technique offers precise control over key material characteristics, including microstructure, particle size, and porosity, allowing for the development of composites with customized properties. The sol-gel method is cost-effective, operates at relatively low processing temperatures, and is well suited for fabricating components with complex geometries, making it an attractive choice for advanced composite manufacturing.

14.5.3 Mechanical Milling

Mechanical milling is a widely used technique for synthesizing and dispersing nano-sized fillers within CMCs. This process employs high-energy ball mills to grind and mix materials into fine powders, ensuring the uniform distribution of fillers within the ceramic matrix [Surakasi et al., 2022]. The high-energy collisions between the milling media and the materials lead to particle size reduction, surface activation, and, in some cases, alloying or phase transformations, depending on the materials involved.

Mechanical milling is favored for its simplicity, scalability, and cost-effectiveness, making it ideal for producing large quantities of composite materials. It is particularly effective in breaking up agglomerates of nano-fillers such as carbon nanotubes, silicon carbide, or boron nitride, enabling their even dispersion throughout the matrix

[Wetzel et al., 2001]. To achieve optimal results, careful control of parameters such as milling time, speed, ball-to-powder ratio, and atmospheric conditions (e.g., inert or reactive gases) is essential to minimize contamination and prevent excessive damage to the fillers.

This technique is instrumental in enhancing interfacial bonding between the matrix and fillers, which is critical for improving tribological performance. Additionally, mechanical milling can induce microstructural modifications that enhance the sintering behavior and densification of composites during subsequent processes like hot pressing or SPS [Anand et al., 2016]. Despite its benefits, challenges such as potential contamination and the need for post-milling treatments to refine particle surfaces must be addressed for optimal outcomes.

14.6 CASE STUDIES

14.6.1 Graphene-Reinforced SiC Composites

Graphene-reinforced SiC composites exhibit significant enhancements in tribological performance, attributed to graphene's exceptional properties, including high strength, thermal conductivity, and self-lubricating capabilities. Research shows that adding graphene improves wear resistance and reduces friction by forming protective tribo-films at the sliding interface. SiC ceramics were positioned between a 100-μm-thick composite filler with varying GNP concentrations. Figure 14.2 shows the schematics for applying braze filler powder and putting the brazing structure together. The technique shown in Figure 14.3(a) was used to apply the AgCuTi/GNPs powder to the SiC ceramic surface, ensuring a 100 μm coating powder thickness. The other SiC ceramic was then positioned on top of the prepared sample, creating the sandwich-like structure seen in Figure 14.3(b) [Song et al., 2019].

For example, graphene-SiC composites have achieved up to 50% lower friction coefficients and substantially reduced wear rates under high-temperature conditions compared to unreinforced SiC. Techniques such as SPS enable the uniform dispersion of graphene within the SiC matrix, facilitating effective load transfer and enhanced mechanical properties. These attributes make graphene-SiC composites well suited for demanding applications in extreme environments.

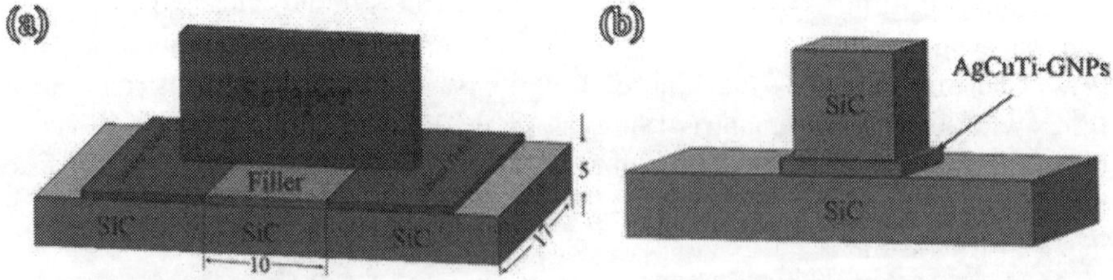

FIGURE 14.3 Schematic diagrams of (a) spreading AgCuTi/GNPs powder on SiC surface, (b) brazing assembly (Song et al., 2019).

14.6.2 Nano-Titania in Alumina Composites

Nano-titania (TiO_2) has been widely investigated as a reinforcing filler in Al_2O_3 composites for its ability to enhance mechanical and tribological performance. Studies indicate that the inclusion of nano-titania increases hardness and wear resistance while reducing friction, primarily through the formation of protective tribo-films during sliding contact. TiO_2 also aids in densification and grain refinement, contributing to the composite's structural integrity [Maddaiah et al., 2024]. Furthermore, its presence enhances thermal stability and oxidation resistance, making it suitable for high-temperature applications. The combination of TiO_2 and Al_2O_3 exhibits significant potential for demanding applications in aerospace, automotive, and cutting tools, where durability and superior performance are essential.

14.6.3 CNT-Reinforced Zirconia

CNT-reinforced zirconia composites exhibit notable enhancements in mechanical and tribological properties. The addition of CNTs improves fracture toughness, wear resistance, and thermal conductivity, making these composites ideal for high-performance applications. Techniques like mechanical milling and sonication ensure uniform dispersion of CNTs within the zirconia matrix, facilitating effective load transfer and reducing defects [Muraliraja et al., 2021]. Research indicates that CNTs lower the friction coefficient by forming self-lubricating tribo-films during sliding contact. Furthermore, their thermal stability enhances zirconia's resistance to high-temperature wear, positioning these composites as promising candidates for aerospace, biomedical, and industrial applications demanding exceptional tribological performance.

14.7 CHALLENGES AND FUTURE DIRECTIONS

Although nano-sized fillers provide significant advantages, several challenges persist.

14.7.1 Agglomeration

The agglomeration of nano-sized fillers, such as CNTs and ceramic nanoparticles, presents a significant challenge in fabricating CMCs. Inadequate dispersion of fillers within the ceramic matrix can result in stress concentrations, structural defects, and diminished composite performance [Shekhawat et al., 2023]. This clustering, driven by van der Waals forces, reduces the fillers' effectiveness in reinforcing the matrix.

To address this issue, future research should prioritize advanced dispersion methods, including ultrasonic processing, high-energy ball milling, and surface functionalization of fillers, to improve their distribution and compatibility with the matrix [Singla and Singh, 2024]. Additionally, optimizing processing conditions and developing innovative stabilizers or dispersants can ensure uniform filler dispersion, thereby enhancing the mechanical and tribological properties of CMCs.

14.7.2 Interface Compatibility

Ensuring interface compatibility between nano-sized fillers and ceramic matrices is a critical challenge in the development of high-performance CMCs. Insufficient interfacial bonding can hinder load transfer, cause filler agglomeration, and lead to structural defects, ultimately degrading the mechanical and tribological properties of the composites. Surface functionalization techniques, such as chemical treatments or coatings, are often required for nanofillers like CNTs and graphene to enhance adhesion with the ceramic matrix [Pun and Siddhartha, 2018].

Future efforts should focus on refining functionalization methods to improve compatibility while preserving the fillers' intrinsic properties. Investigating novel filler-matrix combinations and hybrid reinforcement strategies can further address these compatibility challenges. Additionally, employing advanced characterization tools and computational modeling will be crucial in understanding interfacial interactions, paving the way for the design of durable composites suited for demanding applications [Rodriguez-Suarez et al., 2012].

14.7.3 Cost and Scalability

The high cost of nano-sized fillers, such as CNTs and advanced carbides, presents a major obstacle to the broader implementation of reinforced CMCs. Achieving consistent scalability in production while ensuring uniform filler dispersion and maintaining matrix integrity adds to the complexity [Amir and Mamat, 2011]. Future efforts should prioritize the development of cost-efficient filler synthesis methods, refinement of manufacturing processes, and investigation of alternative, more affordable fillers. These approaches aim to strike a balance between performance and cost-effectiveness, enabling the industrial-scale adoption of CMCs.

14.8 CONCLUSION

The incorporation of nano-sized fillers into CMCs marks a significant breakthrough in material science, driving substantial improvements in tribological performance. By utilizing the exceptional properties of nano-fillers such as CNTs, graphene, and SiC, researchers have effectively tackled key challenges, including enhancing wear resistance, reducing friction, and improving thermal stability. These fillers facilitate the formation of protective tribo-films, minimize surface interactions, and boost the load-bearing capacity of the composites, making them ideal for use in extreme operating conditions.

Advanced processing techniques like SPS, sol-gel methods, and mechanical milling are critical in ensuring uniform dispersion and robust bonding between fillers and the ceramic matrix. Such precision in processing enables composites to maintain superior mechanical integrity even in dynamic and high-temperature environments. Despite these advancements, challenges such as filler agglomeration, interfacial compatibility, and cost limitations remain. Addressing these challenges requires continuous research into innovative surface functionalization, optimized dispersion strategies, and economical filler synthesis methods.

Examples such as graphene-reinforced SiC and CNT-reinforced zirconia highlight the potential of nano-sized fillers in enhancing the mechanical and tribological properties of CMCs. These advancements are particularly promising for aerospace, automotive, biomedical, and energy applications, where materials must withstand rigorous operational demands.

Looking ahead, collaboration among materials scientists, engineers, and industry stakeholders is essential to overcome existing challenges and enable large-scale production. By resolving these limitations and maximizing the capabilities of nano-sized fillers, the development of next-generation CMCs with tailored properties can be accelerated. Such composites have the potential to revolutionize high-performance applications, offering greater durability, efficiency, and reliability in demanding environments.

REFERENCES

Amir, A., & Mamat, O. (2011, November). A review of wear mechanisms of ceramic-reinforced metal-matrix composite: Case study of nano silica reinforced iron. In Kazuo Umemura et al. (Eds.), *Defect and diffusion forum* (Vol. 319, pp. 77–84). Trans Tech Publications Ltd.

Anand, A., Ul Haq, M. I., Raina, A., Vohra, K., Asgar, A., Jha, N., ... & Singh, Y. (2016). Tribological and mechanical aspects of zirconia-reinforced aluminum metal matrix composites. *Materials Focus*, *5*(6), 489–495.

Arunachalam, S. J., & Saravanan, R. (2023). Study on filler reinforcement in polymer matrix composites–A review. *Materials Today: Proceedings*.

Azadi, M., Zolfaghari, M., Rezanezhad, S., & Azadi, M. (2018). Effects of SiO 2 nano-particles on tribological and mechanical properties of aluminum matrix composites by different dispersion methods. *Applied Physics A*, *124*, 1–13.

Babic, M., Stojanovic, B., Dzunic, D., & Pantic, M. (2020). Micro/nanoscale structural, mechanical and tribological characterization of ZA-27/SiC nanocomposites. *Journal of Composite Materials*, *54*(16), 2113–2129.

Chatterjee, A., Roy, P., Birru, A. K., & Sutradhar, G. (2023, June). A comprehensive review on tribological and mechanical properties of metal matrix composites. In *AIP Conference Proceedings* (Vol. 2810, No. 1). AIP Publishing.

Dash, D., Samanta, S., & Rai, R. N. (2018, June). Study on fabrication of magnesium based metal matrix composites and its improvement in mechanical and tribological properties-a review. In *IOP Conference Series: Materials Science and Engineering* (Vol. 377, No. 1, p. 012133). IOP Publishing.

Dass, K., Chauhan, S. R., & Gaur, B. (2017). Study on the effects of nanoparticulates of SiC, Al2O3, and ZnO on the mechanical and tribological performance of epoxy-based nanocomposites. *Particulate Science and Technology*, *35*(5), 589–606.

Erappa Rajj, B., Nagaral, M., Chintakindi, S., Kumar, R., Anqi, A. E., Rajhi, A. A., ... & Chan, C. K. (2024). Nano-sized Al2O3–Gr reinforced Al7075 hybrid composite: Impact of cooling agents on mechanical, wear, and fracture behavior. *ACS Omega*, *9*(16), 17878–17890.

Faisal, N., & Kumar, K. (2018). Mechanical and tribological behaviour of nano scaled silicon carbide reinforced aluminium composites. *Journal of Experimental Nanoscience*, *13*(supl), S1–S13.

Friedrich, K., Zhang, Z., & Schlarb, A. K. (2005). Effects of various fillers on the sliding wear of polymer composites. *Composites Science and Technology*, *65*(15–16), 2329–2343.

Gangwar, S., & Pathak, V. K. (2021). A critical review on tribological properties, thermal behavior, and different applications of industrial waste reinforcement for composites.

Proceedings of the Institution of Mechanical Engineers, Part L: Journal of Materials: Design and Applications, *235*(3), 684–706.

Ghahremani, A., Abdullah, A., & Arezoodar, A. F. (2022). Wear behavior of metal matrix nanocomposites. *Ceramics International*, *48*(24), 35947–35965.

Jayanthi, A., Priya, R. K., Surakasi, R., Vanam, J. P., Puthilibai, G., Santhanakrishnan, M., & Rajkumar, S. (2022). Mechanical and tribological properties of TiC nano particles reinforced polymer matrix composites. *Materials Today: Proceedings*, *59*, 1472–1477.

Kala, H., Mer, K. K. S., & Kumar, S. (2014). A review on mechanical and tribological behaviors of stir cast aluminum matrix composites. *Procedia Materials Science*, *6*, 1951–1960.

Kundie, F., Azhari, C. H., Muchtar, A., & Ahmad, Z. A. (2018). Effects of filler size on the mechanical properties of polymer-filled dental composites: A review of recent developments. *Journal of Physical Science*, *29*(1), 141–165.

Maddaiah, K. C., Kumar, G. V., & Pramod, R. (2024). Studies on the mechanical, strengthening mechanisms and tribological characteristics of AA7150-Al2O3 nano-metal matrix composites. *Journal of Composites Science*, *8*(3), 97.

Mohammadi, S., Montazeri, A., & Urbassek, H. M. (2020). Geometrical aspects of nanofillers influence the tribological performance of Al-based nanocomposites. *Wear*, *444*, 203117.

Muraliraja, R., Tamilarasan, T. R., Udayakumar, S., & Arvinda Pandian, C. K. (2021). The effect of fillers on the tribological properties of composites. *Tribological Applications of Composite Materials*, 243–266.

Naskar, M. K., Basu, K., & Chatterjee, M. (2009). Sol–gel approach to near-net-shape oxide–oxide composites reinforced with short alumina fibres—The effect of crystallization. *Ceramics International*, *35*(8), 3073–3079.

Nilagiri Balasubramanian, K. B., & Ramesh, T. (2018). Role, effect, and influences of micro and nano-fillers on various properties of polymer matrix composites for microelectronics: A review. *Polymers for Advanced Technologies*, *29*(6), 1568–1585.

Padhan, M., Paul, G., & Bijwe, J. (2020). Roles of size, shape, amount, and functionalization of nanoparticles of titania in controlling the tribo-performance of UHMWPE composites. *Frontiers in Materials*, *7*, 205.

Pun, A. K., & Siddhartha. (2018). A comparative study for the leverage of micro and nano SiC fillers on thermo-mechanical and erosion wear peculiarity of woven glass fibre-based vinyl ester composites. *International Journal of Surface Science and Engineering*, *12*(5–6), 467–495.

Richard, S., Selwin Rajadurai, J., & Manikandan, V. (2017). Effects of particle loading and particle size on tribological properties of biochar particulate reinforced polymer composites. *Journal of Tribology*, *139*(1), 012202.

Rodriguez-Suarez, T., Bartolomé, J. F., & Moya, J. S. (2012). Mechanical and tribological properties of ceramic/metal composites: A review of phenomena spanning from the nanometer to the micrometer length scale. *Journal of the European Ceramic Society*, *32*(15), 3887–3898.

Sharma, N., Saxena, T., Alam, S. N., Ray, B. C., Biswas, K., & Jha, S. K. (2022). Ceramic-based nanocomposites: A perspective from carbonaceous nanofillers. *Materials Today Communications*, *31*, 103764.

Shekhawat, D., Agarwal, P., Singh, T., Singh, A., & Patnaik, A. (2023). Effect of control factors on wear rate of the ceramic particulate filled metal alloy composites: A comparative analysis. *Arabian Journal for Science and Engineering*, *48*(9), 11819–11840.

Singh, A. K., Siddhartha, & Yadav, S. (2017). Investigation of sub-micron size cenosphere fillers and filler loading on the mechanical and tribological peculiarity of polyester composites. *Polymers for Advanced Technologies*, *28*(12), 1764–1777.

Singhal, V., Shelly, D., Babbar, A., Lee, S. Y., & Park, S. J. (2024). Review of wear and mechanical characteristics of Al-Si alloy matrix composites reinforced with natural minerals. *Lubricants, 12*(10), 350.

Singla, A., & Singh, Y. (2024). Nanosize Al2O3 reinforcement in Si-rich aluminum MMC for enhanced tribological performance in engine components. *Proceedings of the Institution of Mechanical Engineers, Part E: Journal of Process Mechanical Engineering, 238*(1), 180–191.

Song, Y., Liu, D., Hu, S., Song, X., & Cao, J. (2019). Graphene nanoplatelets reinforced AgCuTi composite filler for brazing SiC ceramic. *Journal of the European Ceramic Society, 39*(4), 696–704.

Sosiati, H., Fahrezy, R. A., Pamasti, A., Hata, S., & Gao, H. (2024). Improvement of the nylon (N)-glass fiber (G) reinforced polyester composite properties by varying oxide ceramic particles and stacking sequences of N and G. *Materials Research Express, 11*(12), 125301.

Sosiati, H., Supatmi, S., Wijayanti, D. A., Widyorini, R., & Soekrisno, S. (2014). Properties of the treated kenaf/polypropylene (PP) composites. *Advanced Materials Research, 896*, 566–569.

Surakasi, R., Prasad, A. R., Pattnaik, B., Rao, M. V., Puthilibai, G., Vibhakar, C., & Rajkumar, S. (2022). Mechanical behaviour of nano ceramic particles reinforced aluminium matrix composites. *Materials Today: Proceedings, 59*, 1452–1456.

Thirupathy, M., Vadivel, M., Subbiah, T., & Pathinettampadian, G. (2024). Experimental investigation of the mechanical and tribological properties of jute fiber composites with nano-sized Al2O3 ceramic particle reinforcement. *Engineering Proceedings, 61*(1), 39.

Wetzel, B., Haupert, F., Friedrich, K., Zhang, M. Q., & Rong, M. Z. (2001). Mechanical and tribological properties of microparticulate and nanoparticulate reinforced polymer composites. *Proceedings of the ICCM-13, Wan Fang Digital Electronic Publisher, Beijing, ID, 1021.*

Xiao, N., Wu, C., Yang, K., & Tang, J. (2024). Progress of multidimensional nano-additives under dry/liquid wear: A review. *Lubricants, 12*(10), 332.

15 Influence of Fillers on High-Temperature Tribology of Ceramic Matrix Composites

Shuhaib Mushtaq, Mohammad Arif Parray, and Mohd Nadeem Bhat

15.1 INTRODUCTION

Filler is defined as a solid substance that has the ability to modify the properties of materials through surface contact or lack thereof, as well as its own physical features [Wypych, 2016]. Based on the requirements and kind of application, fillers are employed in a widespread range of industrial applications. These additives are mixed into a combination in order to improve the qualities and lower the cost of the composites. The fillers are accessible in solid, liquid, and gaseous forms that are sold commercially. In addition to their use in improving thermal as well as dielectric properties, fillers are also employed in improving mechanical and tribological properties [Abenojar et al., 2017]. They are employed in various industrial applications for a range of objectives such as adhesives, farming, space exploration, household appliances, automotive parts, packaging, construction materials, office equipment, electrical wiring, textile coatings, paint products, cosmetics, pharmaceuticals, dental materials, fibers, packaging films, cushioning materials, food packaging, friction materials, geotechnical applications, piping systems, magnetic technology, healthcare products, filtration systems, noise reduction, optical instruments, paper products, railway infrastructure, roofing materials, telecommunications equipment, tire manufacturing, sealant products, sports gear, water-resistant materials, and window components [Wypych, 2016].

Initially fillers were used to minimize material cost, but the size of particle is also a major factor. Several factors influence the efficacy of filler addition, including purity, surface preparation, shape, and dispersal [Chiellini and Solaro, 1996]. Metallic powder fillers can be used to change the color of a material so that the finished polymer has the appearance of a composite metal. Graphite, polytetrafluoroethylene (PTFE), and MoS_2 are examples of fillers that can be used to improve a material's surface properties and lower its coefficient of friction (COF) because they have self-lubricating qualities [Kano and Akiyama, 1996]. Fillers are used in polymer foams to help materials maintain their shape. Furthermore, the best insulation against heat and electricity conduction is achieved by using hollow spherical

DOI: 10.1201/9781003530909-19

particles as a filler. The appropriate selection of fillers can also affect a material's porosity [Srirangan and Paulraj, 2016]. The majority of mechanical properties of a material have been shown to be considerably influenced by fillers [Asadi et al., 2011]. Fillers lower the auto-ignition temperature, which lowers the formation of smoke in dangerous conditions and increases the formation of char. These factors help to reduce the number of fire incidents. In addition, the fillers can stop dripping and lower the rate of heat transmission. Some of the essential characteristics of material can be enhanced by right and proper use of fillers.

15.2 CLASSIFICATION OF FILLERS

Fillers can be classified based on their shape, size, and material. Particles and fibers are frequently utilized fillers in polymer matrix composites (PMCs) and metal matrix composites (MMCs), respectively, which are added to act as fillers or reinforcements. Based on their size, the particle fillers in composites are categorized as nanofillers, microfillers, minifillers, midifillers, and macrofillers. The particle fillers have size limits of 10–100 μm, 1–10 μm, 0.1–1 μm, 0.01–0.1 μm, and –0.005 to 0.01 μm, respectively. There are several ways in which the fillers vary from each other including impurities, shape, size, and dispersion of the particles as well as surface characteristics. According to a review of the literature, fillers with finer particle sizes resulted in composites with superior mechanical properties, while fillers with coarser particles exhibited tendency towards a decline in the mechanical properties of the composite. Impurities may cause major problems, such unintentional chemical interactions with the matrix components at high temperatures [Muraliraja and Arunachalam, 2018; Nassar et al., 2017]. Rubbery fillers can be added to thermoplastics to increase their toughness. To boost the strength, layers of fibrous filler materials are inserted into the polymer or plastic material [Raja and Retnam, 2019]. Natural and artificial fibers are the two basic categories into which they are divided. Natural fibers are derived from plants, animals, and minerals that are found in nature. The synthetic fibers used in composites are both inorganic and synthetically synthesized. Figure 15.1 shows the schematic of classification of fillers [Nassar et al., 2017].

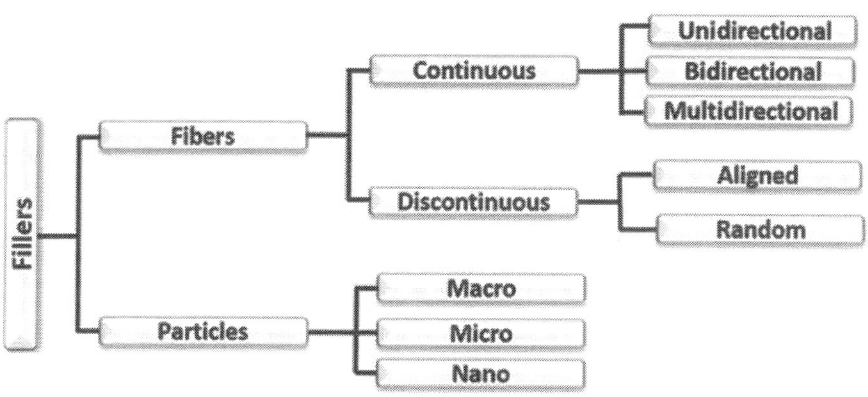

FIGURE 15.1 Classification of fillers [Nassar et al., 2017].

15.3 ROLE OF FILLERS IN CERAMIC MATRIX COMPOSITES

Fillers serve as multifunctional additives in ceramic matrix composites (CMCs), influencing their tribological behavior in several ways. Fillers provide the ceramic matrix extraordinary hardness and wear resistance. Examples of these fillers are silicon carbide (SiC) and alumina (Al_2O_3). As a result, CMCs can endure surface deterioration and abrasive wear, extending component life in high-temperature applications where sliding or abrasive contact occurs. Fillers have the ability to significantly reduce the COF in CMCs, reducing energy losses and increasing the effectiveness of sliding or rotating components. Fillers help reduce wear and promote smoother operation by reducing frictional forces, which is essential for preserving performance and dependability in demanding environments [Gevorkyan et al., 2020]. In order to keep the microstructure of CMCs stable and avoid softening or deformation at high temperatures, fillers are essential. Because of their ability to maintain tribological performance in the face of extreme heat, CMCs with fillers are the perfect choice for high-temperature seals, engine components, gas turbines, and other applications [van Roode et al., 2007].

15.4 PROCESSING METHODS OF FILLERS

The preparation composites with varying fillers are listed in Table 15.1. This section aims to define the various procedures or techniques that are applied when adding fillers to newly created composite materials.

TABLE 15.1
Methods of Using Different Fillers

S. No.	Method	Filler Used	References
1.	Blow molding	Glass fiber	Palutkiewicz et al. (2019)
2.	Compression molding	Carbon fiber sheet	Wulfsberg et al. (2014)
3.	Dip coating	Particulate filler (nano YSZ)	Bakhsheshi-Rad et al. (2016)
4.	Dispersion	Particulate fillers	Li et al. (2012)
5.	Extrusion	Biomass residues as fiber fillers	Bajwa et al. (2011)
6.	Foaming	Wood fiber	Matuana et al. (1998)
7.	Injection molding	Cellulose fiber	Graupner et al. (2016)
8.	Pultrusion	Glass fiber (treated)	Chen et al.(2017)
9.	Reaction injection molding	Silane treated glass fiber	Yoo et al. (2017)]
10.	Resin transfer molding	Natural fiber (Jute)	Ashworth et al. (2016)
11.	Rotational molding	Natural fibers (agave, coir, and pine)	Cisneros-López et al. (2017)
12.	Sheet molding	Chopped carbon fiber	Tang et al. (2019)
13.	Spinning	Multi-walled carbon nanotubes	Lai et al. (2015)
14.	Thermoforming	Hemp fibers	Ciupan et al. (2017)

15.5 FACTORS INFLUENCING THE FUNCTION OF FILLERS

The strength of the composite is typically maintained while its weight is decreased by fillers. Since they replace the specific content of the material, they lower the price of the matrix material. The shape as well as size of the fillers has a major effect on the properties of the composite. Fillers generally have the shapes of spheres and flakes. Due to their increased surface area at the point of contact, fillers with sizes between those of micrometers and nanometers are usually desired. The impact of various fillers on the characteristics and behaviour of composites vary depending on their size and shape. The application for which the composites are designed is specifically taken into consideration during the selection of fillers [Muraliraja et al., 2021]. Polymer-derived CMCs, like C/C composites, often exhibit cracked matrix and small pores during processing. Adding particulate fillers to the matrix can reduce shrinkage and cracking by stiffening the material. Figure 15.2 illustrates the role of fillers in decreasing porosity and cracking in the ultimate product [Chawla, 2012]. Particulate or whisker ceramics employed as fillers in polymeric matrices can serve several functions:

- Disrupt polymer shrinkage-induced matrix cracks.
- Increase ceramic yield by generating reaction products during pyrolysis.
- Improve the composite's interlaminar shear strength by strengthening and toughening its weak amorphous matrix.

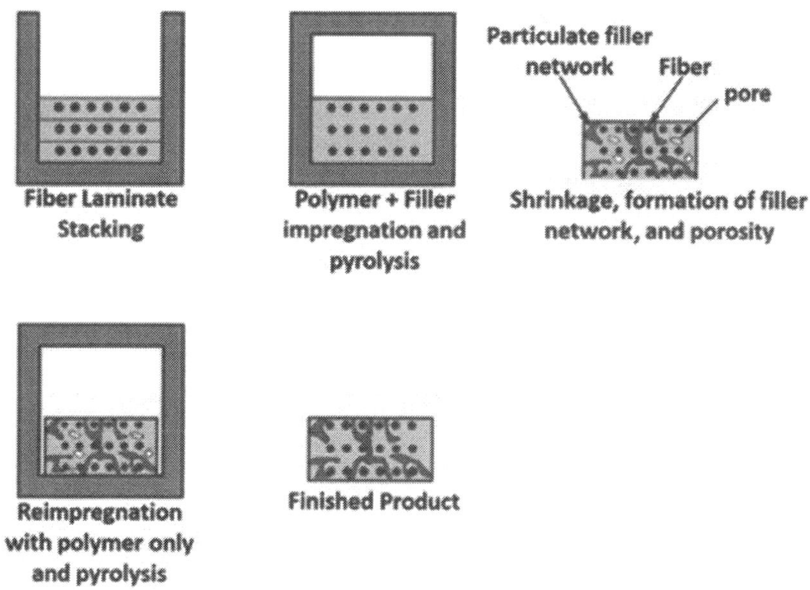

FIGURE 15.2 The role of fillers in decreasing the porosity and cracking in the ultimate carbon matrix composite (CMC) product [Chawla, 2012].

15.6 INFLUENCE OF FILLERS ON MECHANICAL CHARACTERISTICS OF THE COMPOSITES

The inclusion of fillers generally enhances the mechanical characteristics like hardness, ultimate tensile strength and impact strength of the PMCs [Hemanth et al., 2014]. One of the primary benefits of adding fillers into CMCs is the major improvement in strength and toughness. Additionally, these fillers inhibit crack propagation, hence enhancing the overall toughness of the material. This reinforcement mechanism is particularly valuable in the applications where mechanical strength and durability are paramount, such as aerospace components, automotive parts, and structural engineering [Hulugappa et al., 2016]. Fillers improve the hardness as well as wear resistance of CMCs in addition to their strength and toughness. High hardness fillers, such as boron nitride or silicon carbide, give the composite material superior wear resistance when they are incorporated into the matrix. This feature is crucial for applications like cutting tools, wear-resistant coatings, and friction materials that are vulnerable to abrasive wear or surface degradation [Yadav et al., 2023].

The incorporation of certain fillers can increase thermal conductivity an essential characteristic for applications that require efficient heat dissipation and management [Mumtaz et al., 2024]. Fillers with high thermal conductivity, such as boron nitride or aluminum nitride, can be strategically dispersed throughout the ceramic matrix to improve heat transfer capabilities. This is especially useful in high-temperature applications where thermal stability and performance are critical, such as aerospace propulsion systems, thermal barriers, and heat exchangers [Wu et al., 2023]. To further ensure compatibility with other materials and lower the chance of thermal stress and cracking during thermal cycling, fillers can be chosen and added to CMCs to modify their coefficient of thermal expansion (CTE) [Sommers et al., 2010]. Through precise manipulation of filler composition and distribution, engineers can enhance the thermal expansion properties of CMCs, consequently extending their dependability and durability under severe operating circumstances.

15.7 EFFECT OF FILLERS ON HIGH-TEMPERATURE TRIBOLOGICAL PROPERTIES OF CERAMIC MATRIX COMPOSITES

15.7.1 High-Temperature Tribology

High-temperature tribology typically commences around 300–350°C, marking the point where organic base oils and polymers begin to deteriorate, and extends up to 1000°C. Experiments in this field of tribology commonly involve dry or solid-state friction, unless a solid lubricant is employed, as most lubricants degrade or oxidize at high temperatures. Consequently, conducting tribotests in this manner is beneficial for understanding the tribological characteristics of coatings, ceramics, alloys, cermets, and related materials, including wear and friction. Moreover, comprehending high-temperature tribology is essential for gaining insight into the adhesive behavior and frictional interactions of contacts operating under these elevated temperatures [Shah et al., 2020].

15.7.2 Tribology

Tribology is all about understanding how surfaces in motion interact. Tribology mainly focuses on friction, wear, and lubrication. The British government received access to the renowned "Jost Report" in 1966. Ever since, various studies have been done in this area and there is widespread use of the term "tribology" [Gohar and Rahnejat, 2018]. The phenomena studied in tribology are among the most fundamental and common encountered by humans in their contact with a mostly solid world. Tribology affects our lives to a much greater degree than is commonly realized. Everything that a person makes, almost wears out or always wears out because of interaction between the surfaces [Stachowiak and Batchelor, 2013]. Friction is the primary cause of wear and loss of energy as it reduces the lifetime of an asset. Friction and wear, two key tribological properties of mechanical systems, are neither mutually exclusive nor completely inclusive. Tribology has a far-reaching impact on life of people that is often overlooked. Improving friction control can significantly lead to cost savings. The COF can be lowered by adjusting the surface finish of the interacting surfaces, temperature rise, relative speed, operating load, interaction between the surfaces, and lubrication. To reduce friction in a mechanical system, consider using sacrificial bearing surfaces which are designed for high resistance to wear or low shear materials, coating the component, replacing sliding friction with rolling friction, and improving lubricity between sliding surfaces through viscosity changes, additives, or lubricants [Kumar and Srivastava, 2016; Vinayagamoorthy, 2020].

One of the main areas of advanced material development is the use of design and production technology to the processing of composite materials. This is based on the idea that composites are multipurpose materials with previously unheard-of mechanical and physical qualities which can be customized to fulfill the needs of a specific application. Excellent wear resistance, elevated temperatures, corrosion, and oxidation are exhibited by many composite materials. Because of these special qualities, engineers have access to design possibilities that are not achievable with traditional monolithic materials [Ganesh and Dhamejani, 2016]. CMCs are materials with enhanced properties that are specifically made by adding reinforcing ceramic phases inside a ceramic matrix. High-temperature stability, excellent hardness, high resistance to wear and corrosion, non-conductive qualities, high thermal shock resistance, and increased versatility in offering distinctive engineering solutions are among the desired qualities of CMCs [Sommers et al., 2010].

Fillers are essential for adjusting the tribological behavior of CMCs for high-temperature applications [Parikh and Gohil, 2015]. Certain fillers, such as MoS_2 or ceramic particulates, increase the material's hardness and provide additional support when in contact, reducing wear. Fillers with lubricating properties, such as graphite or fluoropolymers, form a low-shear layer at the contact interface, reducing friction at elevated temperatures [Bashandeh et al., 2019]. High-temperature resistant fillers help the CMC remain more thermally stable overall by avoiding softening or degradation at high temperatures, which can cause wear. The predominant wear process may be influenced by particular fillers. As an illustration, certain fillers encourage resistance to abrasive wear, while others may strengthen tribochemical interactions and provide layers of protection on the surface [Yin et al., 2017].

Generally, the role of a filler in composite materials varies depending on the matrix and filler material. Fillers offer several benefits, including increased load-carrying capacity, improved crack arrest, increased thermal or electrical conductivity, and reduced shear stress at the interface. While fillers have superior mechanical and tribological properties, using specific types may affect other properties. To avoid this, proper optimization of the composite's mechanical and tribological properties is necessary. Fillers can also be used as inexpensive additives in polymer applications, lowering material costs while improving tribological properties [Bobby and Samad, 2017; Saba et al., 2014; Senbet, 2008]. Fillers in CMCs have a complex overall effect on high-temperature tribology. To attain the intended tribological performance, filler type, size, and content must be carefully chosen and optimized.

15.7.3 General Testing Methods for Tribology Tests

There are several approaches for characterizing the tribological behavior of a certain material. In order to provide significant analyses of the tribological data, using the appropriate test for the specified objective will be optimum. Tribotest results represent the behavior and qualities of the material as well as the mechanical system. Standardized testing of composites can provide insight into the product's expected life span. Tribological characteristics of composites, including matrix, type and structure of the reinforcement, filler-matrix interface, and internal lubricants, often impact wear and friction measurements. Friction and wear testing varies by tribological system, so composites should be tested using standard procedures before applying the results. Society of Automotive Engineers (SAE) and American Standard for Testing Materials (ASTM) have produced guidelines for testing and assessing material qualities. Following established rules means that similar equipment generate almost comparable findings, allowing for easy comparison and correlation. Some of the most often utilized tribological testing devices include [Bhushan, 2000]:

1. ***ASTM G65 Abrasion Test:*** It uses a dry sand/rubber wheel equipment to test metallic abrasion resistance of a material. This test determines the quality, reliability, and toughness of a sample. Metallic materials are evaluated based on their scratch resistance and abrasion resistance in a controlled setting.
2. ***ASTM A514 Pin-on-Drum:*** It involves high-pressure, abrasive wear between two surfaces. In this process, one end of a cylindrical pin-shaped sample is pushed against abrasive paper, resulting in material removal and compression of the abrasive particles. This type of wear is believed to simulate the abrasive wear experienced during crushing and grinding in ore processing, where abrasive particles are crushed, and thus it is termed high-pressure abrasive wear.
3. ***ASTM G99 Pin-on-Disk:*** This technique describes an experimental process for assessing the wear of materials while sliding with a pin-on-disk equipment. Materials are evaluated in pairs under supposedly non-abrasive circumstances. The appeal of the approach stems from its relative simplicity and number of tribological interactions.

4. ***Linear Tribometer:*** Linear tribometers typically consist of a stationary sample stage and a reciprocating pin or ball. The normal load, frequency, and amplitude of the reciprocating motion can be controlled, allowing researchers to simulate a wide range of real-world conditions.
5. ***ASTM G77, G137-95 Block-on-Ring:*** This technique describes an experimental process for measuring the wear resistance of plastic materials in dry sliding state. The test uses a block-on-ring geometry to rank materials based on their sliding wear properties under a variety of situations.
6. ***ASTM D2266 Four Ball Test:*** This method assesses how well greases prevent wear in situations where steel surfaces slide against each other. Its purpose does not extend to predicting wear properties for metal pairs other than steel, nor does it evaluate the grease's ability to withstand high pressure.

Standard testing procedures can precisely analyze the failure mode and life span of a tribological system; however, they might not accurately act out the experience of a composite in a real system. Since bench model tests are efficient and cost-effective for evaluating materials under controlled conditions, they have limitations. Moving from a field test to a simpler model test reduces the practical representation and realism of surface damage data, as well as the ability to make reliable inferences about its performance and usability for applications. To determine the applicability of a model test, consider how well it replicates the wear mechanisms found in the actual working conditions. Understanding the wear mechanism is crucial for exactly describing the tribological behavior of materials under test.

15.8 DIFFERENT FILLERS USED IN COMPOSITES FOR IMPROVING TRIBOLOGICAL PROPERTIES

Fillers are now used as multifunctional additives in various industries, rather than just cost-cutting measures. Graphite, CuO, silica, $CaCO_3$, PTFE, and molybdenum sulphide (MoS_2) are popular fillers used in composites to increase tribological qualities. Previous research has mainly focused on evaluating the tribological behavior of composite materials. This review focuses on fillers used in composites to enhance tribological properties.

15.8.1 Ceramic Fillers

Ceramics, including metallic nitrides, oxides, silicides, borides, and carbides, can maintain mechanical properties at high temperatures. Ceramic fillers, primarily composed of inorganic compounds, utilize this inherent property of ceramics. Composites or nanocomposites with fillers such as whisker, microfiber, or ceramic particles exhibit a threefold increase in wear resistance compared to those without fillers. Nanometer-sized particles are commonly used as fillers for improving tribological properties.

In a dry sliding environment, research examined the tribological characteristics of polyetheretherketone (PEEK) and polytetrafluoroethylene (10%) PEEK counters loaded with Al_2O_3 particles (5%) in relation to mild carbon steel (AISI 1045) steel rings. The addition of nanometer- and micron-sized Al_2O_3 to PEEK composites significantly reduced the wear rate (without PTFE) but had little effect on the COF. The pure PEEK composite containing 10% PTFE had a significantly lower wear rate and COF. When 10% PTFE was added to Al_2O_3/PEEK composites, the COF decreased but the wear rate increased, which contradicted expectations [Qiao et al., 2007]. A study compared the impact of SiC and Al_2O_3 on the wear behavior of jute/epoxy composites by adjusting the filler weight percentage in relation to the resin content. Jute-epoxy composites were more likely to wear at high normal loads and sliding speeds when fillers were not present. However, using ceramic fillers considerably lowered wear rates. The 15 wt.% filled jute-epoxy composites exhibited the lowest COF, which matched their excellent wear resistance. Al_2O_3-filled jute epoxy composites demonstrated reduced COF and material loss due to wear than SiC-filled composites across all compositions. In-filled composites showed microcracking, pitting, and debris formation, while unfilled composites exhibited fiber breakage and plastic deformation as wear mechanisms [Sabeel Ahmed et al., 2012]. A different study employed the three-body abrasive wear mechanism to explore how changing loads and abrasion distances affect blends of polyamide 66/polypropylene (PA66/PP), nano clay-filled PA66/PP, and short carbon fiber–reinforced PA66/PP nanocomposites. Introducing nano clay/short carbon fiber into PA66/PP had a significant impact on wear rate under various abrading distances and loads. The research revealed that composites containing nano clay/short carbon fiber showed reduced wear rates compared to those with only short carbon fibers in PA66/PP blends [Ravi Kumar et al., 2009]. In bamboo-glass hybrid polymer composites, ceramic fillers such as TiO_2 and ZrO_2 were utilized. These similar filler materials were employed to investigate their influence on the friction and wear behavior of reinforced thermosets containing microparticles. When comparing the wear properties of bamboo-glass-epoxy hybrid composites, it was analyzed that ZrO2 fillers surpassed TiO_2 fillers in terms of strength and wear resistance [Latha and Rao, 2018]. Additionally, ceramic fillers comprising silicon carbide (SiC) and boron carbide (B_4C) were experimented with in three-dimensional needled carbon fiber to fabricate improved ceramic C/C composites. Adding ceramic fillers to C/C composites resulted in homogeneous pore size distribution, as observed through morphology analysis. The ceramic improvement significantly reduced COF fade in marine environments for C/C composites. C/C-SiCf outperformed C/SiC in terms of friction stability and fade-free performance in marine conditions. Lubrication from water and SiO_2 films led to higher COF fading for C/SiC than C/B_4C [Cai et al., 2013]. Nanostructured fillers, including BN and SiO_2 microparticles, were employed to enhance the AK12M2MgN alloy. The filler structure ensured consistent dispersion of the alloy's structural components, resulting in enhanced wear resistance and lower COF [Vityaz' et al., 2013].

15.8.2 Carbon and Organic Fillers

Numerous thermoplastic and thermoset resins have been modified with the inclusion of carbon fibers and black carbon. The layered structure of graphite has a high electrical conductivity and is distinguished by a weak van der Waals force within the layers. Because the graphite structure forms a lubricant film between the mating surfaces, it also contributes to the reduction of friction and wear. This film is used as a filler in graphite applications. When carbon black was added to composites as a filler, it displayed appealing tribological properties. When PTFE composites were combined with carbon black fillers (at a 1%–5% filling ratio), the wear resistance increased dramatically by up to 700 times. In a dry-testing environment on flat metal surfaces, the large dispersal of fillers in the PTFE matrix exacerbated the interaction impact of carbon black.

Additionally, the properties of a highly dispersed filler, such as its specific surface area, significantly impacted the rate of wear in a PTFE composite. On increasing the specific area of filler, the corresponding wear rate increases because of the amplified interphase interaction area between the components per unit mass of the filler and heightened activity at its surface centers [Aderikha and Shapovalov, 2010]. The wear characteristics and dry friction of hybrid fillers containing multiwalled carbon nanotubes and carbon black in natural rubber were also examined. For each fixed sliding speed, it was deduced that a rise in the applied load led to a notable rise in wear resistance and a surprising fall in the COF [Ojha et al., 2014]. The wear mechanism of low-filled PTFE-CB was investigated and the results showed that wear occurred via delamination mode in which variations in wear resistance were related to structural changes in the composite. Researchers have used carbon fibers extensively because of their advantageous mechanical qualities. The tribological properties of surface-improved carbon nanofiber (CNF)-packed PTFE composites were investigated under dry sliding circumstances. CNF was pre-treated with HNO_3 and a coupling agent before being integrated into the PTFE matrix. The appropriate CNF material in PTFE reduced the wear rate by approximately 30% compared to pure CNF-filled PTFE at a load of 200 N [Shi et al., 2008].

Graphene and graphene oxide fillers are essential in any discussion of carbon-based fillers. Several articles have reported on the friction and wear characteristics of graphene-containing metal matrix composites. With the use of graphene fillers, such as multilayer graphene and graphene oxide, can greatly increase composite strength and high temperature tribological characteristics [Li et al., 2017; Llorente et al., 2019]. Carbon fiber composites are commonly applied in several industries, including marine, sports, construction, and aerospace.

15.8.3 Metallic Fillers

Coating substrates of mild steel with metallic fillers, for example micro nickel, aluminum, silver, and zinc powders, improves their tribological performance and surface energy. The success of metallic filler-based composites has prompted researchers to investigate additional metallic powders as a possible alternative. Metallic fillers, including steel fibers, brass fibers, and copper powder, were employed to create non-asbestos organic (NAO) composites. The composites' tribological characteristics were investigated under various loads and speeds. Adding metallic fillers improved the friction performance of composites, while increasing the quantity of filler increased wear resistance. Increasing metallic content led to improved wear resistance and increased thermal conductivity (TC). Copper filler (10%)–based composites outperformed brass in terms of wear and friction qualities, while iron powder–based composites performed moderately [Kumar and Bijwe, 2010].

Researchers investigated how the inclusion of metallic fillers (copper, steel, or aluminum) affects the frictional properties of composite brake pads. They conducted multiple experiments using a small-scale friction tester against two different counter-disks (gray cast iron and aluminum metal matrix composite) across various temperature settings. In tests conducted at room temperature against a gray cast iron disk, composites containing copper fibers exhibited a significant negative correlation between the COF and sliding velocity (μ-v), indicating stick-slip behavior at low speeds. However, at elevated temperatures, copper-fiber composites demonstrated superior wear resistance compared to other composites. Steel fibers exhibited subpar performance when tested against aluminum metal matrix composite disks due to material degradation and inconsistent frictional behavior at high temperatures [Jang et al., 2004]. Copper and its alloys are often used to improve heat diffusivity at friction interfaces. Copper's ability to withstand high temperatures and greater friction coefficients is considered advantageous. Copper oxide generated at the interfacial layer effectively drives away frictional heat even at high temperatures. Copper and its alloys are used as fillers to reduce friction and minimize fierceness against the counterpart. Al-MMC braking rotors use composites with aluminum fibers [Urquhart, 1991; Wilson and Alpas, 1996].

Steel fibers have been proposed as an alternative for fillers with inferior mechanical qualities. Composites with continuous steel fibers were studied for their wear and friction properties. Wear rates increased with sliding directions and depended on the film stability. The COF (0.49–0.54) was determined using fiber alignment directions. At elevated temperatures, the specimen's iron-rich transfer coating increases adhesive contact, leading to steady-state friction [Qu et al., 2004]. Friction materials reinforced with brass fibers exhibited superior wear resistance compared to gray cast iron. The interaction between the components displayed fatigue wear, which was set by the following observations: (i) Application of a copper transfer coating to the frictional surface of the gray cast iron counterpart and (ii) the presence of cracks on the worn-out surface of the counterpart attributed to fatigue. Incorporating brass fibers at a mass fraction exceeding 19% resulted in slightly increased wear loss and COF [Xian and Xiaomei, 2004]. Additionally, the incorporation of carbon fibers and nanofillers like aluminum and zinc into epoxy hybrid composites via compression molding reduced the specific wear rate by up to 0.5 wt.% of Al/Zn filler loading [Divya and Suresha 2018]. Other metallic fillers commonly used in composites include steel wool, nickel, brass fibers, silver, copper powder, and sodium. Mineral silicates [Vasilev et al., 2019], talc [Zhao et al., 2012], feldspar [Cai et al., 2015], nano clays [Bobby and Samad, 2017], and calcium carbonate are commonly utilized as fillers in polymer applications. $CaCO_3$ can be used as a filler in resins such as polyvinyl chloride, polyolefins, phenolics, polyesters, and epoxies due to its economic and performance benefits [Palanikumar et al., 2019].

15.8.4 Polymeric Fillers

Polymers are becoming more widely available and affordable. Combining fillers with resin can provide materials with qualities that are intermediate between the two elements, making the use of fillers less important for cost reduction.

Polymeric fillers can be natural or synthetic. Natural polymers include cellulose, graphite, wood grain, flax, cotton, and starch. Synthetic polymers, such as polyamide, polyethene, polytetrafluoroethylene, polyester, aramid, and polyvinyl alcohol, are customized to specific purposes as they are not found naturally. Polymeric fillers are typically used to control permeability, soften the matrix, dampen vibrations, and impart appropriate tribological properties to composite systems.

Polymer-based fillers like Fluorinated polyether ketone and PTFE have been utilized to improve the characteristics of epoxy-based composites. FPEK's scratch resistance enhances the composite's resistance to abrasion when combined with a matrix [Saba et al., 2014]. Adding PTFE fillers significantly reduces the composite's friction coefficient. PTFE, a commercial filler, can reduce frictional coefficients and thus wear rates in polymeric composites. PTFE's unusual structure provides exceptional tribological characteristics. When a polymer slides on steel, it forms a third-body transfer layer, resulting in a lower frictional coefficient [Sahin, 2018; Zhang et al., 2004].

15.9 CHALLENGES AND FUTURE TRENDS

Fillers combined with CMCs for high-temperature tribological applications are the most effective solution. However, there are obstacles to overcome and exciting new directions to pursue. A very important obstacle is making sure that the filler is compatible with a CMC matrix. This suggests that an ideal filler should not only enhance the properties of the matrix, but also exhibit sliding wear resistance. Furthermore, preserving the filler's attributes at high temperatures guarantees that performance will stay the same during operation. Another issue involves possible cracking arising from a mismatch in CTE between the filler and matrix materials. Significant differences in CTE can make the CMC crack when thermally cycling it. For this reason, researchers are finding ways to achieve homogenous distribution of fillers within CMCs. Thus, even distribution of fillers throughout prevents agglomeration leading towards wear out.

CMCs' future in tribology is full of promising trends. One such direction is the use of nanoparticle fillers. Nanoparticles possess a high surface area and reactivity, giving better wear resistance and even self-lubrication potential. Additionally, the development of in-situ fillers is being investigated by scientists. These fillers are generated directly during CMC's manufacture, which may lead to improved compatibility and dispersion. Furthermore, another trend that is quite promising connotes the integration of friction as well as wear modifiers into these fillers. This approach makes it possible to achieve both reduction of wear rates on one hand and improvement in sliding behavior on the other. Consequently, advances in computational design have become valuable tools for predicting customized filler properties necessary for specific tribological applications.

By overcoming these problems, researchers can produce CMCs with unique performance under high-temperature conditions which will accelerate progress in many fields, including aerospace, power generation, and automobile industry.

15.10 CONCLUSION

In this chapter, various fillers used in CMCs have been discussed broadly. The review identifies the several fillers used in CMCs and their impact on high-temperature tribology. This study would be beneficial for the researchers working on CMCs and other composite materials. The study includes roles, classification, effect on mechanical properties, processing methods, testing methods, and influence of different categories of fillers which are explained clearly. The study also presents the challenges and future trends. Addressing these challenges will be crucial for successful implementation.

REFERENCES

Abenojar, J., Tutor, J., Ballesteros, Y., del Real, J.C. & Martínez, M.A. (2017) Erosion-wear, mechanical and thermal properties of silica filled epoxy nanocomposites. *Composites Part B Engineering*, 120, 42–53. DOI: 10.1016/j.compositesb.2017.03.047.

Aderikha, V.N. & Shapovalov, V.A. (2010) Effect of filler surface properties on structure, mechanical and tribological behavior of PTFE-carbon black composites. *Wear*, 268, 1455–1464. DOI: 10.1016/j.wear.2010.02.022.

Asadi, P., Givi, M.K.B., Abrinia, K., Taherishargh, M. & Salekrostam, R. (2011) Effects of SiC particle size and process parameters on the microstructure and hardness of AZ91/SiC composite layer fabricated by FSP. *Journal of Materials Engineering and Performance*, 20, 1554–1562. DOI: 10.1007/s11665-011-9855-x.

Ashworth, S., Rongong, J., Wilson, P. & Meredith, J. (2016) Mechanical and damping properties of resin transfer moulded jute-carbon hybrid composites. *Composites Part B Engineering*, 105, 60–66. DOI: 10.1016/j.compositesb.2016.08.019.

Bajwa, S.G., Bajwa, D.S., Holt, G., Coffelt, T. & Nakayama, F. (2011) Properties of thermoplastic composites with cotton and guayule biomass residues as fiber fillers. *Industrial Crops and Products*, 33, 747–755. DOI: 10.1016/j.indcrop.2011.01.017.

Bakhsheshi-Rad, H.R., Hamzah, E., Ismail, A.F., Daroonparvar, M., Yajid, M.A.M. & Medraj, M. (2016) Preparation and characterization of NiCrAlY/nano-YSZ/PCL composite coatings obtained by combination of atmospheric plasma spraying and dip coating on Mg-Ca alloy. *Journal of Alloys and Compounds*, 658, 440–452. DOI: 10.1016/j.jallcom.2015.10.196.

Bashandeh, K. et al. (2019) Tribological performance of graphene and PTFE solid lubricants for polymer coatings at elevated temperatures. *Tribology Letters*, 67, 1–14.

Bhushan, B. (2000). *Modern Tribology Handbook: Volume One: Principles of Tribology. Mod Tribol Handb Vol One Princ Tribol 1–1697.*

Bobby, S. & Samad, M.A. (2017) Enhancement of tribological performance of epoxy bulk composites and composite coatings using micro/nano fillers: A review. *Polymers for Advanced Technologies*, 28, 633–644. DOI: 10.1002/pat.3961.

Cai, P., Li, Z., Wang, T. & Wang, Q. (2015) Effect of aspect ratios of aramid fiber on mechanical and tribological behaviors of friction materials. *Tribology International*, 92, 109–116. DOI: 10.1016/j.triboint.2015.05.024.

Cai, Y., Yin, X., Fan, S., Zhang, L. & Cheng, L. (2013) Tribological behavior of three-dimensional needled ceramic modified carbon/carbon composites in seawater conditions. *Composites Science and Technology*, 87, 50–57. DOI: 10.1016/j.compscitech.2013.07.026.

Chawla, K. K. (2012). Ceramic matrix composites. In: *Composite Materials: Science and Engineering*, pp. 249–292. Springer New York.

Chen, K., Jia, B. & Liu, X. (2017) Comparison and analysis dissimilar joint strength of pultrusion GFRP composites. *DEStech Transactions on Engineering and Technology Research*, (), 184–191. DOI: 10.12783/dtetr/icaenm2017/7795.

Chiellini, E. & Solaro, R. (1996) Biodegradable polymeric materials. *Advanced Materials*, 8, 305–313. DOI: 10.1002/adma.19960080406.

Cisneros-López, E.O., González-López, M.E., Pérez-Fonseca, A.A., González-Núñez, R., Rodrigue, D. & Robledo-Ortíz, J.R. (2017) Effect of fiber content and surface treatment on the mechanical properties of natural fiber composites produced by rotomolding. *Composite Interfaces*, 24, 35–53. DOI: 10.1080/09276440.2016.1184556.

Ciupan, E. et al. (2017) Characterization of a thermoforming composite material made from hemp fibers and polypropylene. *MATEC Web of Conferences*, 137.

Divya, G.S. & Suresha, B. (2018) Role of metallic nanofillers on mechanical and tribological behaviour of carbon fabric reinforced epoxy composites. *Materials Sciences and Applications*, 9, 740–750. DOI: 10.4236/msa.2018.99054.

Ganesh, K.G. & Dhamejani, C.L. (2016) Investigation of tribological behaviour of peek composite with glass fiber filled under harsh operating condition. *International Journal of Advance Research and Innovative Ideas in Education (IJARIIE)*, 2, 271–279.

Gevorkyan, E., Rucki, M., Panchenko, S., Sofronov, D., Chałko, L. & Mazur, T. (2020) Effect of SiC addition to Al2O3 ceramics used in cutting tools. *Materials*, 13, 5195. DOI: 10.3390/ma13225195.

Gohar, R. & Rahnejat, H. (2018). *Fundamentals of Tribology*. World Scientific Publishing. DOI: 10.1142/q0152.

Graupner, N., Ziegmann, G., Wilde, F., Beckmann, F. & Müssig, J. (2016) Procedural influences on compression and injection moulded cellulose fibre-reinforced polylactide (PLA) composites: Influence of fibre loading, fibre length, fibre orientation and voids. *Composites – Part A*, 81(October 2015), 158–171. DOI: 10.1016/j.compositesa.2015.10.040.

Hemanth, R., Sekar, M. & Suresha, B. (2014) Effects of fibers and fillers on mechanical properties of thermoplastic composites. *Indian Journal of Advances in Chemical Science*, 2, 28–35.

Hulugappa, B., Achutha, M.V. & Suresha, B. (2016) Effect of fillers on mechanical properties and fracture toughness of glass fabric reinforced epoxy composites. *Journal of Minerals and Materials Characterization and Engineering*, 4, 1–14. DOI: 10.4236/jmmce.2016.41001.

Jang, H., Ko, K., Kim, S.J., Basch, R.H. & Fash, J.W. (2004) The effect of metal fibers on the friction performance of automotive brake friction materials. *Wear*, 256, 406–414. DOI: 10.1016/S0043-1648(03)00445-9.

Kano, Y. & Akiyama, S. (1996) Estimation of surface tension and surface segregation of poly(ethyl acrylate)/poly(vinylidene fluoride-co-hexafluoro acetone) blends. *Polymer (Guildf)*, 37, 4497–4503.

Kumar, M. & Bijwe, J. (2010) Role of different metallic fillers in non-asbestos organic (NAO) friction composites for controlling sensitivity of coefficient of friction to load and speed. *Tribology International*, 43, 965–974. DOI: 10.1016/j.triboint.2009.12.062.

Kumar, P. & Srivastava, V.K. (2016) A review on wear and friction performance of carbon-carbon composites at high temperature. *International Journal of Applied Ceramic Technology*, 13, 702–710. DOI: 10.1111/ijac.12538.

Lai, D., Wei, Y., Zou, L., Xu, Y. & Lu, H. (2015) Wet spinning of PVA composite fibers with a large fraction of multi-walled carbon nanotubes. *Progress in Natural Science: Materials International*, 25, 445–452. DOI: 10.1016/j.pnsc.2015.10.003.

Latha, P.S. & Rao, M.V. (2018) Investigation into effect of ceramic fillers on mechanical and tribological properties of bamboo-glass hybrid fiber reinforced polymer composites. *Silicon*, 10, 1543–1550. DOI: 10.1007/s12633-017-9637-7.

Li, J.F., Chen, B., Shi, Q., Li, C., Chu, Y. & Li, C. (2017) Tribological properties of novel Cu/NbSe2 composites reinforced with reduced graphene oxide filler. *Chalcogenide Letters*, 14, 499–510.

Li, Z., Gao, Y., Moon, K.S., Yao, Y., Tannenbaum, A. & Wong, C.P. (2012) Automatic quantification of filler dispersion in polymer composites. *Polymer (Guildf)*, 53, 1571–1580.

Llorente, J., Ramírez, C. & Belmonte, M. (2019) High graphene fillers content for improving the tribological performance of silicon nitride-based ceramics. *Wear*, 430–431, 183–190. DOI: 10.1016/j.wear.2019.05.004.

Matuana, L.M., Park, C.B. & Balatinecz, J.J. (1998) Cell morphology and property relationships of microcellular foamed pvc/wood-fiber composites. *Polymer Engineering and Science*, 38, 1862–1872. DOI: 10.1002/pen.10356.

Mumtaz, N., Li, Y., Artiaga, R., Farooq, Z., Mumtaz, A., Guo, Q. & Nisa, F.U. (2024) Fillers and methods to improve the effective (out-plane) thermal conductivity of polymeric thermal interface materials–A review. *Heliyon*, 10, e25381. DOI: 10.1016/j.heliyon.2024.e25381, PubMed: 38352797.

Muraliraja, R. & Arunachalam, R. (2018) Al-fori I, al-maharbi M, Piya S. In: *Development of Alumina Reinforced Aluminum Metal Matrix Composite with Enhanced Compressive Strength Through Squeeze Casting Process. Proceedings of the IMechE Part LJ Mater des Appl.*, pp. 1–8.

Muraliraja, R., Tamilarasan, T.R., Udayakumar, S. & Arvinda Pandian, C.K. (2021) The effect of fillers on the tribological properties of composites. In: *Tribological Applications of Composite Materials* (edited by M. T. Hameed Sultan, M. R. Mohd Jamir, M. S. Abdul Majid, A. I. Azmi & N. Saba). Springer Singapore, pp. 243–266. DOI: 10.1007/978-981-15-9635-3_9.

Nassar, M.M.A., Arunachalam, R. & Alzebdeh, K.I. (2017) Machinability of natural fiber reinforced composites: A review. *International Journal of Advanced Manufacturing Technology*, 88, 2985–3004. DOI: 10.1007/s00170-016-9010-9.

Ojha, S., Acharya, S.K. & Gujjala, R. (2014) Characterization and wear behavior of carbon black filled polymer composites. *Procedia Materials Science*, 6(Icmpc), 468–475. DOI: 10.1016/j.mspro.2014.07.060.

Palanikumar, K., AshokGandhi, R., Raghunath, B.K. & Jayaseelan, V. (2019) Role of calcium carbonate(CaCO3) in improving wear resistance of polypropylene(PP) components used in automobiles. *Materials Today: Proceedings*, 16, 1363–1371. DOI: 10.1016/j.matpr.2019.05.237.

Palutkiewicz, P. Trzaskalska, M. & Bociąga, E. (2019) The influence of blowing agent addition, talc filler content, and injection velocity on selected properties, surface state, and structure of polypropylene injection molded parts. *Cellular Polymers*, 35, 159–192.

Parikh, H.H. & Gohil, P.P. (2015) Tribology of fiber reinforced polymer matrix composites—A review. *Journal of Reinforced Plastics and Composites*, 34, 1340–1346. DOI: 10.1177/0731684415591199.

Qiao, H.B., Guo, Q., Tian, A.G., Pan, G.L. & Xu, L.B. (2007) A study on friction and wear characteristics of nanometer Al2O3/PEEK composites under the dry sliding condition. *Tribology International*, 40, 105–110. DOI: 10.1016/j.triboint.2006.02.069.

Qu, X., Zhang, L., Ding, H. & Liu, G. (2004) The effect of steel fiber orientation on frictional properties of asbestos-free friction materials. *Polymer Composites*, 25, 94–101. DOI: 10.1002/pc.20007.

Raja, D.B.P. & Retnam, B.S.J. (2019) Effect of short fibre orientation on the mechanical characterization of a composite material-hybrid fibre reinforced polymer matrix. *Bulletin of Materials Science*, 42, 111.

Ravi Kumar, B.N., Suresha, B. & Venkataramareddy, M. (2009) Effect of particulate fillers on mechanical and abrasive wear behaviour of polyamide 66/polypropylene nanocomposites. *Materials and Design*, 30, 3852–3858. DOI: 10.1016/j.matdes.2009.01.034.

Saba, N., Tahir, P.M. & Jawaid, M. (2014) A review on potentiality of nano filler/natural fiber filled polymer hybrid composites. *Polymers*, 6, 2247–2273. DOI: 10.3390/polym6082247.

Sabeel Ahmed, K., Khalid, S.S., Mallinatha, V. & Amith Kumar, S.J. (2012) Dry sliding wear behavior of SiC/Al2O3 filled jute/epoxy composites. *Materials and Design*, 36, 306–315. DOI: 10.1016/j.matdes.2011.11.010.

Sahin, Y. (2018) Dry wear and metallographic study of PTFE polymer composites. *Mechanics of Composite, Materials*, 54(3), 403–414.

Senbet, D. (2008) Measuring the impact and international transmission of monetary policy: A factoraugmented vector autoregressive (favar) approach. *European Journal of Economics, Finance and Administrative Sciences*, 7, 121–143.

Shah, R., Chen, R., Woydt, M., Baumann, C., Jurs, J. & Iaccarino, P. (2020) High temperature tribology under linear oscillation motion. *Lubricants*, 9, 5. DOI: 10.3390/lubricants9010005.

Shi, Y., Feng, X., Wang, H. & Lu, X. (2008) The effect of surface modification on the friction and wear behavior of carbon nanofiber-filled PTFE composites. *Wear*, 264, 934–939. DOI: 10.1016/j.wear.2007.06.014.

Sommers, A., Wang, Q., Han, X., T'Joen, C., Park, Y. & Jacobi, A. (2010) Ceramics and CMCs for heat exchangers in advanced thermal systems—A review. *Applied Thermal Engineering*, 30, 1277–1291. DOI: 10.1016/j.applthermaleng.2010.02.018.

Srirangan, A.K. & Paulraj, S. (2016) Multi-response optimization of process parameters for TIG welding of Incoloy 800HT by Taguchi grey relational analysis. *Engineering Science and Technology, an International Journal*, 19, 811–817. DOI: 10.1016/j.jestch.2015.10.003.

Stachowiak, G. & Batchelor, A.W. (2013) *Engineering Tribology*. Butterworth-Heinemann.

Tang, H., Chen, Z., Zhou, G., Li, Y., Avery, K., Guo, H., Kang, H., Zeng, D. & Su, X. (2019) Correlation between failure and local material property in chopped carbon fiber chip-reinforced sheet molding compound composites under tensile load. *Polymer Composites*, 40(Supplement 2), E962–E974. DOI: 10.1002/pc.24767.

Urquhart, A.W. (1991) Novel reinforced ceramics and metals: A review of Lanxide's composite technologies. *Materials Science and Engineering*, 144, 75–82. DOI: 10.1016/0921-5093(91)90211-5.

van Roode, M., Price, J., Kimmel, J., Miriyala, N., Leroux, D., Fahme, A. & Smith, K. (2007) Ceramic matrix composite combustor liners: A summary of field evaluations. *Journal of Engineering for Gas Turbines and Power*, 129, 21–30. DOI: 10.1115/1.2181182.

Vasilev, A.P., Struchkova, T.S., Nikiforov, L.A., Okhlopkova, A.A., Grakovich, P.N., Shim, E.L. & Cho, J.H. (2019) Mechanical and tribological properties of polytetrafluoroethylene composites with carbon fiber and layered silicate fillers. *Molecules*, 24, 224. DOI: 10.3390/molecules24020224.

Vinayagamoorthy, R. (2020) Friction and wear characteristics of fibre-reinforced plastic composites. *Journal of Thermoplastic Composite Materials*, 33, 828–850. DOI: 10.1177/0892705718815529.

Vityaz', P.A., Komarov, A.I., Komarova, V.I., Shipko, A.A., Ovchinnikov, V.V. & Kovaleva, S.A. (2013) Effect of phase composition of nanostructured refractory modifier on structure and tribological behavior of AK12M2MgN alloy. *Journal of Friction and Wear*, 34, 329–338. DOI: 10.3103/S1068366613050127.

Wilson, S. & Alpas, A.T. (1996) Effect of temperature on the sliding wear performance of Al alloys and Al matrix composites. *Wear*, 196, 270–278. DOI: 10.1016/0043-1648(96)06923-2.

Wu, W., Zheng, M., Lu, K., Liu, F., Song, Y.H., Liu, M. & Dang, Z.M. (2023) Thermally conductive composites based on hexagonal boron nitride nanosheets for thermal management: Fundamentals to applications. *Composites Part A*, 169, 107533. DOI: 10.1016/j.compositesa.2023.107533.

Wulfsberg, J., Herrmann, A., Ziegmann, G., Lonsdorfer, G., Stöß, N. & Fette, M. (2014) Combination of carbon fibre sheet moulding compound and prepreg compression moulding in aerospace industry. *Procedia Engineering*, 81, 1601–1607. DOI: 10.1016/j.proeng.2014.10.197.

Wypych, G. (2016) Fillers in different products. In: *Handbook of Fillers*. Elsevier, p. 823–872. DOI: 10.1016/B978-1-895198-91-1.50021-X.

Xian, J. & Xiaomei, L. (2004) Friction and wear characteristics of polymer-matrix friction materials reinforced by brass fibers. *Journal of Materials Engineering and Performance*, 13, 642–646. DOI: 10.1361/10599490420061.

Yadav, R., Singh, M., Shekhawat, D., Lee, S.Y. & Park, S.J. (2023) The role of fillers to enhance the mechanical, thermal, and wear characteristics of polymer composite materials: A review. *Composites Part A*, 175, 107775. DOI: 10.1016/j.compositesa.2023.107775.

Yin, X.W., Cheng, L.F., Zhang, L.T., Travitzky, N. & Greil, P. (2017) Fibre-reinforced multifunctional SiC matrix composite materials. *International Materials Reviews*, 62, 117–172. DOI: 10.1080/09506608.2016.1213939.

Yoo, H.M., Kwon, D.J., Park, J.M., Yum, S.H. & Lee, W.I. (2017) Mechanical properties of norbornenebased silane treated glass fiber reinforced polydicyclopentadiene composites manufactured by the S-RIM process. *e-Polymers*, 17, 159–166. DOI: 10.1515/epoly-2016-0257.

Zhang, Z., Breidt, C., Chang, L., Haupert, F. & Friedrich, K. (2004) Enhancement of the wear resistance of epoxy: Short carbon fibre, graphite, PTFE and nano-tio2. *Composites Part A*, 35, 1385–1392. DOI: 10.1016/j.compositesa.2004.05.005.

Zhao, G., Wang, T. & Wang, Q. (2012) Studies on wettability, mechanical and tribological properties of the polyurethane composites filled with talc. *Applied Surface Science*, 258, 3557–3564. DOI: 10.1016/j.apsusc.2011.11.113.

16 Influence of Fillers in the Tribological Performance of Self-Lubricating Ceramic Matrix Coating

A Comprehensive Study

Naveen Kumar and Sujeet Kumar

16.1 INTRODUCTION

Wypych (2016) defines filler as a solid material that alters physical and chemical properties through surface interaction and its qualities. Fillers serve various industrial applications worldwide, depending on their nature and requirements. Adding these components is mixed to improve characteristics and lower the cost of composites. Fillers are supplied commercially as gases, solids, and liquids. Fillers are chosen from numerous categories to meet the material's application needs. These fillers lower composite costs and enhance material processing and mechanical performance. Various industrial applications include adhesives, agriculture, aerospace, appliances, automotive, building components, cables, wires applications, etc.

Initially, fillers were employed to save material costs. However, particle size issues also arose. Purity, surface preparation, form, and distribution affect filler effectiveness. Material density can be changed with fillers. High-density polymer material up to 2 g/cm^3 can be created using fillers. Polymers with a low density and lighter fillers, such as foam, can also be created (Chiellini and Solaro 1996). Using metallic powder fillers can change a material's color, creating a composite metal-like polymer. Graphite, PTFE, and MoS_2 can enhance the surface qualities of a material by reducing friction due to their self-lubricating properties (Kano and Akiyama 1996). Fillers assist by keeping the shape of materials in polymer foams. Hollow spherical particles are the finest filler for electric and thermal insulation. Choosing the appropriate fillers might affect the porosity of a material (Srirangan and Paulraj 2016). Most material mechanical properties are affected by fillers. It is possible for fillers to minimize the number of fire incidents by reducing the temperature of auto-ignition, minimizing smoke generation, and increasing char formation. Fillers can reduce heat transmission and avoid leaks. Proper filler application can enhance important material characteristics.

Machining hard materials like Ni and Ti alloys is problematic, particularly at high rates of cutting. It is necessary to develop appropriate materials for the tools.

DOI: 10.1201/9781003530909-20

Designed for dry cutting of alloys, this technique is environmentally benign and economically viable. Making cuts with ceramic composite inserts, with varying attributes such as volume fraction, size, matrix-reinforcement interface, and self-lubricity, are gaining popularity for their strong mechanical, thermal, and tribological properties. Dry cutting generates excessive heat due to intense contact between the tool and the workpiece. This results in severe tool wear, reducing its life span. Ceramic tools that have self-lubricating properties can create a tribofilm during sliding, improving antifriction qualities. A recent study suggests that the incorporation of solid lubricants into ceramic composites may result in a reduction in the mechanical properties of the material. For dry cutting of complicated materials, improving the design and synthesis of self-lubricating composites that involve a ceramic matrix is of utmost importance. Proper design, processing, characterization, and testing are necessary to create a ceramic cutting material that works exceptionally well. This study aims to investigate the research conducted on self-lubricating ceramic composite cutting tools, as well as their limitations, practical concerns, and possible applications in the cutting industry.

In-depth description research is needed to determine optimal solid lubricant combinations for improved structural, thermal, and tribological performance in ceramics. Select ceramic materials and inclusions for optimal mechanical, thermal, and tribological properties. Researchers have investigated various inclusions such as hexagonal boron nitride (hBN), cBN, graphene platelets, CNTs, MoS_2, metals, oxides, and fluorides in ceramic matrices for self-lubrication. No systematic evaluation of self-lubricating ceramic composites for cutting tools has been documented despite their expanding importance in the industry. A thorough review of this field is expected to aid researchers and the cutting tool industry in producing new ceramic cutting tools. This overview covers research from the past decade, including current and future developments. The following topics are discussed in this review: current research trends, the requirement for self-lubricating ceramic tools, a variety of high-temperature solid lubricants, practical issues, mechanisms of self-lubrication, existing and upcoming synthesis processes, research gaps, and prospects. This review aims to improve the understanding of a cutting device that lubricates itself technology, identify gaps and opportunities for future research, the development of harder-to-cut ceramic cutting tool inserts, and also focus on the influence of fillers in the tribological performance of self-lubricating ceramic matrix coating.

16.2 RECENT RESEARCH HIGHLIGHTS THE NEED FOR SELF-LUBRICATING CERAMIC EQUIPMENT

The cutting insert's geometry and material must match the workpiece material for efficient machining. Milling hard-to-cut, high-temperature, wear-resistant metals requires more cutting tools. Cutting tool manufacturers using these alloys have realized that they will need more than conventional materials to keep them competitive. They must be prepared to offer more cost-effective, environmentally friendly, wear-resistant, and tribologically superior alternatives.

We recommend ceramic-composite tools for cutting such alloys. These composite tool inserts can be made using strong, self-lubricating, and heat-resistant materials.

The cutting tool material may withstand dry cutting conditions while machining hard-to-cut materials like creating superalloys. The increasing demand for materials that are difficult to cut in the machining sector is causing the cutting tools business to struggle. Check out the most recent research on designing and developing new tool materials with excellent performance. Cutting tool prices for machining expensive and difficult-to-cut items greatly impacts manufacturing costs and product pricing. The extensive product productivity research that has been conducted has led to the development of new ceramic cutting tools for high-speed machining, particularly dry cutting. Ceramic composite tools are being used in numerous fields. Reports show that 9%–10% of the materials used in tool construction are ceramic composites, mostly indexable inserts for productivity [Bobzin 2017; Rizzo et al. 2020].

16.2.1 Material Categorization for Ceramic Tools

Composite cutting tools have garnered a lot of interest in recent times because they are quite inexpensive while still providing a high level of efficacy. Ceramics make cutting steel possible due to several important features, including their remarkable abrasion resistance and heat resistance, and low degree of reactivity with ferrous particles. On top of that, because of their excellent hot hardness, these cutting materials are capable of being employed at high cutting speeds without experiencing any distortion [Kumar et al. 2024]. As a result of the fact that ceramics are not particularly resistant to fracture and are extremely sensitive to thermal stress, the utilization of composites based on ceramics in machining applications is severely limited. Cutting tools typically are comprised of the following fundamental categories of ceramic-based cutting tools, as shown in Figure 16.1.

Oxide ceramics are made from aluminum oxide (Al_2O_3), a fundamental ceramic material with high chemical stability. However, its limited thermal conductivity is its biggest drawback. Its shock resilience limits its use in pure cutting instruments. Continuous medium to fine machining of cast irons, carbon steels, and alloy steels is the primary use for this material. Its low cost and impressive resistance to wear at extreme temperatures make it an ideal choice for this application. [Alam et al. 2024] explain that pure Al_2O_3 cutting tools are frequently utilized to machine gray cast iron bits.

Mixed Al_2O_3 ceramics combine Al_2O_3 with carbides and carbo-nitrides that have been strengthened, such as SiC, ZrO_2, TiC, and TiN. Reinforcing materials aim to improve structural and thermal qualities. These tool materials are appropriate for continuous and intermittent cutting operations, including milling, with various harsh

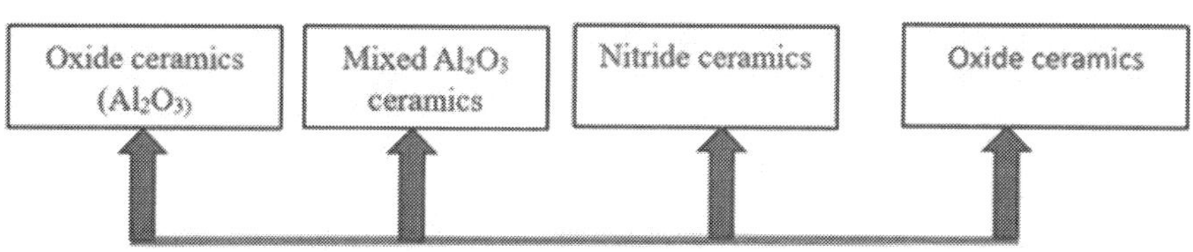

FIGURE 16.1 Categories of ceramic-based cutting tools.

and delicate cutting machining situations. Cutting tools can mill various materials, including cast irons, brittle materials, intermetallic steels, and high-alloyed steels with tungsten [Teng et al. 2007]. Black ceramics, or Al_2O_3-TiC, are frequently used in cutting tools because of their increased toughness and hardness than cemented carbide tools. Incorporating ZrO_2 into Al_2O_3 increases its toughness by causing a stress-induced change from tetragonal to monoclinic [Wang et al. 2001]. Doping Al_2O_3 with oxide or nonoxide compounds enhances its mechanical capabilities, making the blend of these additives appealing. In addition, incorporating nano-sized oxides and nonoxides into Al_2O_3 can enhance its hardness and toughness by causing segregation at grain boundaries. [Katiyar et al. 2024] researched the hybrid addition of oxides and nonoxides in the Al_2O_3 matrix. There is an opportunity for more research to be conducted on this topic. According to Gogotsi [1994], hot-pressed Al_2O_3–TiC–ZrO_2 composites have high hardness and fracture toughness due to increased densification from TiC dispersion in the matrix. The extraordinary hardness, toughness, and wear resistance of nitride ceramics are the primary factors behind the remarkable demand for light ceramic tools, particularly metal cutting. Liu and Huang [2003] found that ceramics based on Si_3N_4 are prevalent due to their strong heat resistance and excellent chemical resistance. However, ceramics like Si_3N_4 have limitations, including being an insulator, making them unsuitable for spark plasma sintering (SPS) processing techniques for cutting tools. To address this issue, silicon nitride can be doped with 10%–12% of oxides (Al_2O_3 and Y_2O_3) [Kumar et al. 2024; Tokita et al. 2011]. Si_3N_4-grade tools are ideal for rough machining and semi-finishing cast irons throughout the turning and milling processes due to their increased toughness and chemical resistance [Munir et al. 2006]. SiAlON cutting tools are gaining favor, combining Si_3N_4 networks' strength and chemical stability.

16.2.2 The Need of Self-Lubricating Ceramic Cutting Tools and Their Limitations

In high-speed machining, ceramic composite materials are increasingly used for cutting tools, leading to increased efficiency and productivity [Belmonte et al. 2006]. The limited usage of bulk Al_2O_3 as a cutting tool is due to its lower fracture toughness and flexural strength, especially in intermittent machining. To boost the hardness and strength of Al_2O_3, secondary particles may be added [Teng et al. 2007], resulting in ceramic matrix composites with enhanced characteristics. For enhanced fracture toughness and flexural strength, the utilization of second phases that are either micro- or nano-sized, such as SiC, WC, TiC, or TiB_2, is common [Wang et al. 2001]. Research suggests that adding micro- and nano-sized phases to Al_2O_3 may enhance its flexural strength and fracture toughness [Fernández et al. 2015]. Considering technology, Al_2O_3 and its composites have great promise for many industrial uses, such as cutting.

Al_2O_3-SiC material is increasingly used for cutting inserts because of its physical and mechanical qualities, including hardness and crack resistance [Kumar et al. 2024]. Composite cutting tools are often used to machine tempered steels and heat-resistant alloys. Research indicates ceramic tools are 3–10 times more robust than sintered carbide tools [Suresha et al. 2006]. Complex machining and dry-cutting techniques were developed using ceramic materials in tool manufacturing. A constant effort minimizes machining time by improving cutting speed and enhancing tool life. High-speed machining improves accuracy and surface quality in difficult-to-cut materials, including titanium alloys, tool steels, and stainless steels used in dies and moulds. In addition, these tools lowered cutting forces and workpiece deformation by improving heat dissipation via chips. These benefits simplify tool manufacturing throughout design and development. Precise and stiff machine tools with regulated spindle speeds and unique tool-holding mechanisms are necessary for cutting tools. The cutting tools' high hardness and wear resistance are crucial because they need less frequent re-grinding and replacement. Ceramic tools have higher fracture toughness and resistance to dynamic stresses, often seen in intermittent milling processes.

16.3 UTILIZATION OF FILLERS IN COMPOSITE MATERIALS

16.3.1 Fillers and Their Function in Composites

Applying additives, fillers, and reinforcements may improve or alter a composite material's characteristics. Various qualities, including wear, friction, corrosion, heat conductivity, and material deterioration, are enhanced using fillers in composites today. Lower the amount of shrinkage and swelling, decrease the amount of thermal expansion, and decrease the material's electrical characteristics. According to Suresha et al. (2006), incorporating fillers like graphite and SiC particles into glass-epoxy composite systems results in exceptional wear-resistant qualities and a considerable improvement in the frictional behavior of the composite. Brief fibers are used as reinforcements in polymer composites for tribological applications. Fillers have been employed in polyetheretherketone (PEEK) composites to reduce wear rates; however, the degree of their impact depends on the composite layer's ability to form transfer films (Bahadur and Gong 1992). In natural rubbers, lignin fillers safeguard the material from the deterioration caused by thermo-oxidative degradation in the atmosphere, as stated by Košíková et al. (2007). In order to boost the breakdown power of epoxy composites (Li et al. 2011), fillers with high heat conductivity are added to the mixture. In order to accomplish several objectives, such as bonding, color, and strength, fillers are also included in the composites used in dental applications.

16.3.2 Filler Categorization

The material, size, and form of fillers are the primary factors that determine their classification. Particles and fibers are two types of fillers often used in producing matrix composites made of polymers and metals, respectively.

Depending on the situation, they are either included to act as reinforcements or fillers. Five different types of fillers are used in composites and categorized according to their size: fillers that are macro, midifillers, mini, micro, and nano. According to the recommended guidelines, the size limitations of the particle fillers are as follows: 10–100 μm, 1–10 μm, 0.1–1 μm, 0.01–0.1 μm, and –0.005 to 0.01 μm, specifically.

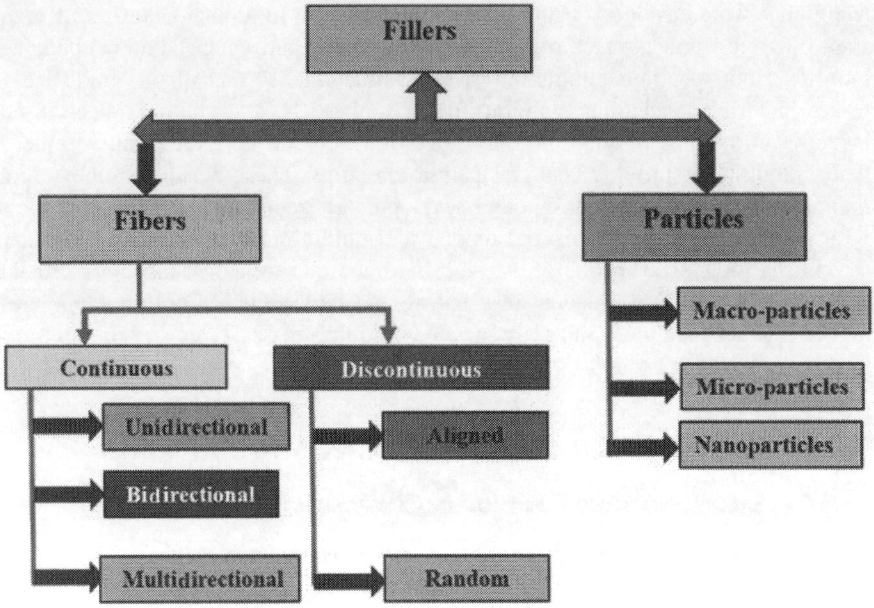

FIGURE 16.2 Detailed filler categorization.

Various features, including surface qualities of the filler, form, size, distribution of the particles, and contaminants, are among the characteristics that distinguish one filler from another. These filters are distinguished from one another. A search of the relevant literature revealed that composites with finer particle sizes created composites with superior mechanical characteristics. On the other hand, composites with coarser particles exhibited a downward trend in terms of their qualities.

According to Nassar et al. (2017), impurities can cause significant issues, such as harmful chemical interactions with the matrix components, when the temperature is raised. By using fillers that have a rubbery consistency, thermoplastics may have their toughness increased. According to Raja and Retnam (2019), increased strength may be achieved by implanting layers of fiber filler materials into the plastic or polymer material. Fibers may be divided into two categories: those that are naturally occurring and those that humans produce. The extraction of natural fibers involves using plants, animals, and minerals found in their natural state (Nassar et al. 2017). Figure 16.2 shows that the artificially synthesized organic and inorganic fibers are the components that are used in composites.

16.4 FACTORS INFLUENCING FILLER FUNCTIONS

Fillers are often used to lower composites' weight without affecting the composite's strength. Changing the matrix's unique composition also brings down the cost of the matrix material. Characteristics of the composites that are mainly influenced by the variables in terms of both the size and the form of the fillings. Most fillers are shaped like spheres or flakes, depending on their use. As a general rule, fillers that have a size that falls somewhere between micrometers and nanometers are chosen since they include a greater surface area to the contact region. There is a tendency for various fillers to have varied impacts on the characteristics and behavior of composites, and these effects are based on the size and form of the filler. More specifically, the fillers' selectivity is determined by the application for which the composites are built.

16.4.1 Fillers' Impact on the Composites' Mechanical Properties

A composite's mechanical properties may be improved by combining fillers with a base matrix. Many factors influence the mechanical behavior of composites, including dispersion, distribution, adhesion, and contact between the filler and the matrix. Polymer composites enhance mechanical properties such as ultimate tensile strength, impact strength, and hardness. Polymer composites also improve the flexibility of the composite. In both natural and synthetic forms, fillers can significantly improve the material properties they apply. Composites that include filler have damage modes that absorb energy and efficiently transfer stress between the particles and the matrix. They combine the advantages of component phases incorporated into the composite.

Nanofillers made of graphitic carbon nitride were investigated by Naidu et al. (2019) to see how they affected the mechanical properties of epoxy and glass fiber composites. The tensile and flexural strengths of the epoxy matrix were enhanced by 11% and 13%, respectively, when various filler quantities were added to the matrix (Naidu et al. 2019). These filler amounts were as follows: 1.5%, 1.5%, 2.5%, and 3%. The influence of silica fumes on the mechanical characteristics of jute-linen composites was contrasted in research conducted by Pandian and Jailani (2019). Composites were manufactured both with and without the presence of 1% silica fumes.

Compared to the composite that did not include any filler, the composite with filler exhibited a tensile strength that was 7% higher and flexural characteristics that were 5.2% better (Pandian and Jailani 2019). All these improvements were seen. Using silica fumes from industrial waste as fillers in polymer composites was the subject of research conducted by researchers. When applied to natural fiber textiles reinforced with polymer composites, industrial waste silica fumes at weight percentages of 1, 2, and 3 were evaluated for their dynamic mechanical properties. The composite's dynamic mechanical properties (storage modulus, loss modulus, and glass transition temperature) were significantly enhanced by adding a modest quantity of silica fumes.

The effects of nano silicon dioxide on fatigue, flexural, and tensile characteristics were investigated using jute-epoxy composites with varying concentrations of nano silicon dioxide (0, 5, and 10%). The maximum load, the ultimate tensile strength, the flexural strength, and the Young's modulus. It was found that the composite with 5% filler had the greatest attributes compared to others.

According to Ashik et al. (2017), the 5% nano silicon dioxide composite outperformed other loaded composites and suffered the most cycles during the fatigue test. The use of molybdenum disulfide, shungite, organic montmorillonite, and graphite nanoplates as fillers in ultra-high molybdenum polyethylene (UHMWPE) was investigated. We investigated the deformation and strength of UHMWPE composites over a range of filler concentrations up to 0.06 vol.%. In this experiment, the inclusion of filler had a substantial influence on the parameters of deformation and strength. Graphite nanoplates and montmorillonite laminates demonstrated the most significantly improved elastic modulus in composites with filler levels as high as 0.06 vol.%, according to research conducted by Grinev et al. (2018). The tensile properties of epoxy composites are altered when alumina nanoparticles, glass fiber, and carbon fiber are included in the composites, as stated by Mohanty et al. (2014). The features of four different composites were studied using variable amounts of the ingredients stated above. Individually, alumina particles, short glass, and carbon fibers were added at weight percentages ranging from 1% to 5% in the first three kinds of composites. On the other hand, the fourth composite type had alumina particles and fibers at a weight percentage of 2%. Compared to all of the developed materials, composites manufactured using alumina particles and fibers in epoxy had greater tensile strength and modulus.

16.4.2 Environmental Benefits and Health Concerns

The environment is being polluted by hundreds of different sources of industrial waste, which has a severe influence on the planet and the people who live on it. More than that, because of "ecological necessities and stringent regulations," the incorporation of natural fibers instead of synthetic fibers has become unavoidable "for the manufacturers to accomplish new composite materials originated from renewable sources" (Arvinda Pandian et al. 2017). Arvinda Pandian and her colleagues reached this conclusion. Arvinda Pandian et al. (2017) state that as a result, the "whole gamut of the engineering sector" has begun to move away from synthetic materials and toward materials that are oriented toward natural fibers.

In order to evaluate the mechanical characteristics of filled glass/epoxy composites while they were being used, many natural fillers, such as coconut coir, rice husk, and wheat husk fillers, were included in the composites as fillers. Dhawan et al. (2013) found that the composite made with coconut coir displayed the best mechanical properties among the other composites. Additionally, in different research, the characteristics of epoxy composites were modified by using pulverized walnut shells as organic waste fillers. Evaluations were conducted on composites that included 20, 30, 40, and 50 wt.% of a walnut shell. Salasinska et al. (2018) found that the impact and tensile properties of the composites containing the natural filler declined. Significant improvements were seen in hardness and stiffness in the composites, including the natural filler. According to Raja and Retnam (2019), the composite material's fatigue life and tensile strength significantly increased when coconut shell powder was used as a filler. Other materials used to produce polyester composites were bamboo, E-glass, and coconut shells.

16.5 THE MOST RECENT DEVELOPMENTS, OBSTACLES, AND PROSPECTIVE TRENDS

Matching cutting insert material and shape to workpiece material is crucial for efficient and productive machining. With the increased demand for wear-resistant hard-to-cut material with the rise of high-temperature alloys, the need for cutting tools to manufacture them also rises. Cutting tool makers working with these alloys see the need for more productive, wear-resistant, tribologically effective, ecologically friendly, and cost-efficient options when competing with conventional materials.

Ceramic-composite cutting tools are ideal for machining alloys. Composite tool inserts may be tailor-made for high-temperature resistance, strength, and self-lubrication. Cutting tool material can endure harsh circumstances, such as dry cutting or developing superalloys and other hard-to-cut materials. The cutting tools sector struggles to meet the demand for machining challenging materials. Explore current research trends for developing new tool materials with more excellent performance than existing ones. Cutting tool costs for machining complex and costly components significantly impact manufacturing and component costs. This project aims to increase product productivity by developing new ceramic cutting tools for high-speed machining, particularly in arid conditions. Composite ceramic tools have several uses and are highly sought after. According to research, ceramic composites are used in 9%–10% of tool materials, mainly in indexable inserts, for higher productivity [Bobzin, 2017; Rizzo et al. 2020].

Composite materials have become a sustainable substitute for conventional materials in the automobile, aerospace, and other industries in recent years. While this technique allows for creating composite materials with desirable qualities, the production procedure for commercializing them presents significant hurdles. Conversely, reinforced composites have grown in technical capabilities and applications. Composites' expanding uses stem from the desire to replace costlier materials, often metals. Despite the lack of metals in metal processing sectors, conventional composites need upgrading to meet the demand for higher characteristics and excellent performance. Nanocomposites are making progress in meeting the demands of contemporary industries. Advanced composites provide the opportunity to enhance many desirable qualities concurrently. Research on tribological nanocomposites, particularly hybrid and polymer nanocomposites, shows promise for both domestic and commercial use.

Classically, fillers were seen as additions, primarily for cost reduction in composites. Due to its wide range of applications, composite technology strongly needs fillers or reinforcing fillers. Fillers have become more diverse, but choosing the most suitable one requires knowing the performance objectives of the composite being formed.

Several commercial polymer composites passed market needs by combining materials to enhance desired properties and minimize losses. Furthermore, filler makers must consider environmental factors including recycling, sustainability, and life cycle effect. Biological and organic fillers are replacing synthetic ones because of their environmental friendliness. Improving the durability and viability of fillers will be a priority in the future. Large-scale adoption of natural fillers has been hindered by unfamiliarity and restricted availability. Significant growth is expected as major corporations join the industry and technology progresses.

Future tribological research on nanocomposites is expected to advance steadily, focusing on developing green methods and modifying the tribological behavior of polymer composites (pristine, hybrid, and nanocomposites).

16.6 CONCLUSION

For the first time, this chapter reviews several fillers used in composites. The significant contribution of this study is finding fillers in composite materials and their impact on tribology. The study should help researchers get suggestions on fillers for producing novel composite materials. Different fillers' functions, classifications, processing techniques, and effects in composites are examined. Highlighted are composite fabrication problems. The article briefly reviews recent achievements in integrating fillers into composites and suggests future research opportunities. Ceramic tools that generate a lubricating layer on their own and are self-lubricating increase the tribological performance of cutting tools. Solid lubricants may reduce the tool's wear resistance and mechanical properties. Solid lubricant selection, applications, synthesis, characterization, and performance evaluation are only some topics that have been intensively researched about self-lubricating ceramic tools. When it comes to material design, modeling, process simulation, and numerical optimization, most of the work published is based on experimental trials, with very little computational work being done. Based on the studies reviewed, creating self-lubricating ceramic tools for various machining settings is conceivable. To better understand the impact of matrix and secondary phases on composite properties, lubrication mechanisms, and synthesis techniques, sophisticated computational models can be developed using extensive experimental data. For specific applications, predicting the structural, thermal, and tribological features of self-lubricating composites is feasible.

REFERENCES

Alam MS, Kumar N, Kumar Das A (2024) Behaviour of thermally sprayed coating for hot corrosion applications. Journal of Electrochemical Science and Engineering 14(4):459–472.

Arvinda Pandian CK, Siddhi Jailani H, Rajadurai A (2017) Natural fabric sandwich laminate composites: Development and investigation. Bulletin of Materials Science 40(1):139–146.

Ashik KP, Sharma RS, Raghavendra N (2017) Effect of filler on mechanical properties of natural fiber reinforced composites. Asian Journal of Chemistry 29(8):1697–1701.

Bahadur S, Gong D (1992) The role of copper compounds as fillers in the transfer and wear behavior of polyetheretherketone. Wear 154(1):151–165.

Belmonte M, Nieto MI, Osendi MI, Miranzo P (2006) Influence of the SiC grain size on the wear behaviour of Al2O3/SiC composites. Journal of the European Ceramic Society 26:1273–1279. https://doi.org/10.1016/j.jeurceramsoc.2005.01.049.

Bobzin K (2017) High-performance coatings for cutting tools. CIRP Journal of Manufacturing Science and Technology 18:1–9. https://doi.org/10.1016/j.cirpj.2016.11.004

Chiellini E, Solaro R (1996) Biodegradable polymeric materials. Advanced Materials 8(4):305–313.

Dhawan V, Singh S, Singh I (2013) Effect of natural fillers on mechanical properties of GFRP composites. Journal of Composites 2013:1–8.

Fernández A, Peretyagin P, Solís W, Torrecillas R, Borrell, P (2015) Functionalization of carbon nanofibres obtained by floating catalyst method. Journal of Nanomaterials 2015: 395014. https://doi.org/10.1155/2015/395014

Gogotsi YG (1994) Particulate silicon nitride-based composites. Journal of Materials Science 29:2541–2556. https://doi.org/10.1007/BF00356803.

Grinev VG et al. (2018) The effect of filler type on the mechanical properties of composite materials based on ultra-high-molecular-weight polyethylene. Polymer Science Series D 11(2):202–208

Kano Y, Akiyama S (1996) Estimation of surface tension and surface segregation of poly(ethyl acrylate)/poly(vinylidene fluoride-co-hexafluoro acetone) blends. Polymer (Guildf) 37(20):4497–4503.

Katiyar PC, Singh BP, Chhabra M, Kumar N, Parle D (2024) Analysis of tensile strength of thin ABS specimens 3-D printed in various build orientations for turbomachinery applications, International Journal on Interactive Design and Manufacturing (IJIDeM) 19(3):1869–1881.

Košíková B, Gregorová A, Osvald A, Krajčovičová J (2007) Role of lignin filler in stabilization of natural rubber–based composites. Journal of Applied Polymer Science 103(2):1226–1231.

Kumar N, Alam MS, Mishra V, Vasudev H, Yadav PC, Kumar Choubey V (2024) A comparative investigation of the effects of temperature on the oxidation resistance of high-velocity oxy-fuel coating on AISI316L. Physica Scripta 99:055031.

Kumar N, Kumar Choubey V (2024) Investigation of microstructure and isothermal oxidation resistance of cermet HVOF coated on AISI316L at 900°C. Results in Surfaces and Interfaces 14:100173.

Kumar S, Madugula NS, kumar R, Kumar N, Giri J, kanan M (2024) An extensive analysis of GTAW process and its influence on the microstructure and mechanical properties of SDSS 2507. Journal of Materials Research and Technology 33:8675–8686.

Li Z, Okamoto K, Ohki Y, Tanaka T (2011) The role of nano and micro particles on partial discharge and breakdown strength in epoxy composites. IEEE Transactions on Dielectrics and Electrical Insulation 18(3):675–681.

Liu CC, Huang JL (2003) Effect of the electrical discharge machining on strength and reliability of TiN/Si3N4 composites. Ceramics International 29:679–687. https://doi.org/10.1016/S0272-8842(02)00217-1

Mohanty A, Srivastava VK, Sastry PU (2014) Investigation of mechanical properties of alumina nanoparticle-loaded hybrid glass/carbon-fiber-reinforced epoxy composites. Journal of Applied Polymer Science 131(1):1–7.

Munir ZA, Anselmi-Tamburini U, Ohyanagi M (2006) The effect of electric field and pressure on the synthesis and consolidation of materials: a review of the spark plasma sintering method. Journal of Materials Science 41:763–777. https://doi.org/10.1007/s10853-006-6555-2

Naidu PP, Raghavendra G, Ojha S, Paplal B (2019) Effect of g-C3N4 nanofiller as filler on mechanical properties of multidirectional glass fiber epoxy hybrid composites. Journal of Applied Polymer Science 48413:1–9.

Nassar MMA, Arunachalam R, Alzebdeh KI (2017) Machinability of natural fiber reinforced composites: A review. The International Journal of Advanced Manufacturing Technology 88(9–12):2985–3004.

Pandian, A., & Jailani, S. (2019). *Development and Investigation of Jute/Linen Fibre Reinforced Polymer Composite* (No. 2019-28-0171). SAE Technical Paper.

Raja DBP, Retnam BSJ (2019) Effect of short fibre orientation on the mechanical characterization of a composite material-hybrid fibre reinforced polymer matrix. Bulletin of Materials Science 42(3):111.

Rizzo A, Goel S, Grilli ML, Iglesias R, Jaworska L, Lapkovskis V, Novak P, Postolnyi BO, Valerini D (2020) The critical raw materials in cutting tools for machining applications: A review. Materials 13(6):1377. https://doi.org/10.3390/ma13061377

Salasinska K, Barczewski M, Górny R, Kloziński A (2018) Evaluation of highly filled epoxy composites modified with walnut shell waste filler. Polymer Bulletin 75(6):2511–2528.

Srirangan AK, Paulraj S (2016) Multi-response optimization of process parameters for TIG welding of Incoloy 800HT by Taguchi grey relational analysis. Engineering Science and Technology, an International Journal 19(2):811–817.

Suresha B, Chandramohan G, Prakash JN, Balusamy V, Sankaranarayanasamy K (2006) The role of fillers on friction and slide wear characteristics in glass-epoxy composite systems. Journal of Minerals and Materials Characterization and Engineering 5(1):87–101.

Teng X, Liu H, Huang C (2007) Effect of Al2O3 particle size on the mechanical properties of alumina-based ceramics. Materials Science and Engineering: A 452–453:545–551. https://doi.org/10.1016/j.msea.2006.10.073

Tokita M, Bldg E, Sc K, Ksp P (2011) Mechanism of spark plasma sintering. Ceramics 044903: 605–608. https://doi.org/10.1063/1.3622584.

Wang L, Shi JL, Gao JH, Yan DS (2001) Influence of tungsten carbide particles on resistance of alumina matrix ceramics to thermal shock. Journal of the European Ceramic Society 21:1213–1217. https://doi.org/10.1016/S0955-2219(00)00325-3.

Wypych G (2016) Fillers in different products. In: Handbook of Fillers, p 851. Emerald Insight.

17 Case Studies Related to the Tribological Performance of Reinforced Ceramic Matrix Composites

Ashokkumar Mohankumar, Vignesh Packkirisamy, and Arunkumar Thirugnanasambandam

17.1 INTRODUCTION

It is for this very reason that ceramic matrix composites (CMCs) were selected as a point of interest in tribological application, as they could resist such an aggressive mechanical and thermal environment where conventional materials crumbled. A CMC is composed mainly of ceramic fiber placed in a ceramic matrix; according to Naslain (2004), Padture (2016) Campbell et al. (2000).

Such materials are relatively lightweight yet strong and have excellent heat resistance in these properties, which makes CMCs suitable for manufacturing such parts subjected to very high wear and corrosion, e.g., turbine blades or structural components in aerospace and automobile markets with high temperatures; also, braking systems can be produced (Krenkel, 2005; Sheehan et al., 1994; Heidenreich, 2014).

One of the most outstanding systems is carbon fiber–reinforced silicon carbide (C/SiC), which encompasses the hardness and thermal stability of SiC as well as the toughness of carbon fiber (Windhorst and Blount, 1997; Stimson and Fisher, 1980; Arai et al., 2019).

Such a combination is useful when metal alloys are either oxidized or are no longer able to provide mechanical integrity, making it possible to keep a high level of tribological performance even at high sliding speeds and high temperatures. The efficiency of CMCs to conduce heat, which also decreases coefficients of thermal expansion, leads to reducing the possibilities of thermal damage and guarantees that the coefficient of friction (COF) remains constant (Fouquet et al., 2008; Almeida et al., 2022; Zhang and Li, 2009). Sliding velocity and applied load are some environmental factors that contribute to the tribological performance of CMCs. In many cases, these materials show more resistance to wear compared to their metallic counterparts (Li et al., 2019; Le et al., 2021; Christin, 2002; Glass, 2008).

DOI: 10.1201/9781003530909-21

Also, contact surface degradation can be reduced and friction more uniform increased with the use of surface treatments and fillers (such as coke) in order to improve wear properties. The latest development proves that even in the conditions of high mechanical load C/SiC rotors and appropriate brake pads are capable of achieving significant values of COF and lower values of wear (Kumar et al., 2017a; Krenkel, 2004; Shi et al., 2021).

Despite their potential for tribological applications, several problems still have to be overcome, such as material costs and manufacture as well as the problem of tailoring CMC composition. This notwithstanding, constant and progressive research and development (R&D) effort is broadening the scope for CMC use in extreme conditions where load and temperature regimes are highly elevated through constant development of the structure and performance parameters of CMCs.

17.2 TRIBOLOGICAL PERFORMANCE OF REINFORCED CERAMIC MATRIX COMPOSITES

17.2.1 Case Studies Based on the Impact of Constituent Inclusion on Tribological Characteristics

C/SiC composites are made from carbon fibers and a SiC matrix; samples prepared via liquid silicon infiltration often contain residual silicon. Investigations show that the latter negatively impacts the tribological properties of the material due to increased wear rates and poor friction stability. High friction conditions cause the silicon phase to soften, causing adhesion on the friction surface that contributes to adhesive wear. Therefore, it is necessary to remove silicon from C/SiC composites for better tribological characteristics (Vaidyaraman et al., 2001; Wang et al., 2021; Toor, 2018).

C/SiC composites experience four main wear mechanisms during friction: grain abrasion, oxidation abrasion, fatigue wear, and adhesive wear, which occur together and are mutually dependent. The tribological properties of these composites are mainly affected by their component ratios and porosity. Densification via chemical vapor infiltration (CVI) affects the friction characteristics, with optimal lubrication observed at 30% porosity and a 10 N load, resulting in an average COF of 0.23. Low densification exposes more carbon fibers, enhancing lubrication, while high porosity reduces the SiC hard phase on the surface, increasing wear rates (Beyer et al., 2006; Xu et al., 2001; Cai et al., 2008).

Hot pressing sintering behaves differently when it comes to wear rates. A high content of carbon fibers reduces the COF but initiates a near-increase in wear. When carbon fibers are introduced as lubrication, wear rates increase with the content of carbon fibers because SiC has better durability. SiC mainly contributes to wear resistance by micro-cutting grain abrasion and fatigue wear, transforming from a tetrahedral form to a smoother periodic surface morphology (Ma et al., 2018; Yang et al., 2015; Kumar et al., 2011). The small size of SiC grains contributes to improved wear resistance by creating an oxide film, allowing transgranular fracture. Conversely, the C/C-SiC composite exhibits higher SiC weight fraction with reduced carbon pyrolytic density. Therefore, there is reduced ploughing effect and friction cutting resistance. Because of this, surface roughness and deformation wear becomes less compared to that of C/SiC composites (Mall and Engesser, 2006; Wang et al., 2016; Tang et al., 2009).

Recently, many phases have been added to the matrix of C/SiC composites in order to improve their properties and broaden their applications. For instance, recently discovered new self-healing phases disappear at high temperatures, and there are continuous efforts to increase their life span. New lubricating phases, such as nanoparticles with intrinsic ferroelectricity in cubic systems, can endure up to 1250°C for lead-free ceramics. For instance, the addition of graphite to 3D C/SiC composites decreases the COF from 0.38 to lower values, with considerably decreased wear rates as well (Fan et al., 2007, 2008, 2010). Graphite also decreases the residual silicon content and stabilizes friction by reacting with Si to form SiC. Similarly, coke, as a solid lubricant, decreases the real contact area of the material, improving COF stability by reducing wear. After 7200 seconds of friction, MoS_2 as a lubricant further decreases the mean COF to 0.161.

Research is still focused on the MAX phase, a class of ternary carbides and nitrides, for the development of C/SiC composites with properties of high-temperature stability, thermal shock resistance, and oxidation resistance. Among these, the representative MAX phase material, Ti_3SiC_2, has emerged to be of significant importance. The application of Ti_3SiC_2 in C/SiC composites enhances the tribological properties by substantially reducing the COF and wear rates compared to the nonmodified composites. However, the effect of Ti_3SiC_2 on the tribological performance of C/SiC composites has yet to be fully studied. Thus, there is ample room for further research to improve the modifications in the matrix of such composites (Opel et al., 2022; Ohnabe et al., 1999; Chen et al., 2019).

17.2.2 Case Studies Based on the Impact of Graphitization Process on Tribological Properties

Preparation and processing of carbon preforms are some of the most significant factors influencing the performance of C/SiC composites. Generally, carbon preforms are prepared from organic precursors like phenolic resin. High-temperature treatment, namely graphitization at 2500°C, optimizes the microstructure of the preforms, increases porosity, reduces density, and therefore improves silicon infiltration, increasing the final SiC content.

The internal stress resulting from gas entrapped in the curing of the material is reduced through this graphitization treatment as the material becomes better compacted through this process. The addition of graphite powder significantly enhances the graphitization process beyond the effect of high-temperature treatment alone. Graphite enhances both thermal conductivity and oxidation resistance, thereby elevating the overall material performance. In addition, graphite helps to distribute the remaining silicon residue, which also helps in reducing wear (Kumar et al., 2017; Yang et al., 2017; He et al., 2001).

The degree of graphitization has a significant influence on the tribological properties of the composite. For example, the addition of graphite reduced the COF of the material from 0.38 to 0.34. At a treatment temperature of 1800°C, the COF was reduced by about 26.8% as compared to the untreated sample. However, an excessively high degree of graphitization may increase the COF due to the generation of additional abrasiveness by the introduction of more SiC particles. For instance, though low level of graphitization could minimize the COF, the overly high level has been observed to have opposing effects.

A higher degree of graphitization leads to the creation of denser, stable friction films, improving the wear resistance. On the other hand, it also brings complexity in affecting COF due to a trade-off between the lubrication aspect of graphite and abrasive property of SiC. Further research is needed to establish a definitive relationship between graphitization and the tribological performance of C/SiC composites to achieve optimal results (Xu et al., 2007; Wang et al., 2018; Deng et al., 1999).

17.2.3 Case Studies Based on the Impact of the Material Structure and Fiber Direction on Tribological Characteristics

The material structure affects the physical and chemical properties of the C/SiC composites. Differences in component distribution, density, and structure cause differences in performance. The current research is carried out on two major configurations: 3D needled and short-cut fiber configurations.

The 3D needled C/SiC composites will have higher SiC content and less carbon content compared to the short-cut fiber composites. Hence, higher density with friction surface asperities increases the shear force required for displacement, thereby increasing the COF. Relatively high values of COF have been observed under braking tests due to the dense nature of the 3D needled C/SiC composites.

New achievements have brought innovative structures for the improvement of tribological properties. For instance, a sandwich structure consisting of a mechanical functional layer and two friction functional layers exhibits high mechanical strength and a uniform phase distribution with low Si content. Such a design minimizes adhesion effects and increases stability in comparison with traditional 3D needled C/SiC composites (Inoue et al., 2013; Fan et al., 2015; Zhao et al., 2015).

The orientation of the fibers in the material also determines its tribological characteristics. The two-dimensional laminate structure of C/SiC composites displays lower COF when the fibers are aligned vertically (0.14 with ZrO_2) than a parallel alignment (0.19 with ZrO_2). This is because, due to the lamination of fibers in a vertical position, shear resistance is much improved and wear debris is minimized, thus lowering COF and wear rates.

During friction, carbon fiber bundles contribute to the material's performance through retention and cutting actions. The best tribological performance is achieved at a 90° fiber orientation because it offers the lowest COF, high friction stability, and minimal wear due to its high retention capacity and low cutting ability. On the other hand, a higher COF is obtained from a 30° fiber orientation.

Studies show that for C/C-SiC composites with short fibers, vertically aligned fibers have the lowest COF when sliding against steel (Li et al., 2008; Yang et al., 2021; Zhang et al., 2021; Kumar et al., 2021). This orientation supports the development of a continuous friction film, resulting in a lower COF, whereas randomly distributed fibers tend to have a higher COF.

17.3 TESTING

17.3.1 Braking Test

During the braking test that targets a proper and stable COF as well as the minimum rate of wear, it was evident that C/SiC's COFs commonly rise in a way and fall for increasing the speed of the break. Under the relatively small speed levels of around 15 m/s or lower level of about 10 m/s, shear asperities in the friction plane are under insufficient impacts and shears by sheer forces. This results in an increased COF due to brittle fracture and the formation of new asperities. Once the speed is above a certain value, debris fills surface grooves, reducing the ploughing effect and thus resulting in a lower COF and wear rate. At higher speeds, more abrasive debris is produced, eventually forming a friction film that stabilizes both the COF and wear rate (Duan et al., 2021; Sziroczak and Smith, 2016).

However, high braking speeds also increase temperature and vibrations, which can cause hot spots and unstable tribological performance. For C/SiC composites, when matched with 30CrSiMoVA counterparts, the COF decreases with an increasing braking speed because of the rise in surface temperature. This temperature rise softens and deforms materials, facilitates friction film formation, and reduces the ploughing effect. At excessively high speeds, the effects of plastic deformation and adhesion at the asperities cause fluctuation in the COF and unstable performance (An et al., 2021).

17.4 CONSTANT SPEED TEST AND IMPACTS ON TRIBOLOGICAL CHARACTERISTICS

Constant speed tests are essential for determining the tribological properties of high-temperature structural parts used in aerospace applications, such as seals, flaps, and bearings. These materials must have low COFs and wear rates to provide reliable and stable operation with increased equipment life (Fournier et al., 2000).

In constant speed friction tests of C/SiC composites with ZrO_2, as the product of velocity and load (FV value) increases, tribological properties first improve and then stabilize. At high FV values, the COF becomes less sensitive to external conditions because the wear debris becomes finer and more uniform. This fine debris helps fill surface defects and forms a lubricating film, thereby reducing the COF and wear rates. However, at a braking speed of more than 20 m/s, the carbon fibers close to the friction surface get oxidized because of high temperature, which decreases their compressive strength and increases wear. Besides, at a velocity above 15 m/s, C/SiC composites combined with 30CrSiMoVA show higher oxidation and wear, thus producing FeO, γ-Fe_2O_3, and α-Fe_2O_3. These compounds increase wear (Paris et al., 2001; Friess et al., 2003).

17.4.1 Tribological Performance in Different Conditions

Most tribological research is done in dry friction conditions, while practical applications often include wet environments. SiC-based ceramics are increasingly applied in water lubrication systems because of their excellent tribological properties. The investigations revealed that the COF in wet environments is usually smaller than in dry conditions for C/SiC composites. This can be explained by the presence of a lubricating film that is determined by material composition and the hydrophilicity of the components.

While the COF is significantly reduced for wet conditions for C/C composites because of the presence of hydrogen bonds, more stable friction behavior is expected for C/SiC composites. Additionally, the presence of debris combined with water may lead to enhanced lubrication that contributes to reducing the COF. Yet, the measured COF is greater than the dry values in the case of seawater and calls for further research into this matter.

C/C-SiC composites typically exhibit a smaller COF in a wet environment as thermal and adhesive stresses are decreased. Water film lubrication also reduces abrasive wear and improves their tribological behavior (Kochendorfer, 2003).

17.5 IMPACT OF TEMPERATURE ON TRIBOLOGICAL CHARACTERISTICS

Limited work has been reported on the high-temperature behavior of C/SiC composites, although such materials are expected to withstand extreme conditions up to 1800°C, for example, in aerospace applications. High-temperature tests show that the average COF of C/SiC composites decreases with increasing temperature, leading to more stable and reliable performance compared to room temperature.

When working at elevated temperatures, it has a surface adsorption film that acts to offer protection and enhance the resistance to wear. This allows for the oxidation of the SiC wear debris in SiO_2 and promotes more lubrication to minimize COF. Research indicates that at high-temperature environments in gases, the COF can be as low as 0.15. Otherwise, at room temperature, this is as low as 0.4.

However, high-temperature tests also disclose complex interactions of oxidation-abrasion with the adhesive wear mechanisms. At higher temperatures, surface roughness drops significantly. Additionally, a smooth glassy friction film contributes to increased performance. However, considering these findings, more investigation is still needed on the tribological properties of C/SiC composites for extreme conditions involving thermal, mechanical, and chemical stresses (Fouquet et al., 2004; Zhang et al., 2005).

17.6 IMPACT OF COUNTERPARTS ON TRIBOLOGICAL CHARACTERISTICS

The choice of friction pairs strongly influences tribological performance. In the case of self-mated C/SiC counterparts, the wear mechanism is essentially mechanical with a high COF at the initial stages of friction tests owing to deformation, fracture, and peeling of surface asperities. In the course of the friction process, the accumulation of wear debris causes an additional increase in the COF.

In contrast, the C/SiC-C/C composite pairs present with decreased wear resistance owing to the intrinsic lower wear resistance of the C/C composites. However, flake carbon chips from C/SiC, formed during friction, substitute for the ploughing action by providing lubrication through the friction film that promotes stability and decreases the COF (Abu El-Hija et al., 2005).

Recently, C/SiC composites paired with various metals and ceramics have been researched. For example, with the pairing with 100Cr6 steel, periodic mechanical and thermal stresses induce microcracks and enhance the COF caused by iron debris adhesion. With the pairing of α-Al_2O_3, low COF values (0.04–0.2) were found due to fine grinding powder lubrication of the C/SiC composites. In addition, C/SiC composites exhibit excellent wear resistance over their counterparts, and wear rates for C/SiC-Ti alloy pairs are only 5%–10% of those for Ti alloys.

When combined with ZrO_2, C/SiC composites show a very low COF of less than 0.2, with abrasive wear and material transfer between the friction surfaces being the primary mechanisms observed (Fan et al., 2007).

17.7 CONCLUSIONS

The tribological performance of C/SiC composites has been researched over such factors as fiber orientation, material structure, and temperature conditions. This extensive review summarizes recent findings in areas such as the considerable contributions of the SiC hard phase to abrasive wear, and the importance of high graphitization for stable friction. The COF increases and then decreases with speed in braking tests, while in constant-speed tests, tribological properties improve and stabilize with increasing FV values. C/SiC composites are not so sensitive to humidity, hence they perform better under wet conditions than dry conditions. The COF is lower and the performance is more stable at high temperatures. C/SiC composites show excellent potential in many applications and it is essential to continue to optimize their properties and performance in extreme environments. There will be a great need for extensive interdisciplinary research in order to advance these materials and improve their practical applications.

REFERENCES

H. Abu El-Hija, W. Krenkel, S. Hugel, Development of C/C-SiC brake pads for high-performance elevators. Int. J. Appl. Ceram. Technol. 2 (2005) 105–113. https://doi.org/10.1111/j.1744-7402.2005.02012.x.

R.S. Almeida, B. Besser, K. Tushtev, Y. Li, K. Rezwan, Fatigue behavior and damage analysis of PIP C/SiC composite. J. Eur. Ceram. Soc 42 (2022) 5391–5398. https://doi.org/10.1016/j.jeurceramsoc.2022.06.053.

Q. An, J. Chen, W. Ming, M. Chen, Machining of SiC ceramic matrix composites: A review. Chin. J. Aeronaut. 34 (2021) 540–567. https://doi.org/10.1016/j.cja.2020.08.001.

Y. Arai, R. Inoue, K. Goto, Y. Kogo, Carbon fiber reinforced ultra-high temperature ceramic matrix composites: A review. Ceram. Int. 45 (2019) 14481–14489. https://doi.org/10.1016/j.ceramint.2019.05.065.

S. Beyer, S. Schmidt, F. Maidl, R. Meistring, M. Bouchez, P. Peres, Advanced composite materials for current and future propulsion and industrial applications. Adv, Sci. Technol 50 (2006) 174–181. https://doi.org/10.4028/www.scientific.net/AST.50.174.

Y. Cai, Y. Xu, B. Li, S. Fan, L. Zhang, L. Cheng, B. Dong, J. Jiang, Microstructures and mechanical properties of a low-cost three-dimensional needled carbon/silicon carbide composite. Mat. Sci. Eng. A-Struct. 497 (2008) 278–282. https://doi.org/10.1016/j.msea.2008.07.011.

T. Campbell, J. Ting, P. Avitabile, J. Min, Dynamic properties of 3–D reinforced C/SiC for the RS-2200 linear aerospike engine, in: Todd Jessen, Ersan Ustundag (Eds.), 24th Annual conference on composites, advanced Ceramics, materials, and structures: A: Ceramic Engineering and Science Proceedings, Wiley Online Library, Florida, 2000, pp. 517–524, https://doi.org/10.1002/9780470294628.ch61.

M. Chen, H. Qiu, W. Xie, B. Zhang, S. Liu, W. Luo, X. Ma, Research progress of continuous fiber reinforced ceramic matrix composite in hot section components of aero engine, in: The 11th International Conference on High- Performance Ceramics, 2019. https://doi.org/10.1088/1757-899X/678/1/012043.

F. Christin, Design, fabrication, and application of thermostructural composites (TSC) like C/C, C/SiC, and SiC/SiC composites. Adv. Eng. Mater. 4 (2002) 903–912. https://doi.org/10.1002/adem.200290001.

J.Y. Deng, Y.L. Wei, W.C. Liu, Carbon-fiber-reinforced composites with graded carbon-silicon carbide matrix composition. J. Am. Ceram. Soc. 82 (1999) 1629–1632. https://doi.org/10.1111/j.1151-2916.1999.tb01976.x.

J. Duan, M. Zhang, P. Chen, Z. Li, L. Pang, P. Xiao, Y. Li, Tribological behavior and applications of carbon fiber reinforced ceramic composites as highperformance frictional materials. Ceram. Int. 47 (2021) 19271–19281. https://doi.org/10.1016/j.ceramint.2021.02.187.

S. Fan, L. Zhang, L. Cheng, G. Tian, S. Yang, Effect of braking pressure and braking speed on the tribological properties of C/SiC aircraft brake materials. Compos. Sci. Technol. 70 (2010) 959–965. https://doi.org/10.1016/j.compscitech.2010.02.012.

S. Fan, L. Zhang, Y. Xu, L. Cheng, G. Tian, S. Ke, F. Xu, H. Liu, Microstructure and tribological properties of advanced carbon/silicon carbide aircraft brake materials. Compos. Sci. Technol. 68 (2008) 3002–3009. https://doi.org/10.1016/j.compscitech.2008.06.013.

S.S.W. Fan, Y.D. Xu, L.T. Zhang, L.F. Cheng, L. Yu, Y.D. Yuan, F.K. Zhang, G.L. Tian, Z.J. Chen, J.J. Lou, Three-dimensional needled carbon/silicon carbide composites with high friction performance. Mat. Sci. Eng. A-Struct. 467 (2007) 53–58. https://doi.org/10.1016/j.msea.2007.02.053.

X. Fan, X. Yin, X. Cao, L. Chen, L. Cheng, L. Zhang, Improvement of the mechanical and thermophysical properties of C/SiC composites fabricated by liquid silicon infiltration. Compos. Sci. Technol. 115 (2015) 21–27. https://doi.org/10.1016/j.compscitech.2015.04.019.

S. Fouquet, S. Jouannigot, J. Alexis, R. Pailler, X. Bourrat, A. Guette, R. Naslain, Nanoindentation, microscratch, friction and wear studies of carbon-fibre reinforced SiC-Si matrix composite. Key Eng. Mater. 264–268 (2004) 917–920. https://doi.org/10.4028/www.scientific.net/KEM.264-268.917.

S. Fouquet, M. Rollin, R. Pailler, X. Bourrat, Tribological behaviour of composites made of carbon fibres and ceramic matrix in the Si-C system. Wear 264 (2008) 850–856. https://doi.org/10.1016/j.wear.2006.12.081.

P. Fournier, P. Reynaud, F. Platon, J. Absi, Tribological behaviour of carbon-fibrereinforced SiC matrix composites. Proc. Inst. Mech. Eng., Part J: J. Eng. Tribol. 214 (2000) 291–306. https://doi.org/10.1243/1350650001543188.

M. Friess, X.D. Huang, W. Krenkel, Diamond and other hard materials in fibre reinforced ceramic matrix composites, in: Waltraud M. Kriven, H.-T. Lin (Eds.), 27th International Cocoa Beach Conference on Advanced Ceramics Composites, Cocoa Beach, Fl, 2003, pp. 605–611, https://doi.org/10.1002/9780470294826.ch88.

D. Glass, Ceramic matrix composite (CMC) thermal protection systems (TPS) and hot structures for hypersonic vehicles, in: 15th AIAA international space planes and hypersonic systems and technologies conference, 2008. https://doi.org/10.2514/6.2008-2682.

X.-B. He, X.-M. Zhang, C.-R. Zhang, X.-G. Zhou, A.-C. Zhou, Microstructures of a carbon-fiber-reinforced silicon-carbide composite produced by precursor pyrolysis and hot pressing. Compos. Sci. Technol. 61 (2001) 117–123. https://doi.org/10.1016/S0266-3538(00)00202-5.

B. Heidenreich, C/SiC and C/C-SiC composites, in: N.P. Bansal, J. Lamon (Eds.), Ceramic Matrix Composites: Materials, Modeling and Technology. Wiley. (2014) pp. 147–216, https://doi.org/10.1002/9781118832998.ch6.

R. Inoue, J.M. Yang, H. Kakisawa, Y. Kagawa, Mode I fracture toughness of short carbon fiber-dispersed SiC matrix composite fabricated by melt infiltration process. Ceram. Int. 39 (2013) 8341–8346. https://doi.org/10.1016/j.ceramint.2013.04.014.

R. Kochendorfer, High friction, low wear composites based on fibre reinforced ceramics, in: E.E. Gdoutos, Z.P. Marioli-Riga (Eds.), Recent advances in composite materials: In honor of S.A. Paipetis, Springer, Netherlands, Dordrecht, (2003), pp. 313–323. https://doi.org/10.1007/978-94-017-2852-2_26.

W. Krenkel, Carbon fiber reinforced CMC for high-performance structures. Int. J. Appl. Ceram. Technol. 1 (2004) 188–200. https://doi.org/10.1111/j.1744-7402.2004.tb00169.x.

W. Krenkel, Carbon fibre reinforced silicon carbide composites (C/SiC, C/C-SiC). in: N. P. Bansal (Ed.), Handbook of ceramic composites, Springer, (2005), pp. 117–148. https://doi.org/10.1007/0-387-23986-3_6.

S. Kumar, M. Bablu, M.K. Mishra, A. Ranjan, N.E. Prasad, Fabrication and characterization of PIP based C/SiC composites having improved mechanical properties using high modulus M40J carbon fiber as reinforcement. Ceram. Int. 43 (2017a) 8153–8162, https://doi.org/10.1016/j.ceramint.2017.03.141.

S. Kumar, A. Kumar, K. Sampath, V.B. Prasad, J. Chaudhary, A. Gupta, G.R. Devi, Fabrication and erosion studies of C–SiC composite Jet Vanes in solid rocket motor exhaust. J. Eur. Ceram. Soc. 31 (2011) 2425–2431. https://doi.org/10.1016/j.jeurceramsoc.2011.06.007.

S. Kumar, K.C. Shekar, B. Jana, L. Manocha, N.E. Prasad, C/C and C/SiC composites for aerospace applications, in: N. Eswara Prasad, R.J.H. Wanhill (Eds.), Aerospace materials and material technologies, Springer Singapore, (2017b), pp. 343–369, https://doi.org/10.1007/978-981-10-2134-3_15.

V.T. Le, N. San Ha, N.S. Goo, Advanced sandwich structures for thermal protection systems in hypersonic vehicles: A review. Compos. Part B. 226 (2021) 109301. https://doi.org/10.1016/j.compositesb.2021.109301.

Y. Li, L. Zhang, R. He, Y. Ma, K. Zhang, X. Bai, B. Xu, Y. Chen, Integrated thermal protection system based on C/SiC composite corrugated core sandwich plane structure. Aerosp. Sci. Technol. 91 (2019) 607–616. https://doi.org/10.1016/j.ast.2019.05.048.

Z. Li, P. Xiao, X. Xiong, S.-h Zhu, Tribological characteristics of C/C-SiC braking composites under dry and wet conditions. Trans. Nonferrous Met. Soc. China 18 (2008) 1071–1075. https://doi.org/10.1016/s1003-6326(08)60183-1.

Y. Ma, X. Yin, X. Fan, P. Ju, X. Dang, Modification and toughening of 3D needled C/SiC composite by deformable MAX phase-based matrix. Mater. Sci. Eng. A. 712 (2018) 397–405. https://doi.org/10.1016/j.msea.2017.11.110.

S. Mall, J.M. Engesser, Effects of frequency on fatigue behavior of CVI C/SiC at elevated temperature. Compos. Sci. Technol. 66 (2006) 863–874. https://doi.org/10.1016/j.compscitech.2005.06.020.

R. Naslain, Design, preparation and properties of non-oxide CMCs for application in engines and nuclear reactors: An overview. Compos. Sci. Technol. 64 (2004) 155–170, https://doi.org/10.1016/S0266-3538(03)00230-6.

H. Ohnabe, S. Masaki, M. Onozuka, K. Miyahara, T. Sasa, Potential application of ceramic matrix composites to aero-engine components. Compos. Part A. 30 (1999) 489–496. https://doi.org/10.1016/S1359-835X(98)00139-0.

T. Opel, N. Langhof, W. Krenkel, Development and tribological studies of a novel metal-ceramic hybrid brake disc. Int. J. Appl. Ceram. Technol. 19 (2022) 62–74. https://doi.org/10.1111/ijac.13826.

N.P. Padture, Advanced structural ceramics in aerospace propulsion. Nat. Mater. 15 (2016) 804. https://doi.org/10.1038/nmat4687.

J.Y. Paris, L. Vincent, J. Denape, High-speed tribological behaviour of a carbon/silicon-carbide composite. Compos. Sci. Technol. 61 (2001) 417–423. https://doi.org/10.1016/s0266-3538(00)00124-x.

J.E. Sheehan, K. Buesking, B. Sullivan, Carbon-carbon composites. Annu. Rev. Mater. Sci 24 (1994) 19–44. https://doi.org/10.1146/annurev.ms.24.080194.000315.

K. Shi, Y. Yan, H. Mei, L. Cheng, Improving the thermal expansion behavior and damping capacity of C/SiC composites by heat treatment. J. Alloy. Compd. 859 (2021). https://doi.org/10.1016/j.jallcom.2020.157849.

I. Stimson, R. Fisher, Design and engineering of carbon brakes. Philos. Trans. R. Soc. Lond. Ser. A. Math. Geophys. Sci. 294 (1980) 583–590. https://doi.org/10.1098/rsta.1980.0068.

D. Sziroczak, H. Smith, A review of design issues specific to hypersonic flight vehicles. Prog. Aerosp. Sci 84 (2016) 1–28. https://doi.org/10.1016/j.paerosci.2016.04.001.

S. Tang, J. Deng, S. Wang, W. Liu, Comparison of thermal and ablation behaviors of C/SiC composites and C/ZrB2–SiC composites. Corros. Sci. 51 (2009) 54–61. https://doi.org/10.1016/j.corsci.2008.09.037.

Z.S. Toor, Space applications of composite materials. J. Space Technol. 8 (2018) 65–70.

S. Vaidyaraman, M. Purdy, T. Walker, S. Horst, C/SiC material evaluation for aircraft brake applications, in: W. Krenkel, R. Naslain and H. Schneider (Eds.), High Temperature Ceramic Matrix Composites. Wiley, (2001), pp. 802–808. https://doi.org/10.1002/3527605622.ch122.

J. Wang, X. Chen, K. Guan, L. Cheng, L. Zhang, Y. Liu, Effects of channel modification on microstructure and mechanical properties of C/SiC composites prepared by LA-CVI process. Ceram. Int 44 (2018) 16414–16420. https://doi.org/10.1016/j.ceramint.2018.06.052.

X. Wang, X. Gao, Z. Zhang, L. Cheng, H. Ma, W. Yang, Advances in modifications and high-temperature applications of silicon carbide ceramic matrix composites in aerospace: A focused review. J. Eur. Ceram. Soc. 41 (2021) 4671–4688. https://doi.org/10.1016/j.jeurceramsoc.2021.03.051.

Y. Wang, Z. Chen, S. Yu, Ablation behavior and mechanism analysis of C/SiC composites. J. Mater. Res. Technol. 5 (2016) 170–182. https://doi.org/10.1016/j.jmrt.2015.10.004.

T. Windhorst, G. Blount, Carbon-carbon composites: A summary of recent developments and applications. Mater. Des. 18 (1997) 11–15. https://doi.org/10.1016/s0261-3069(97)00024-1.

Y. Xu, L. Cheng, L. Zhang, H. Yin, X. Yin, Mechanical properties of 3D fiber reinforced C/SiC composites. Mater. Sci. Eng. A. 300 (2001) 196–202. https://doi.org/10.1016/S0921-5093(00)01533-1.

Y. Xu, Y. Zhang, L. Cheng, L. Zhang, J. Lou, J. Zhang, Preparation and friction behavior of carbon fiber reinforced silicon carbide matrix composites. Ceram. Int. 33 (2007) 439–445. https://doi.org/10.1016/j.ceramint.2005.10.008.

B. Yang, X. Zhou, J. Yu, The properties of C-f/SiC composites prepared from different precursors. Ceram. Int. 41 (2015) 4207–4213. https://doi.org/10.1016/j.ceramint.2014.12.111.

X. Yang, C. Feng, Z.-h Peng, W. Yi, G.-d Li, Evolution of microstructure and mechanical properties of PIP-C/SiC composites after high-temperature oxidation. J. Asian Ceram. Soc. 5 (2017) 370–376. https://doi.org/10.1016/j.jascer.2017.07.001.

Y. Yang, Y. Zhao, H. Mei, L. Cheng, L. Zhang, 3DN C/SiC-MoS2 self-lubricating composites with high friction stability and excellent elevated-temperature lubrication. J. Eur. Ceram. Soc. 41 (2021) 6815–6823. https://doi.org/10.1016/j.jeurceramsoc.2021.07.030.

C. Zhang, Y. Zhou, H. Liu, J. Li, K. Lu, Y. Li, Influence of pyrolytic carbon interphase and pyrolysis temperature on the mechanical and tribological properties of Cf/SiC after heat treatment at 1800°C. Ceram. Int. 47 (2021) 16282–16293. https://doi.org/10.1016/j.ceramint.2021.02.207.

Q. Zhang, G. Li, A review of the application of C/SiC composite in thermal protection system. Multidiscip. Model. Mater. Struct. 5 (2009), 199–203. https://doi.org/10.1163/157361109787959903.

Y. Zhang, Y.D. Xu, J.J. Lou, L.T. Zhang, L.F. Cheng, J.J. Lou, Z.J. Chen, Braking behavior of C/SiC composites prepared by chemical vapor infiltration. Int. J. Appl. Ceram. Technol. 2 (2005) 114–121. https://doi.org/10.1111/j.1744-7402.2005.02013.x.

Y. Zhao, H. Xia, R. Tang, Z. Shi, J. Yang, J. Wang, A low cost preparation of C/SiC composites by infiltrating molten Si into gelcasted pure porous carbon preform. Ceram. Int. 41 (2015) 6478–6487. https://doi.org/10.1016/j.ceramint.2015.01.087.

Index

K

L

M

N

O

P

R

S

T

Made in the USA
Las Vegas, NV
07 May 2026

59cd350b-457f-47e8-99ce-7c8c6b1a2620R01